WITHD

UTSA LIBRARIES

REINVENTING WATER AND WASTEWATER SYSTEMS

Global Lessons for Improving Water Management

Edited by

**Paul Seidenstat,
David Haarmeyer,
and Simon Hakim**

John Wiley & Sons, Inc.

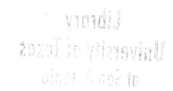

This publication is designed to provide accurate and authoritative information in regard
to the subject matter covered. It is sold with the understanding that the publisher is not
engaged in rendering professional services. If professional advice or other expert assistance is required, the services of a competent professional person should be sought.

Wiley also publishes its books in a variety of electronic formats. Some content that appears
in print may not be available in electronic books. For more information about Wiley products, visit our web site at www.wiley.com.

Library of Congress Cataloging-in-Publication Data:

Reinventing water and wastewater systems : global lessons for improving water
management / by Paul Seidenstat . . . [et al.].
    p. cm.
  ISBN 0-471-06422-X (cloth : alk. paper)
  1. Water—Distribution—Management. 2. Water-supply—Management. 3. Sewage
  disposal—Management. 4. Infrastructure (Economics)—Management.
  5. Privatization. I. Seidenstat, Paul.

TD481 .R45 2002
363.6′1—dc21
                                                                    2001057369

Printed in the United States of America.

10 9 8 7 6 5 4 3 2

Contents

Editors and Contributors

EDITORS

Paul Seidenstat is an associate professor of economics at Temple University. A specialist in the fields of public finance, public management, and water resources, he has written or coedited eight books and several articles on these subjects, including *Contracting Out Government Services* (Praeger). Seidenstat and Simon Hakim are coeditors of *America's Water and Wastewater Industries* (Public Utility Reports), *Privatizing the United States Justice System* (McFarland), *Privatizing Correctional Institutions* (Rutgers), *Privatizing Education and Educational Choice* (Praeger), and *Privatizing Transportation Systems* (Praeger). He has conducted funded research projects for the federal government in the fields of manpower and water resources. His current research is in applying economic analysis to the management of public agencies.

David Haarmeyer is an Associate Director at Cambridge Energy Research Associates (CERA) and an expert in corporate and power industry restructuring, competitive strategy, and economics. He was formerly an economist and executive consultant at Stone and Webster Consultants, where he provided competitive and strategic analysis for clients in the power and water industries worldwide. He also worked with the World Bank; the private equity group Global Environmental Fund; Reason Foundation; the New Zealand government; and the U.S. Federal Trade Commission. He is the author of numerous articles and reports on energy and water market restructuring and privatization and wrote *World Water Privatization: Managing Risks in Water and Sanitation.*

Simon Hakim is professor of economics at Temple University. He is also codirector of the Center for Competitive Government at Temple University and coeditor of the series *Privatizing Government Assets and Services*

(Praeger). He has published over 40 scientific articles in leading economics, criminology, and public policy journals. He has published 50 articles in professional journals for the security industry and has edited 15 books. He has conducted funded research projects for government agencies and private companies and has written extensively on the analysis of criminal behavior, police operations, and the privatization of various types of institutions.

CONTRIBUTORS

Peter Alviti, Jr., Laborers International Union of North America

Maarten Blokland, International Institute for Infrastructure, Hydraulic and Environmental Engineering, Netherlands

H. Richard Borer, Jr., Mayor of West Haven, Connecticut

Penelope J. Brook, The World Bank

Lee P. Brown, Mayor of Houston, Texas

Tyler Cowen, George Mason University and the Center for the Study of Public Choice

Debra G. Coy, Washington Research Group

Michael J. Crean, Chief Operating Officer, Washington Suburban Sanitary Commission, Laurel, Maryland

George Day, Office of Water Services (Ofwat)

Mark Dumol, Consultant and former advisor to Philippines

Jane Eddy, Standard & Poors Corporation

Rodney Eng, City of Seattle, Washington

Vivien Foster, The World Bank

James S. Harris, Lovells, Singapore

Scott Haskins, Seattle Public Utilities

Elizabeth S. Kelly, Seattle Public Utilities

Kristin Komives, Institute for Social Studies, Netherlands

Frank J. Mangravite, Public Works Management

Paul J. Matacera, GluckShaw Group

Patrick J. Meyers, Resource Conservation and Management Consultants

Yogita U. Mumssen, Stone and Webster

Stephen Ramsey, WRC Group, London

Rebecca Reehal, Lehman Brothers

Skip Rimsza, Mayor of Phoenix, Arizona

Lillian Saade Hazin, International Institute for Infrastructure, Hydraulic and Environmental Engineering

Klaas Schwartz, International Institute for Infrastructure, Hydraulic and Environmental Engineering

Neel A. Shah, Six Open Systems, Inc.

James H. Sills, Jr., Former mayor of Wilmington, Delaware, and Associate Professor Emeritus, University of Delaware

Edward Sondey, PSEG Power Development Corporation

Michael A. Traficante, Laborers International Union of North America and former Mayor of Cranston, Rhode Island

Nicola Tynan, Dickinson College

Michael Wilkins, Standard & Poors Corporation

Brian Williamson, National Economic Research Associates, Great Britain

Part I

Introduction

Introduction

Securing safe, reliable, and reasonably priced water has been one of the leading economic and social challenges faced by countries worldwide as their populations, economies, and urban areas grew. Especially as urban centers grew and technology advanced, piping of water became economically feasible. With the greater challenge of scale and larger number of customers, organization of water supply became necessary. An entrepreneur or organizational entity had to take responsibility for planning and financing the construction of piping facilities and for operation of the resultant distribution system. Eventually, an organized commercial water business evolved that was responsible for delivering safer and more reliable water.

Today, across the globe, the range of the level and quality of service is very broad. While most water and wastewater systems are publicly owned, there is a large and growing industry of private service providers who compete for the right to finance, build, and operate facilities worldwide. Where governments are committed and capable of securing private sector participation, this industry has made substantial progress in upgrading and expanding service. Both in developed and developing countries, contractual and regulatory issues involved in ensuring reasonable performance and remuneration have been significant challenges. This book offers a rich and insightful look into many of these challenges, and into the lessons learned by different countries in applying different approaches to improving water and wastewater services.

An overview of the private sector's role in the global water and wastewater industry is presented by David Haarmeyer and Debra G. Coy in the opening chapter. Particular emphasis is given to the sector's unique attributes, and to the risks that increase the costs of private capital and management participation, and thus slow the industry's efforts to modernize, commercialize, and innovate. Nevertheless, while there have been setbacks, the private sector's role is expanding in both the international and the US water and wastewater markets, and this expansion has led to improved operations, greater capital availability for infrastructure expansion, and the creation of commercially viable enterprises.

A key challenge the industry faces is to follow the path laid down by other infrastructure sectors, such as gas, electric, and telecom, who have introduced competition through fundamental industry restructuring, and thus mitigated political and regulatory risk. By opening the sector up to greater competition, the industry can expect to benefit from an increasing flow of private capital, greater technological innovation, better services, and lower prices. As noted by Haarmeyer and Coy, restructuring efforts are especially evident in the United Kingdom. In the United States, the industry's private service providers are consolidating to provide better service in what is a fragmented industry. Finally, in the area of desalination, competition and technological change are starting to make substantial gains in providing lower cost service.

The regulatory framework presents the background against which water and wastewater systems operate. Yogita Mumssen and Brian Williamson review the economics of network industries in Chapter 2, and analyze the major issues involved in regulating them. They compare the regulatory schemes in the United Kingdom, the United States, and Australia in light of the common objectives of protecting consumer interests, meeting customer demand, ensuring adequate financing, and, where possible, promoting competition. An evaluation of the regulatory systems in the three nations then follows.

During the 1990s, most countries in Latin America undertook major reforms to their water supply industries. Vivian Foster dissects this reform movement in Chapter 3, with special emphasis on the changes that were wrought in the regulatory framework.

A more intensive view of the regulatory system of Great Britain is undertaken in Chapter 4 by Stephen Ramsey, who provides in-depth analysis of how the regulatory system utilizes a comparison of the costs and operational benchmarks of regulated companies to assist in adjusting water rates for each company. One of the methods that the regulatory agency, The Office of Water Services (Ofwat), uses to establish standards of performance is metric benchmarking, which uses various quantitative measures to compare companies and to establish operational standards. Also, to aid in improving performance, regulated companies are encouraged, or in a few cases required, to undertake a review of their operations through the methodology of process benchmarking. George Day gives a broader view of the system in Chapter 5. He explores the role of regulation in water and wastewater and ends with the consideration of structural issues and how competition may be introduced in the future.

The all-important issue of financing the water and wastewater sector is discussed in Part III of the book. In the context of the pressure to increase resources and diversify capital structures, Michael Wilkins and Jane Eddy (Chapter 6) explain the criteria used by Standard & Poor's, a provider of objective credit information, in providing ratings and risk analysis to the global financial community. Water industry structures are evolving differently around the world, and Standard & Poor's rating criteria address the diversity of credit risks inherent within each structure. The chapter provides a description of the rating methodology for non- or limited-recourse water and wastewater projects.

Kristin Komives is concerned with the problems of connecting low-income residents to water systems in developing countries. In Chapter 7, she focuses upon features of concession contracts with water utility operators that bear on this issue. The emphasis is on provisions offering financial incentives and on allowing new providers or service alternatives to emerge.

Edward Sondey and Neel Shah elaborate, in Chapter 8, on the problem of attracting investment in developing countries. They point to the success of Mexico's state-owned oil company in structuring its wastewater treatment outsourcing contract to provide risk-sharing with service providers. The risk-sharing feature encourages qualified private project developers to submit efficient and competitive proposals. The challenge for municipal governments in Mexico is to follow this contract model.

How water systems have attempted to make improvements and reinvent themselves is the subject of Part IV. Public–private partnerships have become a valuable tool for improving the operation of publicly owned water systems. In Chapter 9, Lillian Saade Hazin examines the use of "private sector participation" in Mexico's Federal District. She details how the contracting out of specific functions was used with the awarding of service contracts to four private partners to implement universal water metering, to rehabilitate the distribution system, and to carry out a loss-detection program.

Mark Dumol and Paul Seidenstat spotlight, in Chapter 10, how concessions granted to two private companies were used to revamp the Manila water system. The methods used and the obstacles faced in developing a viable concession contract are examined in this chapter.

The use of government-owned stock corporations in water and wastewater is discussed by Klaas Schwartz and Maarten Blokland in Chapter 11. They look in detail at how the Dutch use this organizational device to achieve the production of water services.

Part V examines the restructuring efforts for water and wastewater utilities of municipal governments, both large and small, across the United States. In Chapters 12–18, mayors and public works officials document these efforts in Seattle; Wilmington, Delaware; Houston; Cranston, Rhode Island; North Brunswick, New Jersey; and Phoenix.

The final section, Part VI, looks at some history of how water systems were owned and operated in the past, and then fast-forwards to the twenty-first century to focus on how the future might look.

In assessing the role of the private sector in the production of water services, the observer should be aware that in some countries private provision of piped water has a long history. Nicola Tynan presents London's experience with private water operations in Chapter 19, and evaluates the accomplishments of the privatized sector.

In analyzing water privatization options for developing countries in Chapter 20, Penelope Brook and Tyler Cowen argue that, since the major problem in developing countries is connecting citizens to a safe and reliable water supply, unregulated monopoly may be a viable choice. Although there may be significant downsides to unfettered monopoly, the authors suggest that it may be the best option if the objective is to maximize the number of water hookups.

The future of the structure of regulation of the water sector in Great Britain is discussed by Rebecca Reehal in Chapter 21. She peers into the future through the prism of new legislation that offers to inject competition into British water markets.

The future of business and the Internet are closely connected. In Chapter 22, Patrick Meyers shows that the Internet may be an efficiency-maximizing device to improve the management of the water industry. Meyers illustrates how the new technology can be used in the southwest region of the United States. Another technology, desalination, is examined by James Harris in Chapter 23. He details the recent history of the technology and discusses the financing of desalination projects.

In Chapter 24, Michael Crean examines how water and wastewater companies operating in the public sector can reengineer to improve performance. The experience of the Washington Suburban Sanitary Commission may be used as a model for public sector companies in the twenty-first century.

Finally, Paul Seidenstat looks at the sector's expanding global menu of operations and ownership arrangements in Chapter 25. As a result of growing global experience, owners and operators can take advantage of the lessons learned to better control costs, improve operations and service, and create more financially sustainable systems.

Chapter 1

An Overview of Private Sector Participation in the Global and US Water and Wastewater Sector

David Haarmeyer and Debra G. Coy

1. INTRODUCTION: RISKS OVERSHADOW COMMERCIAL OPPORTUNITIES

The global water and wastewater sector has the distinction of being the infrastructure sector with greatest promise—steady long-term cash flows from what many describe as the "last monopoly utility business"—and the sector that has shown the least amount of progress in terms of attracting

David Haarmeyer is an Associate Director at Cambridge Energy Research Associates (CERA) and an expert in corporate and power industry restructuring, competitive strategy, and economics. He was formerly an economist and executive consultant at Stone and Webster Consultants, where he provided competitive and strategic analysis for clients in the power and water industries worldwide. He also worked with the World Bank; the private equity group Global Environmental Fund; Reason Foundation; the New Zealand government; and the U.S. Federal Trade Commission. He is the author of numerous articles and reports on energy and water market restructuring and privatization and wrote World Water Privatization: Managing Risks in Water and Sanitation. *Debra G. Coy is Senior Analyst for Water Industry and Environmental Policy at the Washington Research Group, a part of Schwab Capital Markets, L.P. Formerly she was a Equity Research Analyst with HSBC Securities and with NatWest Securities. She has appeared on* Wall Street Week *and has been quoted in numerous financial and industry publications on environmental and water industry issues.*

Reinventing Water and Wastewater Systems: Global Lessons for Improving Water Management, edited by Paul Seidenstat, David Haarmeyer, and Simon Hakim.
0-471-06422-X   Copyright © 2002 by John Wiley & Sons, Inc.

7

private investment. Year after year, figures are released indicating that both developing and developed countries require tremendous capital investment to meet the basic needs of their populations. Estimates from the World Bank put the capital "needs" for developing countries at $60 billion over the next ten years. In the United States, the Environmental Protection Agency (EPA) estimates that $275 billion is "needed" for investment in water and wastewater services over the next 20 years. And in the European Community, the figure is $220 billion.

Yet despite the tremendous needs, the amount of private capital flowing to the sector has been limited, especially when compared to other infrastructure sectors. As indicated in Figure 1.1, investment in water and sanitation has trailed transport, energy, and telecom investment throughout the 1990s during the emerging markets infrastructure revolution. Of the nearly $580 billion in total infrastructure investment directed to these sectors during this time, water and wastewater represented only $30 billion, or about 5%.

Also noteworthy is the water industry's slow metamorpohsis to introduce competition and deregulate potential competitive segments of the sector's value chain, changes already made in the gas, electric, and telecom industries. As a result of this restructuring, consumers in these industries are experiencing significant benefits in terms of lower prices, a broader array of services, and greater innovation.*

Why does the flow of private capital to the water and wastewater sector worldwide appear constrained relative to the tremendous needs for investment in new facilities, rehabilitation of leaking pipes, and system upgrades? The following five characteristics of the sector and their associated risks are a helpful starting place to answering this question:

- *Significant Interaction with Government.* Health, environmental, and monopoly concerns have meant that governments, often at multiple levels, have tended to be heavily involved in regulating water and wastewater services. This level of interaction with the government exposes the sector to significant regulatory and political uncertainty and risk.

---

*See the Crandall and Ellig study (1997), which provides estimates of the annual value of consumer benefits due to deregulation for the gas, long-distance telecom, airlines, trucking, and railroad industries.

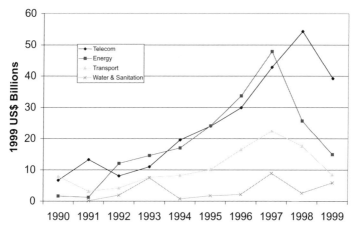

Figure 1.1   Investment in developing country infrastructure with private participation. *Source:* Izaguirre and Rao, 2000.

- *Local and Small Project Size.* Water supply and wastewater systems tend to be small and local, given the relatively low value of the service and high capital cost required for long-distance transmission. Consequently, financiers are relatively less attracted to the sector's fragmented structure, which, compared to gas, electric, and telecom, severely limits its ability to generate efficiencies from regional integration.
- *Political Risks.* As sunk, highly specific, and non-redeployable investments, private investors, especially in developing countries with limited independent regulatory and judicial frameworks, are often exposed to government opportunism and expropriation. These forms of political as well as regulatory risks are costly and difficult to cover.
- *Currency Risk.* Water and wastewater projects in countries with less developed capital markets depend on foreign capital for investment. The mismatch between the foreign currency and the domestic currency that consumers use to pay their bills means that international lenders are exposed to the risk that they may not receive back the full value of their investment because of fluctuations in the domestic currency.
- *Inadequate Tariffs.* Given the high level of capital investment and the history of government-subsidized services, full-cost pricing of

water and wastewater service has yet to take hold fully in many markets, even in the United States. The public perception in many countries is that water is "free," which means that it remains broadly underpriced, which discourages private investment, or forces investors to rely on subsidies to obtain sufficiently attractive rates of return on invested capital.

Taken together, these unique features of the water industry indicate that, despite its potentially attractive cash profile in terms of offering investors steady long-term revenue streams, the sector as it is presently structured and regulated has a relatively unattractive risk profile. Indeed, to adequately address these risks—especially political and regulatory risks—most privately financed water and wastewater projects in developing countries have tended to require some form of political risk insurance or guarantee, and/or the participation of multilateral organizations, (such as the World Bank and the Inter-American Development Bank (Haarmeyer and Mody, 1997). Absent fundamental restructuring to better separate risks and open up the sector to more competition and more accurate market pricing, the monopoly integrated structure of the industry is likely to continue to discourage private capital flows, and thus restrict the improvements and innovations that it tends to bring.

As indicated in the following summaries on international and US developments, over the past several years there have been a number of noteworthy advances, as well as setbacks, in the water and wastewater sector's ability to attract private capital and initiative. In some cases, it is clear that the accumulated experience that the public and private sector has in designing workable contracts and organizing competition can successfully overcome noncommercial risks. In the case of the United Kingdom, the combination of significant regulatory pressures, high capital investment needs, and highly transparent capital market participation is encouraging the industry to go beyond its present organizational and financial structure to consider alternatives such as splitting asset ownership from operations and injecting competition into the sector. Taken together, there are substantial lessons to be learned by governments and private participants in how to better mitigate risks and structure new organizations that properly reward investors and, at the same time, provide the incentive to increase the efficiency and affordability of water and wastewater services.

2. MAJOR INTERNATIONAL DEVELOPMENTS: THE TREND IN PRIVATE PARTICIPATION EXPANDS TO NEW COUNTRIES

2.1 England and Wales

More than a decade after the United Kingdom privatized its water sector by floating 10 English and Welsh water and sewerage companies, its experiment remains controversial as ever. In response to tight price controls and a push to inject competition into the sector, water companies and regulators are fundamentally rethinking the organization of their industry.

In the early years postprivatization, the high returns available through a combination of pricing and efficiencies enabled a number of UK private water companies to build up substantial nonregulated international businesses that today are successfully competing and spreading technology and management practices around the world. The international water business is attractive, both from the perspective of the size of the business opportunities it represents and from the fact that it allows for more growth than the domestic market.

By nearly every measure, there have been substantial improvements in water and wastewater quality, making the United Kingdom one of the leaders in meeting the European Union (EU) Directives. These improvements have been the result of significant capital investment within the allowed price-cap structure. Since then, however, the Office of Water Services (Ofwat), the industry regulator, has dramatically squeezed prices and returns. In its third price review, which determined prices for 2000–2005, Ofwat stipulated average cuts to consumer bills of 12.4%, driving projected returns down to about cost-of-capital levels. In response, water companies announced cuts in dividends and workforce, and equity investors fled the sector.

A more far-reaching response to regulatory pressures has been for water companies to fundamentally rethink their company's financial and organizational structure. In 2000, Kelda, the renamed holding company for Yorkshire Water, proposed the most novel approach, which involves spinning the water and wastewater assets into a nonprofit mutual company financed completely by debt. Equity investors would be left with Kelda's unregulated business, which would be transformed into a contract operation service company, and thus no longer exposed to undue regulatory risk.

Ofwat objected to the scheme on the grounds that a customer-owned company with a nonprofit structure would provide weak investment incentives and make it difficult for Ofwat to take enforcement actions.

In 2001, Ofwat did give the go-ahead to a similarly structured deal involving Glas Cymru, the renamed Welsh Water. The acquisition of the regulated water company is funded through a \$2.7 billion securitization and, as a result of the new capital structure, the utility's cost of capital fell to just over 4%, compared with 6.5% industry threshold set by the regulator (Metcalfe, 2001). Two key differences in the deals are: 1) there was strong political support in the Welsh Assembly for community ownership of the water business, which serves all of Wales, and 2) Glas Cymru created a company that is legally liable for the nonprofit company's debt, rather than customers holding this obligation. To reduce risks for lenders and provide discipline for efficient operations, two four-year contracts were put out to competition: one for operations and the other for customer billing. While it is not clear that this model can be replicated, it does offer an attractive way of financing assets with heavily regulated revenue streams and fairly predictable capital expenditures, while at the same time injecting competition into operations.

Another innovative approach to restructuring the UK water industry has involved the promotion of cross-boundary competition and common carriage. With an Ofwat-approved "inset appointment," water and sewerage companies are able to compete to supply large customers in each other's territory. Similar to the gas industry practice, common carriage in the water sector requires water companies to make their pipes available to competitors. Although proposed many years earlier, in March 2000, the UK Competition Act 1998 was implemented, which gives Ofwat and the Office of Fair Trading legal powers to remove barriers to competition in the water sector and to promote common carriage. In addition to cost savings and better service to customers, the success of these approaches to advancing competition may also serve to minimize the need for regulation.

2.2 Europe

The integration of the European Union (EU), the significant capital demands to meet EU environmental directives, and liberalization are together increasing the attractiveness of private sector participation in the water and wastewater sector. A recent Standard & Poors (S&P) report expects that more than 250 billion Euros (about US \$220 billion) will be required to meet EU environmental standards across Europe by 2005 (Standard & Poors, 2001). While the United Kingdom is the only country to privatize its water industry fully, the French and Spanish governments are relatively supportive of private sector participation compared to other European governments.

In France, under long-term contract, about 75% of water distribution, and 45% of wastewater treatment is provided by private companies. In Spain, almost 45% of the water supply is provided by private companies under long-term concession contracts. In Italy, the 1994 Galli Law is the main driver for making the sector more attractive to private investment by facilitating consolidation of the thousands of water utilities. As a result, international private companies, such as Severn Trent of the United Kingdom, have entered the market. In 1999, Portugal awarded its first major water and wastewater concession with a project financing structure. The Santa Maria da Feira Municipality granted Indaqua Feira a 35-year concession to operate and extend the town and outlying districts' water and wastewater systems (Swann, 2000).

The Netherlands and Germany have been less open to private sector participation, claiming, without substantiation, that involvement of private sector will compromise water quality. However, Nuon NV, the Netherlands' largest utility, is pushing for at least partial privatization within the next two years, in order to help fund its global expansion plans in the water and energy sectors. Germany, with a more mixed system than the Netherlands, has made some modest progress in opening up its industry to private investment. In 2000, Berlin sold a 49% stake in Berlin Water to a consortium that included the German energy and industrial conglomerate RWE and Vivendi.

In Eastern Europe, capital needs are significantly greater, as infrastructure is both less developed and less well maintained. An important obstacle for attracting investment is the absence of supportive and comprehensive legal and regulatory frameworks. Nevertheless, the goal of joining the EU is driving countries to raise their environmental standards in line with the EU, and thus look for near-term solutions to modernize their infrastructure. One of the strategies for making the greatest progress has been privatization.

In 1999, the municipality of Sofia in Bulgaria signed a 25-year concession for water and wastewater services with International Water. The concession serves 1.2 million customers and includes investment of $150 million for upgrade and expansion over 15 years, which is expected to enable the municipality to meet EU standards. The concession is noteworthy not only in that it is Bulgaria's first major municipal infrastructure project that is privately financed, but that the project includes no municipal subsidies or state guarantees (Temple, 2000). The European Bank for Reconstruction and Development (EBRD) is providing about $42 million in debt and equity to the project.

The concession generated substantial competition, with eight international water companies seeking prequalification. The three prequalified firms submitted technical and financial proposals with the lowest proposed tariff submitted weighted as the key decision variable. With no independent regulatory authority in Bulgaria, the city will regulate the project company through the terms of the concession agreement, such as performance targets on coverage, leakage, water and wastewater quality, and other variables. A potential troublesome aspect of the concession, one that raises the potential for conflicts of interests, is the granting of the municipality a minority shareholding in the project company.

In March 2000, the city of Bucharest in Romania signed a 25-year concession with Vivendi Water for the operation of the city's water and wastewater services (D'Amato, 2000). The city faced significant problems: half of its water was lost in leakage, there were serious tariff collection problems, necessary maintenance and expansion required significant capital investment, and drinking and wastewater standards were below EU levels.

To address potential concerns about corruption and endless debates about the fairness of the selection process, the municipality and the International Finance Corp. (IFC), which assisted the city, designed a novel bidding scheme. Instead of awarding the concession based on a technical and a price proposal, Bucharest's scheme was based solely on the lowest price bid by the three firms who met the rigorous prequalification procedure. Consequently, the selection process was highly transparent and could not be criticized by any party for being subjective. The result: the winning bidder submitted a tariff rate that required over the first five years an initial increase in real terms of about 15% on average over the existing tariff, but over the life of the concession, the tariff is expected to be 35% lower than the existing tariff in real terms.

2.3   Asia

Asia has been another market where significant demands for water and wastewater infrastructure have created opportunities to introduce private sector participation. What is noteworthy is the adoption of private sector participation by governments, such as those in China, Vietnam, and South Korea, who do not have a long history of transferring responsibility for utility services to the private sector. Also noteworthy is the progress made in countries such as Thailand and the Philippines, which have been hit hard by the Asian currency crisis.

In August 1999, the Chengdu No. 6 water concession became the first official build, operate, and transfer (BOT) water project in China to reach financial close (Silk and Black, 2000). The consortia of Vivendi and Marubeni were awarded the concession under an open bidding system sponsored by the State Development and Planning Commission. The project cost is $106 million and involves two sets of raw water intake facilities, a water treatment plant, and a transmission pipeline.

The project is unique in that the performance and financial obligations of the Chengdu city government, which is the local offtaker for the project (i.e., is contracted to buy bulk water from the project company), are not guaranteed by either the Provincial, or the Central Government. The commercial lenders take comfort in the fact that political and project risks will be covered in part by the Asian Development Bank and the European Investment Bank, which are participating in the project. Another BOT water project will be getting underway in the summer of 2001 when the winning bidder for the $300-million Beijing No. 10 water treatment project is selected. The 20-year BOT concession involves building and operating a 500,000-cubic-meter capacity water treatment plant in Beijing.

In Vietnam, the Thu Duc water project marks the country's first BOT water project. Suez Lyonnaise des Eaux and Pilecon of Malaysia are the sponsors of the $150-million, 25-year concession for a water treatment plant in Ho Chi Minh City. In addition to commercial lenders, debt finance is also provided by the Asian Development Bank and the Malaysian Export Import Bank. The Malaysia Export Credit Insurance Company is providing political risk cover. All the obligations of the city's water supply company, the offtaker, are guaranteed by the Ministry of Finance.

Another new emerging water market is South Korea, which, as a result of fundamental institutional and political changes, has shown a greater interest in private participation in water and wastewater. Suez Lyonnaise des Eaux teamed with Samsung Engineering to win the contract to manage the Municipality of Pusan's wastewater treatment installations under a 18-year BOT concession. With two water service contracts, Vivendi Water has also been successful in entering the South Korean market.

The two side-by-side concessions for the two halves of the Manila (the

Philippines) Metropolitan Water Works and Sewerage System (MWSS) service territory was the largest water privatization transaction in the last five years. Awarded in 1998, the concessions benefited from a transparent bid process and strong government commitment. Also pushing the process forward were a 60% rate of unaccounted for water (leakage and uncollected tariffs) and a significant need for rehabilitation. As a result of intense competition, the government received a range of bids, all under the existing tariff rate. Though not without problems, the two concessions have weathered the Asian financial crisis, which caused both project companies to suffer serious foreign exchange losses.

In September 2001, Thames Water and CH Karnchang of Thailand signed a 30-year build, own, operate, and transfer (BOOT) concession with the Thai Government. The project, which includes the construction of a 320 million liter per day water treatment plant, trunk mains, reservoirs, and local main distribution system in West Bangkok, may be the largest in Asia. Funded by domestic Thai banks, the project has a take-or-pay agreement with the Provincial Waterworks Authority of Thailand. The country has a number of other bulk water supply and treatment projects in the pipeline in addition to plans to fully privatize both the Metropolitan Waterworks Authority and Provincial Waterworks Authority (Lau, 2001).

A major, but not unanticipated, setback in Asia occurred in early 2000, when the Malaysian government took over the project company that was granted a concession for the country's entire sewerage services in 1993. The fundamental tariff collection and asset valuation problems encountered by the 30-year negotiated concession highlight the value of competition, particularly in countries with not fully developed regulatory institutions. With an open and competitive tender, there would have been a greater likelihood that a number of key project risks, such as the magnitude of the investment requirements, the unique tariff issues, and the financial strength of the prospective operators, would have been revealed (Haarmeyer and Mody, 1997).

2.4   Latin America

Argentina is the undisputed leader in water privatization in Latin America. The country's program started in 1993 with the award of a 30-year concession for operation and expansion of the Buenos Aires water and wastewater systems to a consortium led by Suez Lyonnaise. The success of this contract provided support for similar concessions in the cities of Santa Fe and Cordoba, and the provinces of Mendoza, Buenos Aires, Corrientes, Formosa, and others. By one estimate, private companies today provide 70% of the population with drinking water (Abadie, 2001).

The only major setback in Argentina was in the province of Tucuman. After the concession was awarded in 1992, water tariffs doubled and a new provincial government was voted into office. Following a dispute over the quality of water supplied by the concessionaire, the provincial executive took control of the water regulatory body and revoked the concession.

Another instance demonstrating the highly political and high-risk nature of water infrastructure occurred in the Bolivian city of Cochambaba. In April 2000, after a concession for the city's water system was awarded, a violent riot broke out when the government approved a 35% rate increase. The riot resulted in six deaths, temporary imposition of a nation-wide emergency, and the private concession company's removal from the $200 million, 40-year project.

While Brazil, Mexico, and Bolivia have been successful in attracting private capital and management to their water and wastewater sectors, after Argentina, Chile's program stands out for putting in place the proper legislation and process to make privatization work. Like Argentina, Chile has had a strong privatization program in its other infrastructure sectors including energy, telecom, and ports. Over the past four years, Chile has sold interests in a number of its water and wastewater utilities including: ESVAL, provider of wastewater services for Valparaiso; ESSAL, supplier of water to Puerto Montt; EMOS, supplier of water to Santiago; and ESS-BIO, a utility for Concepcion.

3. NORTH AMERICAN HIGHLIGHTS: CONSOLIDATION INTEGRATES A FRAGMENTED MARKETPLACE

As is true around the world, the North American water industry is being driven by large capital spending requirements. Various industry and government estimates indicate that close to $1 trillion will be needed in the United States over the next 20 years to upgrade, expand, and replace aging municipal water and wastewater infrastructure ("Water Infrastructure Now," 2001). While such estimates are imprecise, there is little doubt that water supply is a capital-intensive and rising cost industry. In addition to the steady rise of basic replacement costs, tighter environmental standards and advances in water treatment technology—such as membrane filtration and ultraviolet (UV) disinfection—are increasing the complexity and cost of operating these systems.

The issue of how to obtain the money needed to meet these needs across a still fragmented industry has begun to move into more prominence on the agenda of the US Congress, which has not addressed water infrastructure funding since it established a federal grants program in the 1970s. Because of their fears of politically unpopular rate increases, municipal officials have pushed for another large dose of federal assistance—recommending a new program of more than $50 billion in water grant funding over five years.

While Congress will likely consider some additional funding for water systems, though not $50 billion, the private sector water industry opposes a broad subsidy program. Water and sewer rates remain low in the United States relative to other utility services, typically in the range of $15 to $30 per household per month, leaving room for prices to rise significantly without undue economic impact for most consumers. Economies of scale resulting from consolidation should also help to mitigate the need for rate increases (Beecher, 1999).

Executives of investor-owned water utilities consider infrastructure replacement a routine cost of doing business, one that can be efficiently managed within the existing capital and rate-making structures without federal intervention. Privately funded investment by the largest investor-owned water utilities has resulted in up-to-date infrastructure and an average net utility plant per customer over the life of the system of about $2,000 (Coy and Uspenski, 2001). While comparable figures for municipal utility investment are not available, they are likely much lower in most areas. As a result, rates for investor-owned utilities are at the higher end of the range noted above, but are still widely affordable.

With only about 15% of the US population receiving water from private sector providers, and less than 5% in the wastewater segment, there appears to be significant scope for expanding the private sector market as municipalities look for economic solutions to their water infrastructure spending needs. However, opposition from local utility workers and some politicians has continued to limit growth in the private sector market.

As the sale of utility water and wastewater assets remains politically controversial at the local level and, in the view of some municipalities, also less economically attractive, given the availability of low-cost tax-exempt financing received by municipally owned systems, much of the growth in private participation in the United States has been focused on the market for contract operations. The 17 largest firms competing in the utility out-sourcing services market reported 2000 revenues of $1.68 billion, 16% higher than in 1999 (Public Works Financing, 2001a). This market includes government and industrial market operations and maintenance (O&M) revenues, as well as the newer design-build contracting market. However, the market share for private vendors is still estimated at less than 5% of the more than $35 billion spent by government authorities for water and wastewater services, excluding capital costs. The level of outsourcing in the industrial market thus far is even smaller.

Table 1.1 highlights a number of the long-term contracts for larger-sized cities and the extent of cost savings offered through competitive contract-ing. The $400+ million Atlanta water privatization contract, which was signed in 1998, was fiercely competed given the high profile and large size of the contract. The 20-year O&M contract was won by Suez–United Water after the company dropped its final bid price by 21% in an open auction, with resulting estimated cost savings for the city of $400 million over the term of the contract, 45% below the city's operating budget. A reflection of the intense competition for contracts ("buyers market") has been a push by some municipal owners to have private contractors accept unlimited li-ability clauses. Fortunately, as a result of the growing experience, local gov-ernments and private contractors appear to be gradually getting more com-fortable with contract structures that will be mutually beneficial. By the summer of 2001, a number of large new privatization deals were in vari-ous stages of evaluation and bidding in various cities, including Phoenix, AZ; Tampa, FL; Stockton, CA; and New Orleans, LA; among others.

The shift toward the design-build method of project delivery in the US civil infrastructure market is also beginning to have a significant effect on the water and wastewater industry. The Design-Build Institute of Amer-ica (DBIA) is tracking over $4 billion worth of water-related design-build projects in the US municipal market scheduled for completion in 2001 and in the future. This compares to less than $1 billion worth of design-build projects in the DBIA database from 1999–2000 (Design-Build Institute of America, 2001).

The historic approach of using different vendors for design and construc-tion is increasingly shifting to an integrated approach, with a single entity taking responsibility for the contract, as municipal owners are seeing the

Table 1.1 Water Privatization Scorecard: Notable Long-Term Public–Private Partnerships

City	System Type	Plant Size (MGD[a])	Contract Term (years)	Winner, Year Won	Estimated Cost Savings for Customer
Atlanta, GA	Water	100	20	Suez, 1998	$400 million (45%)
Augusta, GA	Wastewater	46	10	OMI, 1999	$5 million
Cranston, RI	DBO[b] wastewater	23	25	Poseidon/US Filter, 1996	$35 million
Evansville, IN	Water	60	10	American Water Works, 1997	$8.1 million
Franklin, OH	BOT water	5	20	Earth Tech, 1997	20%
Fulton Co., GA	Wastewater	24	10	OMI, 2000	$4 million
Indianapolis, IN	Wastewater	250	14	Suez, 1994	$250 million +
Milwaukee, WI	Wastewater	550	10	Suez, 1998	$145 million (30%)
New Haven, CT	Wastewater	45	15	OMI, 1997	$53 million (30%)
Newport, RI	Wastewater	10	20	Earth Tech, 2000	$22 million (24%)
Rahway, NJ	Water	6	20	Suez, 1999	$32 million
Seattle, WA	DBO water	120	25	CDM/Azurix, 1997	$70 million (40%)
Springfield, MA	Wastewater	67	20	US Water, 2000	10%
Tampa, FL	DBO water	66	15	US Filter, 2000	$85 million (21%)
Tampa, FL	BOT desalination	25	30	Poseidon/Ogden, 1999	50%
Taunton, MA	Wastewater	8.3	20	US Filter, 1998	$62 million
Wilmington, DE	Wastewater	105	20	US Filter, 1997	$60 million
Woonsocket, RI	DBO wastewater	16	20	US Filter, 1999	$45 million

Source: Public Works Financing, 2000, 2001b.

[a]MGD = million gallons per day.

[b]DBO = design, build, operate.

increased speed of delivery and cost savings that can result. Taking the process a step further, some cities are beginning to add a long-term operations component to the design-build contract, hence the DBO designation.

One of the most prominent DBO contracts was the Tolt River water filtration plant contract awarded by Seattle Public Utilities to a consortium led by Camp Dresser McKee and Azurix in 1997. With a capital cost of about $65 million and a 15-year operating contract, plus two five-year renewal periods, the utility expects to save about $70 million in life cycle costs, about 40% below its benchmark estimate.

Consolidation remains a theme in the water sector, as US companies look to broaden their base of service capabilities and international firms look to enter the US market. After a booming $15 billion in acquisitions of US water companies announced in 1999, 2000 and early 2001 were relatively quiet. (Table 1.2 details transactions in the regulated utility sector, excluding the purchases of large equipment and services firms such as US Filter and Nalco Chemical.) In the past 18 months, the leading French water companies, Vivendi and Suez, have been busy integrating and digesting the earlier deals.

Suez's US operations include Nalco Chemical, United Water, and its equipment subsidiary Degremont, which are now being integrated under the ONDEO organization, while Vivendi has already integrated all of its US water operations under the US Filter umbrella, except for its minority ownership in Philadelphia Suburban. Vivendi Environnement, including waste, energy, and transport operations as well as water, completed its initial public offering (IPO) spin-off from parent Vivendi Universal in July 2000, and it is now planning a US share listing for the fall of 2001. Thames Water, the third largest global player, was itself acquired by German multiutility RWE in 2000, just before it closed on its own acquisition of the small New Jersey water utility E'town Corp., adding to its established presence in operations and water treatment equipment in North America.

While the industry at large remains fragmented, with some 54,000 municipal utilities and hundreds of local equipment and services providers, the top-tier private sector participants have grown both in size and breadth and now are capable of offering a wide range of products and services for municipal, industrial, and residential customers. As a result, the lines between traditionally separate segments in the water industry—regulated utilities, operating services, engineering and design, equipment and systems manufacturing, bottled water, and wholesale water management—have become increasingly blurred as the largest firms have moved to in-

Table 1.2 Major Water Utility Merger and Acquisition Transactions

Target	Status	Acquirer	Total Consideration	Number of Customers	Price per Customer
Dominguez	Completed	CWT	$66.4 million	40,000	$1,660
NEI	Completed	AWK	$700 million	513,000	$1,364
Aquarion	Completed	Kelda (UK)	$600 million	141,000	$4,248
United Water	Completed	Suez Lyonnaise (France)	$2.2 billion	605,000	$3,580
Citizens Water	Pending	AWK[a]	$835 million	305,000	N/A
San Jose Water	Cancelled	AWK	$480 million	216,000	$2,222
E'town Corp.	Completed	Thames (UK)	$957 million	238,000	$4,021
Utilities Inc.	Pending	Nuon NV (Netherlands)	About $400 million	230,000 (est.)	$1,739

Sources: Company data and Schwab Capital Markets L.P. estimates.

[a]AWK assumed no liabilities with the Citizens' assets, and the valuations are based on its assumptions of equivalent fair value; AWK estimates equivalent fair value at $550 million.

tegrate their service offerings to take advantage of cross-selling opportunities.

The newest segment to see some emerging private sector involvement is management of the water resource itself, offering the potential for commodity trading in the water business. While ownership and trading of water rights has established legal roots in the arid American West, pending water shortages and the need for better allocation and management of existing resources is generating more investor interest. Enron's Azurix subsidiary was the most well-known company to try a start-up water trading operation—an effort since abandoned—but many others are exploring the possibilities, and a viable market could develop in the next five years. Smaller firms, such as Cadiz Inc. in Southern California, are developing public–private partnerships in water storage and sales, and this segment should become more visible for investors in coming years.

4. CONCLUSION: THE FUTURE DIRECTION OF INDUSTRY TRENDS

Significant progress has been made worldwide in attracting private capital and management initiative to the water and wastewater sector. This progress has been made despite the serious political and regulatory risks associated with the sector. A highly competitive local and global industry of private developers, operators, and financiers has evolved that is capable of providing a range of services for projects of all sizes. The existence of multiple service providers ensures that, when governments organize open and transparent tenders for new greenfield facilities, or for the operation of existing facilities, they will attract highly qualified bidders. By facilitating risk identification and allocation, competition and strong contractual performance agreements have provided a solid foundation for achieving operations improvements, cost savings, and new infrastructure investment.

As significant as this progress has been, there is still tremendous work to be done in providing basic water and wastewater infrastructure where there is none and in putting existing operations on a commercial and sustainable basis. A major obstacle to achieving these ends more rapidly and comprehensively is the industry's high capital investment needs, generally low or subsidized tariffs, integrated monopoly structure, and significant exposure to regulatory and political risks. Together, these have limited private sector participation, competition, and innovation, all which could result in

significant improvements in the level, quality, and price of the sectors services. Consequently, further and significant progress in the sector will likely
come with the adoption of innovative options that restructure the sector
and more efficiently allocate risks and expose the sector to greater competition. Current efforts in the United Kingdom to split up asset ownership
from operations and to promote cross-boundary competition as well as common carriage are important beginnings of this industry transformation.

In addition to the potential for fundamental structural changes, the
water and wastewater sector will continue to be impacted by a number of
ongoing technological, competitive, regulatory, organizational, strategic,
and financial trends. Indeed, these may all serve to push industry structural changes. Two of these—technology and corporate strategy trends—
are worth raising as important signposts to the industry's future.

Like other industries, the water sector has been looking at the potential
for harnessing the power of the Internet to lower procurement costs and
increase the transfer of knowledge between buyers and sellers of products
and services. Although still at the experimentation stage, the promise is
significant, as e-commerce has the potential to lower transaction costs,
bring more efficient pricing, and consolidate the equipment and services
vendors market. Moreover, particularly in the western United States, the
Internet is already helping to broaden and deepen the market for water
rights, and thus provide water users and sellers with better information
and more options (Fillion, 2000; see also Chapter 22 in this volume).

The other important area where technology is beginning to have a visible impact on the sector, and will likely continue to do so, is desalination—
converting salt or brackish water to fresh water. Already a key supply
source in many wealthy water-scarce areas around the world, primarily
the Middle East and Caribbean, desalination is likely to become an increasingly affordable option for areas like the South and Western regions
of the United States, which suffer from growing scarcity of water supplies.
Significant advances in membrane technology and companies competing
for tenders to provide desalination equipment and operations have pushed
down the price of desalted water to $2–3 per thousand gallons, compared
with $4–7 only a few years ago. A growing market should provide for even
greater opportunity for innovation and cost reduction.

Company strategy provides a good indicator of industry trends, given
that companies will want to have the right strategy, structure, and service
offerings that best match the direction of the industry. Presently, two
broad models can be discerned across the worldwide water and waste-

water market. Under the first model, which is practiced primarily by European utility companies such as Vivendi, Suez Lyonnaise, RWE, Electricidad de Portugal, Endesa (Spain), and Enel (Italy), firms provide services across multiple utility sectors—water, gas, electric, waste, and telecom. Often referred to as the multiutility model, this approach appears to be especially attractive in developing country markets (e.g., there are a number of multiutility concessions in Africa that bundle water, gas, and electric) and in markets where these utility sectors have not been fully opened up to competition. In addition, these firms are beginning to have success in offering multiutility outsourcing services to industrial customers. The relatively faster, and potentially more profitable, development of industrial utility outsourcing contracts compared to municipal contracts is helping to draw more competition and expertise into the overall market.

In the other model, which is more prevalent in the United States, and to some extent in the United Kingdom, the strategy of companies is to focus on one utility sector and either focus on one segment of the value chain or provide an integrated suite of services within that sector. Driving this strategy is the fact that, in these two countries, the gas, electric, and telecom sectors are deregulated and competition is forcing companies to specialize and focus on core businesses where they have a competitive advantage. Shareholder pressures are also pushing this trend, as investors reward "purer plays" whose business they can better understand, and discount conglomerate-type companies whose "sum of the parts" often exceed the market value of the company.

The evolving structure of the US power market and the breakup of the integrated utility structure is one example of how market forces can reshape an industry (Haarmeyer, 2000). The company transformation of Vivendi provides an important example of restructuring within the water industry. Earlier in its corporate life, the company, named Compagnie General des Eaux, focused on providing infrastructure services, primarily in water. After a troubled expansion into and exit from the real estate and construction markets in the 1980s and 1990s, today, many acquisitions later, the renamed Vivendi Universal is the third largest media company in the world. The company's subsidiary, Vivendi Environnement, remains the leader in the global water industry, though it is now much larger, with $12 billion in sales and the broadest portfolio of products and services in the water industry. The parent company, Vivendi Universal, has recently spun off 28% of Vivendi Environnement and is expected to further reduce its stake given the limited synergies between the infrastructure and the media businesses. Thus, the leading water service provider appears to be transforming itself into a more focused and less conglomerate-like company. This structure should become particularly advantageous as the utility industry worldwide continues down the path of greater exposure to competitive forces.

REFERENCES

Abadie, Fernando E. (2001). "Privatizations in South America: The Challenge of Water," *Project Finance International,* July 25.

Beecher, Janice A. (1999). "Consolidated Water Rates: Issues and Practices in Single-Tariff Pricing," A Joint Publication of the U.S. Environmental Protection Agency and the National Association of Regulatory Utility Commissioners, Washington, DC, September.

Coy, Debra G., and Christine Uspenski (2001). *Water Utility Quarterly,* Schwab Capital Markets L.P., May 15.

Crandall, Robert, and Jerry Ellig (1997). *Economic Deregulation and Customer Choice: Lessons for the Electric Industry.* Fairfax, VA: Center for Market Processes.

D'Amato, Erik (2000). "Transparent Waters," *Impact,* International Finance Corp., Vol. 4, No. 2, Spring-Summer.

Design-Build Institute of America (2001). "Representative List of U.S. Municipal Water and Wastewater Design-Build Projects," Washington, DC, January 25.

Fillion, Roger (2000). "Water, Water Everywhere: A Trickle of Websites Promotes Swapping Water Rights," *Business 2.0 Magazine,* July 25 (business2.com).

Haarmeyer, David (2000). "The U.S. Power Market in Transition: The Search for Structure," *Infrastructure Journal,* March/April.

Haarmeyer, David, and Ashoka Mody (1997). "World Water Privatization: Managing Risks in Water and Sanitation," *Financial Times Energy Asia Pacific.*

Izaguirre, Ada Karina, and Geetha Rao (2000). "Private Infrastructure," *Public Policy for the Private Sector,* World Bank, Washington, DC, September.

Lau, Minerva (2001). "Tapping Water in Thailand," *Project Finance International, Asia Pacific Review,* July.

Metcalfe, Steve (2001). "Lock Out: Glas Cymru Structured Finance," *Project Finance,* May.

Public Works Financing (2000). PWF's 2000 International Major Projects Survey, October.

Public Works Financing (2001a). PWF's Fifth Annual Outsourcing Survey, March.

Public Works Financing (2001b). June.

Silk, Mitchell, and Simon Black (2000). "Case Study: Chengdu No. 6 Water Plant," *Project Finance International,* February 23.

Standard & Poors (2001). "Market Liberalization and EU Directives Channel European Water Industry Development," New York, July 5.

Swann, Gilbert (2000). "Water Win Financing for Santa Maria," *Project Finance International,* February 23.

Temple, Richard (2000). "Sofia Water: A Signed and Sealed Water Project," *Project Finance International,* August 23.

"Water Infrastructure Now" (2001). Report from the Water Infrastructure Network, February.

ADDITIONAL READING

Haarmeyer, David, and Ashoka Mody, "Financing Water and Sanitation Projects: The Unique Risks," *Viewpoint,* World Bank, Washington, DC, September 1998.

Part II

The Regulatory Framework

Chapter 2

Comparative Analysis of Regulation

Water and Other Network Industries

Yogita U. Mumssen and Brian Williamson

1. INTRODUCTION

We consider the regulation of network industries—in particular, the regulation of the water and wastewater sector—in two parts. In the first, we discuss the principles of good regulation, drawing on selective examples of experience from the United Kingdom, the United States, and Australia. The second part reviews existing high level or "governance" arrangements in the United Kingdom, the United States, and Australia. Our analysis starts from the basic fact that network industries are characterized by increasing returns to scale and long-lived investments. These characteristics introduce two related concerns: protection of consumers from monopoly abuse—in particular, high prices and/or low quality services—and

This paper draws on the report "Economic Regulation of Network Industries" for the New Zealand Treasury, which was published as Working Paper 00-5 (http://www.treasury.govt.nz/workingpapers/2000/00-5.asp).

Yogita U. Mumssen is Senior Consultant, specializing in economics, with Stone and Webster, an international consulting firm. She was formerly with National Economic Research Associates and PricewaterhouseCoopers in London. Brian Williamson is Senior Consultant at National Economic Research Associates, a British consulting firm, specializing in utility regulation. Formerly he was at the New Zealand Treasury as Chief Economist.

protection of investors from opportunistic behavior by governments and regulators (recognizing the sunk nature of infrastructure assets).

However, meeting these concerns involves trade-offs that any system of regulation must address: first, maintaining incentives for cost reduction while ensuring that prices are not too out of line with costs; second, ensuring that firms have a reasonable assurance of cost recovery—which requires regulatory commitment—while allowing regulation to respond to new information.

UK-style price cap regulation set out with the aim of promoting cost reduction and innovation with some success, but has failed to deliver the degree of commitment and stability necessary to sustain innovation and incentives for investment. In the United States, the Fifth and Fourteenth Amendments to the Constitution have led to commitment and "due process," the *Federal Power Commission Hope Natural Gas Company* (1944) decision provided investors with a reasonable assurance of cost recovery, and the Administrative Procedures Act established rules that must be followed in making determinations. More recently, "incentive-based" mechanisms, known as Alternative Regulatory Plans, have been introduced; these plans build on the existing institutional base. Australian regulation of the network industries has parallels with the approach in the United Kingdom, particularly reliance on price-cap methodology. However, lessons were learned from the UK experience. In particular, in contrast to the high degree of discretion accorded regulators in the United Kingdom, in Australia, legislation and industry codes are highly prescriptive, narrowing the scope for discretion and opportunism. Australia may therefore have "short-circuited" the evolution of institutional arrangements seen in the United States, ensuring greater regulatory commitment and due process than in the United Kingdom. However, time and further privatization will provide a clearer test for the Australian model.

2. MONOPOLY AND REGULATION

Network industries are characterized by increasing returns to scale and long-lived assets. Increasing returns to scale make monopoly provision efficient, while long-lived assets make repeated competition for the market (as a substitute for competition within the market) difficult to implement efficiently, particularly in relation to long-run investment incentives. Traditionally, the problem of monopoly has been seen as one of excessive prices, allocative inefficiency, and lack of incentives for provision of appropriate service quality. However, regulation cannot resolve all of the problems of monopoly, and in fact, may undermine incentives for efficient operations. Regulation can also be expected to introduce incentives to skimp on quality, since a reduction in quality represents a profit opportunity for a monopoly whose prices are capped (Vickers and Yarrow, 1988). Regulation may nevertheless be implemented to limit potential monopoly abuse. Achieving the objectives of regulation at least cost requires recognition of the inherent constraints on achieving good outcomes, and recognition of the problems regulation itself introduces.

2.1   Alternative Forms of Regulation

Discussion of alternative forms of regulation is sometimes characterized by a false dichotomy between rate of return (cost plus) regulation and "price-cap" (incentive) regulation. In practice, the alternatives are not as distinct as they may appear in theory. Price caps do not remain in place indefinitely—and incentives for cost reduction are blunted to the extent that price caps are reset in line with costs. In addition, neither form of price control in and of itself provides investors with an assurance that they will recover their costs over the life of an asset. An alternative to explicit regulation is reliance on general competition law and the threat of regulation (the New Zealand's system of "light-handed" regulation is an example of this approach). However, the threat of regulation (if successful in constraining monopoly abuse) can be expected to introduce incentives similar to those of explicit regulation. A threat of regulation may also prove unsustainable as a long-term means of control (explicit price controls are now proposed for some services in New Zealand).

2.2   Problems of Regulation

Regulation of monopoly cannot achieve first-best outcomes, due to information constraints and the lack of regulators to credibly commit not to expropriate sunk investments (though some regimes are more successful at providing a credible commitment than others). Recognition of the following trade-offs, which such constraints introduce, is essential to the design and operation of sound regulatory institutions:

1. Maintaining incentives for cost reduction while ensuring that prices are not too far out of line with costs (recognizing that firms will earn "information rents").

2. Ensuring that firms have a reasonable assurance of cost recovery, while limiting the scope for "gold plating" (due to guaranteed returns on investment).
3. Responding flexibly to new information while limiting the scope for regulatory opportunism.

In addressing these fundamental trade-offs, careful attention should be paid to incentives for efficiency, and institutional design that promotes the predictable and sound evolution of regulatory methods.

2.2.1   Long-Term Incentives for Efficiency

In the long run, the costs of network utilities are determined to a large extent by the efficiency of investment, and its cumulative effect on the capital stock of the industry. Regulation (or the threat of regulation) introduces the risk of opportunism whereby regulators may ignore sunk costs in setting prices (Greenwald, 1994). Experience in the United Kingdom illustrating the problem of opportunism is summarized in Box 1.

In commercial relationships involving sunk assets, the potential problem of opportunism is often addressed via long-term contracts. However, it is more difficult for governments and regulators to commit since they are in a position to change the rules and/or act retrospectively. Institutional arrangements laying down rules and "rules for changing the rules" are therefore required, which facilitate commitment. For example, in the United States, regulatory commitment ultimately derives from the written Constitution.

Both theory and experience in the United Kingdom and elsewhere point to the fact that the solution to the problem of commitment cannot rely on giving discretion to individuals—irrespective of how well intentioned the individuals may be (Kydland and Prescott, 1997). Indeed, the problem of achieving commitment has long been recognized.

Various institutional approaches have been implemented to improve commitment and predictability in the regulation of network utilities. Box 2 gives examples from the United States, the United Kingdom, and Australia.

2.2.2   Short-Term Incentives for Efficiency

To provide incentives for short-term efficiency, prices may be fixed so that the regulated firm profits from cost reduction. However, prices and costs will inevitably diverge, leading to pressure to reset prices in line with costs, thereby undermining incentive to reduce costs (Williamson, 1997).

Box 1
Experience of Regulatory Opportunism
in the United Kingdom

The following figure shows the impact on the present value of UK utilities from a series of "rule" changes. Initial reviews, up to the 1995 Regional Electricity Company (REC) price review, more or less validated the net present value (NPV) of the price controls set at privatization. The first significant rule change was in 1995, when the proposed price caps for regional electricity companies were revised following a take-over offer. A series of "rule" changes followed, including the windfall tax in 1997, which extracted £5.2 billion from the privatized utilities. The impact on the water industry is based on the 1999 draft price determination. (Note that the values given are calculated changes in the NPV of the business, not observed changes in market value that reflect many factors including anticipated changes.)

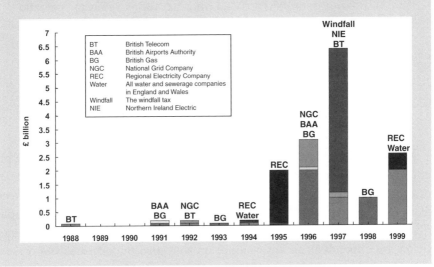

In practice, price caps are periodically reset (typically every five years in the United Kingdom). Periodic adjustment means that incentives are strongest immediately following a price control review. Recognizing this problem of periodicity, the Office of Water Services (Ofwat, the UK water regulator) introduced a rolling five-year mechanism for the retention of savings in 1999 to ensure that incentives for cost reduction were constant.

**Box 2
Commitment in the United States,
the United Kingdom, and Australia**

In the United States, the *Federal Power Commission v. Hope Natural Gas Company* case of 1944 established that utilities should earn the opportunity cost of capital applied in other comparable uses—in order to attract investment. In addition, the Administrative Procedures Act (1946) and other legislation establish administrative rules and procedures that must be followed in making regulatory determinations.

In contrast, in the United Kingdom, commitment mechanisms are relatively underdeveloped. However, even in the United Kingdom, legal precedent may result in the narrowing of regulatory discretion. A decision by the Northern Ireland electricity regulator to disregard part of the decision by the Monopolies and Mergers Commission (MMC) recently went to judicial review. The Appeal Court upheld the MMC decision in full.

In Australia, lessons were learned from the UK experience. Legislation and industry codes are highly prescriptive, narrowing the scope for regulatory discretion compared with the United Kingdom. The approach in Australia may have effectively "short-circuited" the evolution of institutional arrangements seen in the United States.

Attempts have been made to get around a trade-off between passing on efficiency gains and incentives for efficiency via yardstick competition. In practice, it has proved difficult to isolate genuine efficiency differences from cost differences due to varying circumstances of individual companies (Jones, 1999). In theory, the efficiency properties of yardstick competition are dependent on an assumption that also undermines the practical value of the technique, namely, as Shleifer (1985) noted:

> The regulator commits himself not to pay attention to the firms' complaints and to be prepared to let the firms go bankrupt if they choose inefficient cost levels. . . . Unless the regulatory can credibly threaten to make inefficient firms lose money (or, alternatively, can prove in court that firms chose to be inefficient and that their practices were imprudent), cost reduction cannot be enforced.

Clear rules are required so that companies know what incentives are on offer, and can respond to them with a degree of confidence. Objective data and replicable formulas may be preferred, even if the formulas are not "ideal." However, some aspects of regulation will nevertheless be reliant on less robust methods, and evolution of methodology over time may be desirable. To ensure that regulation is transparent and predictable where judgment and evolution are involved, efficient, clear and agreed-upon procedures are necessary. Clear objectives of regulation are also required.

2.3   Principles of "Good" Regulation

Actual objectives of regulation in different parts of the world may diverge from what we consider "good objectives." However, in most cases, we think that governments and their regulatory regimes do attempt to meet the "long-term" and "efficiency" criteria, although some are more successful at doing so than others.

When we discuss *long-term* objectives, this implies that regulators should not be able to forfeit consumers' long-term interests (e.g., investment to secure a reliable supply in the future) in favor of short-term interests (e.g., price reductions during the regulator's term of office). When we discuss *efficiency,* this implies that the achievement of efficiency in operations and investment is a key objective that overlaps with other objectives (e.g., competition, where introduced, should be efficient).

Regulation needs to provide monopolies with proper *incentives* in order for them to aspire to the objectives of meeting long-term customer needs through efficient investment and operations. Companies will only have proper incentives if regulators can demonstrate some degree of *commitment* and *stability.* Commitment and stability, in turn, come from processes that require openness, transparency, consistency, and accountability. Thus, regulation should be:

- *Open.* A process that allows interested parties to put forward their views and be challenged by others, providing maximum access to relevant information;
- *Transparent.* The demonstrable use of available information where a regulator reaches decisions on the basis of observable data sources and a replicable (mechanistic) formula, in order to minimize the scope for discretion;

- *Consistent.* Whereby the regulator uses a stable set of decision criteria to ensure that when change in regulatory methodology/practice is required, this can be done in a manner that is as acceptable and predictable as possible; and
- *Accountable.* Decisions should be reasoned and justified by reference to defined criteria (such as a list of regulatory objectives), so that they can be effectively challenged. This provides an incentive to reach good decisions.

The *independence* of a regulatory body is often cited as one of the most important characteristics of a regulator for ensuring long-term consumer interests and efficient business decisions. Where there is a perceived risk that there will be political pressure to tighten the deal for private investors (which would increase the cost of capital), good regulatory governance is critical; a regulatory agency must not be able to be dismissed or deprived of funds simply for making politically unpopular decisions.

However, independence is difficult to achieve. Regulators must be appointed, and therefore regulation is inherently tied to the political process. It is also questionable whether absolute independence is desirable. Ultimately, the regulatory agency's decisions must be *legitimate* in a democratic society. The requirement to balance the need for regulatory independence with the need to demonstrate legitimacy is critical. Also, by specifying the regulator's duties and powers in law, the regulator can more easily be held accountable for its actions. This will also help ensure some degree of independence from political pressure.

3. INTERNATIONAL EXPERIENCE

The following sections examine experience of economic regulation in the United Kingdom, the United States, and Australia in relation to the principles of good regulation discussed above. Specific experience in the water sector is mentioned where appropriate. However, experience in other sectors, such as electricity and gas distribution, is also relevant to the future development of regulatory regimes in the water and wastewater sectors.

3.1   UK Experience

In the United Kingdom, network industries underwent wide-scale privatization in the 1980s and 1990s. In conjunction with the privatizations, a regulatory regime was established. This regulatory regime is therefore relatively young (e.g., compared to that of the United States). For the majority of the network industries in England and Wales, the regulatory regimes are characterized by:

- Single-sector, national regulatory agencies (although a multisector regulator has been appointed for the gas and electricity sectors);
- Regulatory bodies headed by a single director (although the Utilities Act 2000 established a board for the new Gas and Electricity Authority);
- Price-cap regulation for networks (known as "RPI–X"); and
- Periodic (three- to five-year) price reviews.

Sector-specific Acts of Parliament (e.g., the Electricity Act 1989) are the formal contract between the privatized companies, their regulators, and consumers. In addition to the sector-specific Acts and the individual company licences, there are other acts that legislate the behavior of the privatized companies. These include the Competition Act 1998 and various European Union (EU) Directives (e.g., Bathing Waters Directive).

3.1.1 Evolution of Water and Wastewater Sector

Prior to the Water Industry Act 1973, the water and sewerage industry was highly fragmented, and mainly in public hands. The Water Act 1973 reorganized the water and wastewater sector in England and Wales into ten publicly owned water and sewerage authorities and 29 statutory water companies. However, the industry suffered from underinvestment and required substantial new investment to bring water quality up to the standards prescribed by European Commission (EC) directives.

The Water Industry Act 1989 restructured the industry into separate utility and regulatory functions and provided for the privatization of the industry as vertically integrated regional monopoly companies. Water and sewerage companies were floated on the stock market and economic price-cap regulation was applied by the Office of Water Services (Ofwat).

Over time, a number of mergers have led to consolidation in the industry. Following the five-year price determination announced in 1999, the value of publicly listed water companies declined significantly. Companies in the industry have sought restructuring opportunities. Two examples may point to the future evolution of the industry away from the publicly listed company model that is currently dominant. In November 2000, RWE acquired Thames Water PLC, and in January 2001, Ofwat agreed to allow Glas (a company limited by guarantee and owned by a body of members who have no financial interest, rather than shareholders, and which is wholly debt-financed) to acquire the assets of Dwr Cymru, the Welsh water company. In addition, proposals for the extension of (limited) competition in the sector have received impetus from the introduction of the Competition Act 1998, which came into effect on March 1, 2000.

3.1.2   Evolution to Current Regulatory Regime for the Water Sector

The main objectives of *privatization* were to raise finance, to improve the efficiency of the enterprises (through competition where possible and through incentive-driven regulation where not), to shift control to the private sector (which had a stronger bargaining position with the unions and could impose more credible budgetary limits), and to widen share ownership. Economic regulation was introduced to limit potential monopoly abuse. In 1986, Stephen Littlechild reported on the regulation of the water industry and noted that the industry is largely a natural monopoly and would require ongoing regulation. He proposed price-cap regulation based on the retail price index plus an adjustment factor (RPI–X), with "Periodic revisions in X, based on the yardstick performance in the whole water industry" with a uniform industry X factor, and incentives for efficiency via "fostering competition in the capital market—in particular, by maintaining the threat of takeover." In practice, individual X factors were set on a company-by-company basis, and while significant gains in terms of efficiency and quality have been achieved in the water industry, failures of the current regulatory regime have increasingly come to light:

- Lack of specification of regulatory rules and procedures, leaving the regulators with wide discretion on how they adjust the basis on which companies are rewarded, leading to regulatory opportunism and ongoing uncertainty;
- A lack of clear precedent from the appeals process via the Monopolies and Mergers Commission (MMC, now known as the Competition Commission); and,
- Lack of forethought of the complexities of price-cap adjustment at periodic reviews, especially in relation to incentives and requirement to consider cost-levels.

The rules and procedures of regulation are not well defined for elements of network industries that remain regulated. While the obligations of the regulator are stated (although vague), the regulators are not required to set out the basis upon which they arrive at a decision. As a consequence, regulation in the United Kingdom is less subject to judicial review, and decisions are left unexplained.

Discretionary behavior is manifested by the lack of clear precedents provided by past MMC decisions. However, UK legal precedent may be narrowing in terms of regulatory discretion. A decision by the Northern Ireland electricity regulator to disregard part of the decision by the MMC recently went to judicial review. The Appeal Court upheld the MMC decision in full. In addition, two water only companies, Mid Kent Water and Sutton and East Surrey Water, successfully appealed the 1999 price determinations. In the reports on the appeals the Competition Commission (2000) noted, for example, "We are not persuaded that the methods adopted by either the Director or MKW/SESW for assessing the level of funding required are satisfactory." Further development of regulatory methodology is an anticipated outcome of the appeal.

The Department of Trade and Industry (DTI) released in July 1998 a consultation paper *Modernising the Framework for Regulating Utilities,* which, among other things, proposed a more transparent framework, with greater emphasis on ensuring predictability and consistency of regulatory decisions. However, the Utilities Act introduced in July 2000 arguably increased rather than reduced regulatory discretion (while water was removed from the coverage of the Act, similar provisions may be introduced via the proposed Water Bill).

3.1.3   Objectives of Regulation

The duties and functions of each of the utility regulators in England and Wales are set out in the primary legislation relating to that sector. In all cases, the duties of the regulator are shared with the Secretary of State. The following key regulatory duties—or objectives—apply to most of the network industries in England and Wales: protect consumer interests; meet reasonable demands; ensure adequate financing; and promote ("secure" or "facilitate") competition. The duties of each sector regulator in England and Wales are outlined in their respective Acts of Parliament, including the Electricity Act 1989 and the Utilities Act 2000, which covers electricity and gas.

Regulators in the United Kingdom have sought to provide incentives for efficiency by:

- Setting of objectives for efficiency (e.g., through efficiency comparisons, productivity comparisons, and price caps);
- Systematic regulatory treatment of costs (e.g., through systematic classification of costs, Regulatory Accounting Guidelines in the water sector, investment approval procedures);
- Some retention of savings (e.g., capital expenditures [capex] and operating expenditures [opex] savings); and
- Relating of expenditure to defined outputs and service standards.

However, these building blocks are still neither well defined nor predictable, and a clear and consistent methodology by which regulators provide incentives for efficiency has not been established across the sectors.

3.1.4   Process: Rate-Making, Appeals, and Arbitration

Under the system of price-cap regulation in the United Kingdom, the control of tariffs usually operates by limiting the weighted average annual increase in charges for a basket of principal outputs to the annual change in the retail price index (RPI) plus an adjustment factor, often known as X (or K for the water and sewerage sector). Each company's X factor is set for three to five years.

Additionally, there are, in some cases, cost pass-through provisions that allow X to be altered each year in certain circumstances. The cost pass-through arrangements were designed to cater for the significant uncertainties that prevailed at the time of the UK privatization and, in the case of certain industries, relate to the large capital expenditure programs being implemented to meet higher quality (and environmental) standards being introduced over the next decade.

The sector regulators each follow their own process and timetable for their respective price reviews. While many view the water industry regulator as having followed the most open and well-managed price review procedure of any of the sector regulators in England and Wales, the approach arguably falls well short of international best practice (Williamson, 2001). For example, the previous water sector regulator, Director General Ian Byatt of Ofwat, removed the "glide path" intended to pass on past efficiency gains to consumers instituted in the 1994 price review. In addition, the approach to final determinations at the 1999 review was far from transparent, as the following quote from the Director General Ian Byatt ("Byatt rejects," (1998, p. 3) illustrates: "Ofwat will not be releasing the software to companies. It is for the companies to decide whether to accept our price limits, not to replicate our work."

There are limited requirements on regulators in terms of openness and in terms of the need to justify decisions with reasoned arguments. The review process does not involve open hearings or cross-examination of expert evidence. In addition, while a company can appeal the regulator determinations to the Competition Commission, appeals open up the entire determination and are not limited to a single aspect of methodology.

3.1.5   Regulatory Characteristics

Independence: The sector-specific regulators of network industries in the United Kingdom are set up as independent regulatory bodies, and are, in fact, often referred to as "independent regulators" in the popular press. To ensure a degree of independence from politics, regulators are not part of a Government Ministry, and are not elected but appointed for fixed terms. They can only be removed from office in the case of incapacity or misbehavior. Invariably, the network industry regulators are not wholly independent. They are appointed (and in some cases reappointed) by their respective sector Ministers. They must operate under relevant laws, which are established by Parliament. They are also subject to political pressure, and due to such pressure, may consider issues that they may not otherwise have considered (e.g., increased emphasis on leakage issues after the Labour Government's Water Summit declared it an important topic).

"Transparency, Consistency, and Predictability": In announcing the Government's review of utility regulation on June 30, 1997, Margaret Beckett, then President of the Board of Trade, set out objectives for the review that included the following: "Guiding principles must be transparency, consistency, and predictability of regulation." However, some of the original proposals were dropped, and the Utilities Act 2000 arguably reduced the transparency, consistency, and predictability of regulation.

Commitment: Since the establishment of the regulatory regime for network industries in the late 1980s, commitment has not been demonstrated, and regulatory rule changes have been prevalent. The first instance that marked a regulatory rule change was in 1995, when the regional electricity companies' 1994 price adjustments were revisited by the regulator immediately after they had been announced, though not yet implemented, because ensuing mergers suggested further scope for savings. Rule changes also took place in the UK gas industry. Houston (1999) estimates that the impact on the NPV of regulated business from regulatory rule changes in the United Kingdom since privatization amounts to over £12 billion. Note that the concern is not that regulation has necessarily become tougher, but that there have been instances where regulators have reneged on prior commitments and/or have made retrospective rule changes, demonstrating opportunistic behavior that damages company incentives to invest (and thus damages long-term consumer interests).

Accountability: In future, regulation in the United Kingdom may provide greater scope for judicial review. For example, the Human Rights Act 1998 came into full effect in October 2000 and includes the right to a fair hearing under Article 6, and the right not to be arbitrarily deprived of one's possessions under Article 1. This may open up new avenues to challenge regulators.

3.2   US Experience

Throughout the history of the United States, there has always been both public and private ownership of network industries: gas and electric utilities are, for the most part, investor-owned, while municipal ownership is more common in the water sector. Of course, there are exceptions: there are still large numbers (i.e., hundreds) of small, municipal utilities in the gas and electricity distribution businesses, and private sector participation is on the rise in the water and wastewater sector.

For the most part, regulation of the network industries is divided between the federal government and the individual states. The federal regulators, following Article I, Section 8 of the US Constitution, deal with those network operations that can be called "interstate commerce" (e.g., interstate gas transportation and high-voltage electricity transmission). The states regulate the local distribution of network services for electricity and gas. In the water sector, there is also the local/municipal dimension. For government-owned water and wastewater utilities, the local government usually determines water rates, while investor-owned utilities are mostly regulated by the state Regulatory Commissions.

The main body of law in the United States by which all institutions must abide is the US Constitution. The Constitution, in the form of the Interstate Commerce Clause, enumerates specifically the economic powers delegated to the federal government. All other government powers related to regulation are reserved for the fifty individual states under the police powers delegated to them. Most state commissions regulate the health and safety of the state's citizens and have the authority to issue licences, franchises, or permits for the initiation of services, and to prescribe or change rates, and prescribe uniform systems of accounting and publication of annual reports.

3.2.1   Evolution of the Water and Wastewater Sector

Private companies originally dominated the water sector in the United States. Throughout the eighteenth and nineteenth centuries, many growing cities chartered private water companies to own and operate their systems. Today, government-owned entities furnish over 85% of publicly supplied water, and possibly 95% of publicly supplied wastewater services (Hyman et al., 1998). Most of these entities are small, municipally owned and operated entities.

One of the main reasons for the rapid increase in municipal ownership of water and wastewater utilities was private companies' unwillingness to expand into low-income areas and sparsely populated areas. Therefore, many cities felt that they had to provide the services themselves. Increased municipal ownership led to rapid expansion of systems to unserved and underserved areas. Improvements in terms of water quality and wastewater services were made. These expansions and improvements provided by government entities had a price: non-cost-recovering tariffs. However, unlike private water companies, municipalities sought to internalize the positive externalities of improved water and wastewater services, for example, improved health and hygiene, and were able to subsidize water services from other municipal, state, and federal revenue sources.

However, some private water companies have thrived in the United States for over 100 years. Two large holding companies, American Water Works and United Water Resources, account for approximately 60% of the revenue of the top 10 private water companies (Hyman et al., 1998). It is very likely that the United States will see increased private sector participation in water and wastewater supply in the coming years. Recent initiatives to make it more attractive for municipalities to sell off their water and wastewater utilities, as well as the increased pressure on government-owned entities to reduce capital spending, cut operating costs while increasing maintenance, and meet new environmental standards may make increased privatization a reality.

3.2.2   Evolution of Utility Regulation in the United States

There was little utility sector regulation after the Revolutionary War and War of 1812. Competition, through franchises granted locally, was relied on to protect consumer interests. The majority of utilities—across all sectors including water and wastewater—were privately owned and operated. It was quickly apparent that some form of regulation was needed. Therefore, common law was adapted from England. Monopolies were seen to be against the public interest, and certain occupations were labeled "common callings." Court decisions declared how the firms should be regulated.

Common law was expensive, and "regulation" was inflexible and slow. In reaction, federal and state regulatory commissions started to develop in the period immediately before the 1860s. The first US federal regulatory commission, the Interstate Commerce Commission, was established in 1887 and the first two state commissions, New York and Wisconsin, were started in 1907. However, the regulatory commissions basically followed the rate-making "formula" suggested by court rulings.

The early history of regulation in the United States was characterized by notorious accounting abuses, including overstated expenses, unverifiable investments in plant and equipment, a lack of separation between utility and nonutility businesses, and overcapitalization (Phillips, 1993). Such abuses were effectively ended with the adoption by the federal government in 1938 of the Uniform System of Accounts for the regulation of the natural gas industry. Since that time, various forms and amendments of the Uniform System of Accounts have been the basis for all federal and state utility regulation in the United States.

The *Hope Natural Gas* case in 1944 marked a clear shift away from court-dictated regulation and complicated questions of asset valuation toward a clearer set of standards for the state regulatory commissions—and hence less reliance on the courts. The role of the courts, it declared, was to interpret whether federal and state statutes were unconstitutional—not to determine how regulators should regulate. This became known as the "end result" criteria. In addition, in 1946, the Administrative Procedures Act (APA) was passed. The APA set out the administrative rules and procedures (*not* pricing methodology) that would need to be followed in making regulatory determinations.

Today, all 50 states and the District of Columbia have regulatory commissions. In addition to the state commissions, the Federal Energy Regulatory Commission (FERC) regulates interstate commerce by gas and electricity utilities, and the Federal Communications Commission (FCC) regulates interstate telecommunications. The present regulatory structure in the United States, both on a federal and on a state level, is the end product of a myriad of statutes and court decisions, such as the *Hope Natural Gas* decision and the APA.

3.2.3 Regulation of the Water and Wastewater Sector

The government—at all levels—plays a large role in the regulation of the water and wastewater sector in the United States. At the federal level, the government sets health and environmental standards for the industry. The two main pieces of federal legislation are the Safe Drinking Water Act, which sets maximum levels of pollutants for drinking water, and the Clean Water Act, which sets the allowable level of effluent discharged into waterways. The Environmental Protection Agency (EPA) is the federal body responsible for enforcing these laws.

State regulatory commissions often regulate the investor-owned water and wastewater utilities. This includes monitoring standards of performance and rate setting. Municipal governments "regulate" most publicly owned water and wastewater utilities. Regulation by municipalities, often the mayor's office or the city board/council, largely involves approving/disapproving rate increases proposed by the utility.

Thus, there is a clear distinction in the way that investor-owned water utilities are treated and government-owned water utilities are treated. Investor-owned utilities must follow a structured, regular regulatory program enforced by the state regulatory commissions, whereas government-owned entities (except for when they serve customers outside of their own jurisdiction) are not regulated *per se,* but instead are monitored by the local government. The local government sets customer tariffs per the guidance and appeals of the utility itself.

3.2.4 Objectives of Utility Regulation in the United States

The basic regulatory objectives that have been employed by the state and federal regulatory commissions in the United States (and observable through various precedents) include (Phillips, 1993): preventing excessive (monopoly) profits and unreasonable (inequitable) price discrimination; assuring adequate earnings to the utility (including allowing cost recovery unless unreasonably excessive); ensuring service provision to the maximum number of customers; promoting development and industry in certain regions; ensuring maximum safety; and ensuring management efficiency.

Regulators in the United States have attempted to provide incentives for investment and operational efficiency by:

- Setting unbiased objectives for total efficiency, for example, through total factor productivity trends, and increasingly, through price-cap regulation;
- Establishing systematic regulatory treatment of costs (e.g., National Association of Regulatory Utility Commissions [NARUC] standardized accounting);
- Relating expenditure to defined outputs and service standards; and,
- Assuring a reasonable certainty of cost recovery.

In addition, each state generally has a public service law, and possibly energy, water, and other sector laws. However, the public service laws are often much broader than the respective sector acts of Parliament in the United Kingdom (e.g., Water Industry Act 1991), and therefore, leave more room for interpretation. For example, under New York State Public Service Law, Article 1, Section 5(2), of "Jurisdiction, Powers and Duties of Public Service Commission" for the state regulatory agency in New York State, it simply states:

> The commission shall encourage all persons and corporations subject to its jurisdiction to formulate and carry out long-range programs, individually or co-operatively, for the performance of their public service responsibilities with economy, efficiency, and care for the public safety, the preservation of environmental values and the conservation of natural resources.

It is accepted and expected that court precedents provide the guidance for regulators in interpreting the vague federal and state laws, which often date back several decades. The *Hope Natural Gas* ruling set a precedent on the determination of the appropriate rate of return network industries should earn. *Hope Natural Gas* implies that only in exceptional cases will costs not be allowed to be recovered. One such example is the nationwide ruling on recovery of nuclear generating costs, where companies are able to recover their principal (depreciation), but not the interest (rate of return).

Although *Hope* says that regulators must allow investors a chance to earn the opportunity cost of capital by giving them a revenue that covers both the return *on* capital and the return *of* capital, *Hope* does not say *how* regulators should set revenues in order to meet the opportunity cost criterion. This, in turn, has left much room for discretion on the part of state regulators in determining pricing methodology.

To remedy the uncertainty that can ensue from year-to-year rate changes (characteristic of "rate of return" regulation) and the lack of a more detailed regulatory regime (in terms of regulatory discretion in determining what pricing *methodology* to use in determining "fair returns"), several states have begun to implement alternative rate plans (ARPs). ARPs:

- Are voluntary agreements between regulated utilities and the state regulatory commissions that set out in contract the obligations of the utility over several years, and the allowed returns, often in the form of a price cap;
- Set down procedures for dealing with foreseeable changes;
- Must operate under the legal framework (US and state constitutions and related precedents, as well as federal and state sector-specific laws), and all obligations required therein from both the regulators and the regulated utilities; and,
- Give regulators the ability to revert to "normal" regulatory procedures.

ARPs have been most common in the energy sector, and are not as common in the water sector. The ARP between the state of Maine's regulatory commission and Central Maine Power (CMP) stressed the importance of a stable contractual relationship to create appropriate incentives:

> [The ARP] represents a very positive step for regulation in Maine. [It] provides, under a very broad set of assumptions, a high degree of stability and predictability in electric rates for CMP customers. In light of the substantial and often unpredictable rate increases of recent years, these benefits are worth achieving. As a Commission, we have an obligation to mirror the effects of genuine competition to the extent consistent with our broader commitment to serve the public interest. . . . [The] price cap provisions . . . together with the virtual elimination of the fuel clause, give incentives and create risks for CMP's management much closer to those found in less regulated companies.
>
> —Maine, State of, 1995

The plans often include price caps and exogenous cost pass-through provisions. The plans may also include profit-sharing arrangements, designed to keep profits within acceptable bounds. For example:

> The price cap has a profit-sharing component that adjusts the subsequent year's earnings if the earnings are outside a +/− 350-basis-point bandwidth around the authorized cost of equity (currently 7.05% to 14.05%). CMP's current authorized cost of equity is 10.55%. The profit-sharing component will be in effect for each price change taking place on or after July 1, 1996.
> —Maine, State of, 1995, p. 8

It is important to note how arbitrary it is to label regulation in the United Kingdom as "price cap" and regulation in the United States as "rate of return." Both regimes use various forms of price-adjustment mechanisms, and price-cap regulation has been used in the telecom industry in the United States since the late 1980s. Increasingly, network industries in the US are turning to some form of price-cap regulation and away from "rate of return" or profit sharing.

3.2.5 Process: Rate-Making

There are two principal aspects of rate setting in a regulatory proceeding in the United States. First, based on the regulated company's costs, the commission determines the total revenue to which the company is entitled for provision of service. Second, this total revenue is translated into a rate pattern yielding individual prices for each segment of service.

The state regulatory commissions fix prices by setting the "fair rate of return" (see the *Hope Natural Gas* decision) that companies can earn on their assets. The rates set by the commission remain in effect until the commission deems the rates to be too high or too low, which means that the review can take place at any time. The guiding principle for setting rates is cost recovery. Of course, variations on this exist, as different sectors and states are turning to price-cap oriented regulation.

ARPs have been developed by state regulatory commissions to provide greater incentives through a stable contractual framework. ARPs provide potentially greater gains for both customers and the regulated utilities, normally by stipulating the following:

- The period between rate and/or price reviews (often five or more years)
- What forms of cost will be passed through (normally only exogenous costs)

- Penalties for failure to meet quality standards
- Reporting requirements
- Review procedures
- Dispute resolution and arbitration procedures

It will be interesting to see if ARPs are able to prosper in the water and wastewater sector as well.

3.2.6   Process: Appeals and Arbitration

When investor-owned utilities wish to increase prices, they have to present a case to the state regulatory commissions. On receipt of a request for a rate increase, the regulatory commission typically requires the company to submit proof that such a rise can be justified. The evidence will then be examined in a public, quasi-judicial arena, often presided over by a hearing examiner or an administrative law judge. Companies have the right to appeal regulatory decisions to the State Supreme Court, but these appeals are limited to procedure and it is rare that the Supreme Court does not support the state regulatory commission's decision on matters of substance.

Judicial review plays a larger role in arbitration in US regulation than it does in regulating utilities in most other countries. In the United States, all acts of legislatures and decisions of administrators are subject to judicial review. State courts can rule on the unconstitutionality of state statutes, and federal courts on federal statutes. The final authority is always the US Supreme Court, which is the final interpreter of the Constitution. In particular, the Supreme Court must ensure that due process requirements have been met, per the Fifth and Fourteenth Amendments of the Constitution. The limitations on control of business by the federal government are found in the Fifth Amendment: "No person shall . . . be deprived of life, liberty or property without due process of law, nor shall private property be taken for public use, without just compensation." The Fourteenth Amendment states: "No state shall make or enforce any law which shall abridge the privileges or immunities of citizens of the United States; nor shall any state deprive any person of life, liberty, or property, without due process of law. . . ." Note that neither of these amendments rules out the taking of private property. However, due process plays a vital role in regulation and judicial review.

Through time, judicial review has seldom been used to rule against regulatory commissions. Since a series of Supreme Court decisions in the 1930s (including the *Hope Natural Gas* case), appeals are largely limited to matters of *procedure:* if a commission makes a decision without following its stipulated procedures for examining evidence, calculating rates of return, measuring costs, and so on, then the aggrieved party may appeal that decision to the courts, which may in turn instruct the commission to remake its decision following the specified procedure. The Court's view is that the commission and the commission's expert staff are presumed to be more qualified in making technical (i.e., substantive) decisions than the court.

3.2.7   Regulatory Characteristics

Independence and Accountability: The regulatory commissions in the United States are often called "independent" commissions. They are seen as independent because they do not officially fall within any one of the three branches of government (executive, legislative, and judiciary). Both at the federal and state levels, several factors contribute to the characterization of regulatory "independence": appointments are for defined periods; no more than half (rounded up if an odd number) of the commissioners can be from the same political party; and the removal from office is difficult, unless some form of malfeasance can be proven.

However, "pure" independence, if there is such a thing for a regulatory body, does not exist in the United States. The executive branch sets appointments. The regulatory commissions are created by legislative authority and depend on legislation for their powers (although regulatory laws are left deliberately vague so that regulatory commissions can apply the statutes as they think best, subject to judicial review). Furthermore, the concept of independence of the regulatory commissions has often been challenged on the grounds of legitimacy—if the regulatory authority is not accountable to the (elected) executive and legislature, this can pose problems of its own. In addition, under the US Constitution, the legislature cannot divest itself of its powers or duties. The regulatory commission can be delegated power from the state and federal legislatures—but the extent of that power will ultimately be determined by the legislatures. In this way, the executive and legislatures hold the commissions accountable. The commissions are, in the end, accountable to the judiciary, as their determinations are subject to judicial review.

Transparency: Regulatory decision-making in relation to the network industries in the United States can be considered transparent. The APA sets out the basic requirements for the decision-making procedures of federal agencies and has served as a model for the states, which have adopted similar statutes. The manifest purpose of the APA was to give regulators—who before 1946 operated under what some legal scholars in the United States believed was dubious constitutional legitimacy—a firm legislative, due process underpinning. That underpinning, however, required that regulators follow meticulous steps in terms of notifying parties of impending actions, calling for and examining evidence, providing for testimony and cross-examination of evidence, and formulating their rulings.

The transparency of the process opens up opportunities for appeal when the regulator's decision does not meet the required standards. The desire to avoid appeal means that individual decision-makers take more care to ensure that their decisions are transparent and justified by the most reliable evidence.

Commitment: The US Constitution safeguards private property through the "due process" clauses of the Fifth and Fourteenth Amendments. The "due process" clause and the potential for judicial review have ensured "commitment" on the part of regulators. This can be demonstrated by the recent debate on stranded assets, as markets—particularly energy—open up to competition. For example, New Hampshire was intending to introduce competition into the energy sector without 100% recovery for stranded assets. This has led to a drawn-out legal battle, whereby the liberalization process has been stalled. In California, the debate on stranded assets lasted for several years after liberalization was contemplated. It was decided that all customers must pay a stranded cost, or Competition Transition Charge, to their utility distribution company for specified assets. Clearly, California's deregulation process has been fraught with problems. However, the decision taken on stranded assets demonstrated the commitment to due process and caution against regulatory "takings." Whether these objectives were effectively implemented is an altogether different story, however.

4. AUSTRALIAN EXPERIENCE

4.1   Current Regulatory Regime

Australia, like the United States, is separated into federal and state governments. There are six states and two territories. A significant proportion of Australian network industries are government owned, generally at the state level. Privatization is a relatively recent phenomenon, and has developed to the greatest extent in Victoria. Most utility services are also characterized by uniform prices across vast geographic areas. Almost all water utilities are government-owned entities.

Regulation of network industries has only been considered a priority over the last decade. The first state-based utility regulator was set up in New South Wales (NSW) in 1992. The IPART Act established the Independent Pricing and Regulatory Tribunal (IPART). IPART functions as a tribunal, holds public hearings, and makes final decisions for a range of publicly and privately owned utilities. With the development of privatization, the Office of the Regulator-General Act (1994) established the Office of the Regulator-General (ORG) for Victoria. The Act sets out the ORG's objectives and its approach, and establishes it as a single-person, multi-sector regulator. More recently, the 1996 National Competition Principles Agreement between all Australian state governments requires independent economic regulation (primarily through the presumption of third-party access) of all network businesses, whether state or privately owned.

The Australian Competition and Consumer Commission (ACCC) is the federal competition policy agency, and is now also the main federal regulatory authority for network industries. In the past, the ACCC acted primarily as a competition policy enforcement agency. However, the Competition Principles Agreement and the associated legislative amendments implemented at both state and federal levels have significantly extended the ACCC's role to include arrangements for third-party access to networks, including the pricing of network and other monopoly services.

The introduction of the National Electricity Code and the National Gas Access Code (Appendix to the National Gas Pipelines Agreement) has resulted in an increased role for the ACCC as a national regulator of both gas and electricity transmission, as well as providing a more consistent framework for the regulation of distribution assets by the individual state regulators. However, individual state regulators still retain a significant degree of discretion, and are subject to state-based legislation, in addition to the national codes. While there has been significant movement toward a consistent national framework, both codes give a degree of discretion to the state regulators so that differences in approach between (for example) IPART and the ORG can still have a material impact.

There is currently no national regulatory framework applying to the water sector, although the development of state-based arrangements for water sector regulation is a requirement of the Competition Principles Agreement. In New South Wales, IPART regulates urban water services under the IPART Act 1992. Sydney Water Corporation, Hunter Water Corporation, Gosford City Council, and Wyong Council are all regulated by the IPART Act for water supply services provided under the Water Supply Authorities Act, 1987. In urban Victoria, the Melbourne Metropolitan water industry is divided into one wholesale water and sewerage company, the Melbourne Water Corporation, and three retail water and sewerage companies, City West Water, South East Water, and Yarra Valley Water. Each retail company is government owned and has its own customer base allocated by geographic area. The companies currently do not compete directly for each other's customers, although the regulator, the ORG, uses some form of comparative competition.

4.2   Objectives of Regulation

4.2.1   State Regulators

For New South Wales, Section 15.1 of the IPART Act sets out the factors that have to be taken into account by the Tribunal in conducting its investigation, and includes things such as "the protection of consumers from the abuse of monopoly power," "the appropriate rate of return on assets," and "the need to promote competition." Under Section 15.2, IPART is required to indicate which of the issues in Section 15.1 it has considered in reaching its decision.

The specific objectives of the ORG under the Water Industry Act 1994 are:

- To ensure the maintenance of an efficient and economic water industry;
- To protect the interests of customers vis-à-vis water industry charges and terms of conditions of services supplied;
- To protect customers vis-à-vis quality of water supplied to them; and
- To facilitate the maintenance of a financially viable water industry.

4.2.2   Federal Regulation

The ACCC's objectives (1999) are to:

Improve competition and efficiency in markets;
foster adherence to fair trading practices in well-informed markets;
promote competitive pricing wherever possible and to restrain price rises
    in markets where competition is less than effective;

inform the community at large about the Trade Practices Act and the
Prices Surveillance Act and their specific implications for business and
consumers; and,
use resources efficiently and effectively.

As discussed, the national framework applies mainly to the electricity
and gas sectors, and is not yet relevant to the water and wastewater sector.

The National Electricity Code and the National Gas Code specify asset
valuation principles and other mechanisms to enable the service provider
to earn revenue that will recover its efficient costs, replicate the outcome
of a competitive market, not distort investment decisions, and provide an
incentive for the service provider to reduce costs. For example, to enable a
service provider to earn revenue that will recover its efficient costs, the
Gas Code sets out a range of principles that can be adopted in setting the
initial asset base, including the optimized depreciated replacement cost
(ODRC), or depreciated actual cost (DAC). The Electricity Code is more prescriptive in requiring the use of optimized deprival value (ODV) in valuing
assets (ODRC is the cost of replacing the asset with the cheapest asset that
does the same job, whereas ODV attempts to take into account the economic
value of assets as well as their replacement costs). In determining the cost
of equity, both codes refer to the Capital Asset Pricing Model (CAPM), although the use of alternative approaches is not prevented.

The Electricity Code has been criticized as being too prescriptive and
for placing too great an emphasis on the interests of the utilities themselves, at the expense of customers and related aspects of public policy.
This may be due largely to the fact that the Electricity Code was drafted
by the representatives of the industry in each state. The Gas Code establishes a much more detailed set of economic and regulatory principles than
exist for any equivalent utility in the United Kingdom or Europe, but is
much less prescriptive than the Electricity Code. It also strikes a better
balance between the interests of consumers and service providers, probably because representatives of the federal government, as opposed to industry representatives (of the state-run enterprises), were responsible for
drafting the Gas Code.

Regulation of network industries in Australia is relatively new, although thinking and experience is developing quickly, with fairly rapid
transfer of experience and knowledge from both the United Kingdom and
the United States. While it is too early to conclude that all the purported
objectives have actually been met, the requirement for all significant network businesses to offer third-party access on economic terms has trans-

formed the environment under which state-owned water, gas, and electricity companies operate. Emerging contentious issues have included: the price distortions brought about through a history of uniform tariffs, and the associated incentives for network bypass; the relationship between asset values, depreciation, and the cost of capital; and the appropriate degree of benefit-sharing across review periods.

4.3   Institutional Design/Characteristics

In general, Australian regulation of network industries is more legalistic compared to the UK system, and therefore more closely resembles the US regulatory model. We look in more detail at some of the institutional design and characteristics of Australian regulation below.

4.3.1   Rate-Making, Appeals, and Arbitration

For declared government monopoly services, IPART is empowered to determine maximum prices [Sections 11(1a) and 12(1a)], and/or carry out a periodic review of pricing policies [Sections 11(1b) and 12(1b)]. IPART may fix maximum prices in either of two ways:

1. Maximum prices may be determined in any way the Tribunal considers appropriate (Section 14), including: setting specific prices for individual services, increasing or decreasing prices for individual services or groups of services, setting a rate of return on assets, referring to the CPI; or
2. A methodology may be established for determining maximum prices.

Medium-term price paths are set for urban water and wastewater services supplied by Sydney Water Corporation and Hunter Water Corporation. These price paths are usually set for a period of four years.

If the Tribunal considers that it is impracticable to make a determination of maximum prices as in method (1) above, it may determine under Section 14A a methodology to be used by the agency for fixing the maximum price.

The IPART Act contains a number of provisions to ensure that the Tribunal's activities are carried out through a public process. The main requirement is that the Tribunal must hold at least one public hearing for each investigation. The Tribunal may seek public participation by: advertising public hearings [Section 21(3)]; seeking public comments on terms of reference [Section 13(2)]; providing public access to submissions [Section 22A(1)]; inviting public comment on issues and submissions; holding public seminars and workshops [Section 21(2)]; releasing reports and determinations to the public [Section 19(1)]; and reporting publicly on compliance by agencies [Section 18(4)]. The general assumption of the legislation (Section 22A) is that the public will have access to information provided to IPART for its investigations. That Section also extends the possibility of public access by allowing the Tribunal to approve the release of information that would not otherwise be available under the Freedom of Information Act, following consultation with the supplier of the information. While most Tribunal activities are public, the Tribunal may direct that evidence be considered in private and may restrict access to confidential documents.

The Tribunal is required to consider a range of issues when making determinations and recommendations. The factors can be grouped as follows:

- Consumer Protection. Prices, pricing policies, and standards of service, general price inflation, social impact of decisions;
- Economic Efficiency. Greater efficiency in the supply of services, impact of exercise of functions by some other body, the need to promote competition;
- Financial Stability. Rate of return on assets, impact of borrowing, capital and dividend requirements; and,
- Environmental and Other Standards. Protection of environment by appropriate pricing policies, considerations of demand management, standards of quality, reliability, and safety.

There is no mechanism for businesses to appeal IPART's decisions under the IPART Act. However, IPART's decisions under the National Gas and Electricity Codes are appealable. IPART's decisions in the water sector are not able to be appealed at present. However, a New South Wales' (NSW) Energy and Water Ombudsman has been set up by the electricity industry and Sydney Water Corporation to provide an independent way of resolving customer complaints. The Ombudsman works with a Council made up of representatives of consumer, small business, and industry interests, with an independent chairperson. There is a published Constitution for the Ombudsman. The Ombudsman is possibly a mechanism that can be used for appeals, although its ability to successfully appeal IPART decisions vis-à-vis the water sector has yet to be clearly seen.

The ORG has responsibility for three retail water and sewerage licenses operating in the Melbourne metropolitan area only. The Office of the Regulator-General Act (1994) sets out the ORG's objectives and its approach, and specifies that a three-person appeal body be appointed by the Minister, with 14 days to decide appeals. The basis for appeal is limited to bias or misinterpretation of facts. However, the ORG does not set prices in the electricity, gas, or water industries. Prices remain controlled by the Victorian Government.

Because the state government reserves the right to set prices, regulation of water utilities in Victoria has focused on nonprice issues, such as:

- The development of a Benchmark Customer Contract and Customer Charter;
- The review of the performance standards set in Schedules of the water and sewerage licenses;
- The introduction of public reporting on customer service performance and a comparative competition regime;
- The development of operational audits; and
- The participation in the development of a licensing regime.

ORG decisions are subject to judicial review, under the Administrative and Judicial Review Act (federal legislation). Case law for this act says that administrative bodies must take account of material that is "relevant," and not take account of material that is "irrelevant" when making decisions. ORG decisions are also subject to appeal on grounds that they go beyond the ORG's power or do not comply with various legal requirements.

4.3.2 Regulatory Characteristics

Independence: The issue of independence is more complicated in Australia than in the United States and the United Kingdom. Regulatory independence is difficult to achieve when government-owned entities are being reviewed. Nevertheless, an explicit objective of the Competition Principles Agreement was to ensure a competitively neutral environment for state-owned enterprises. Relative to the situation five years ago, significant progress has been made in this regard.

Transparency and Accountability: As described above, in several cases regulator's are required to give reasons for their determinations. Rule-making procedures are also relatively explicit (as compared to the United Kingdom). Thus, the Australian system exhibits a greater degree of transparency and accountability than presently exists in the United Kingdom.

Some aspects of regulatory design owe much to perceived problems in the United Kingdom, for example, fixed review periods, prescribed approaches to asset valuation, and requirements to give reasons. It has been questioned whether public consultative processes have been compromised or their effectiveness reduced due to the lack of cross-examination. There are also inconsistencies in the appeal arrangements and in the ability to challenge decisions. There are also questions as to the sustainability of the informal consultative processes currently in place, particularly in the face of a growing inclination for litigation (including by third parties). Nevertheless, appeals and judicial review are realities in Australia, and are generally believed to be an important element of framework as well as a significant influence on the decision-making of regulators.

Commitment: Because of the legalistic nature of the Australian system, and the focus on rules and procedures, as in the United States, commitment has so far been less problematic in Australia as compared to the United Kingdom. Laws and regulations, more detailed in content and assuring cost recovery in detail, prevent regulators from more opportunistic behavior. The Gas Code, for example, states that once an asset is included in the asset base, in cannot be taken out without just cause. Decisions that are believed to extend beyond the principles set out in the Code are sure to face either a merits or a judicial review. Indeed, the recent decision of Great Southern Networks to appeal IPART's recent Final Decision on its gas access arrangements confirms this trend.

5. CONCLUSION

In the United Kingdom, regulation has been characterized by reliance on individuals who were given significant discretion to develop the system of price-cap regulation following privatization. Over a decade of experience has seen little, if any, convergence on a sound regulatory "contract." UK regulation is characterized by regulatory discretion and opportunism, and a higher cost of capital than in the United States.

In the United States, the Fifth and Fourteenth Amendments to the Constitution have led to commitment and "due process," the *Hope Natural Gas* (1944) decision provided investors with a reasonable assurance of cost recovery, and the Administrative Procedures Act established rules that must be followed in making determinations. While the US system has achieved a high degree of regulatory commitment, it has been characterized by cost plus regulation, reducing incentives for short-term efficiency. More recently, "incentive-based" mechanisms have been introduced.

In Australia, lessons were learned from the UK experience. Legislation and industry codes are highly prescriptive, narrowing the scope for regulatory discretion and opportunism. Australia's approach may have "short-circuited" the evolution of institutional arrangements seen in the United States, although time and further privatization will provide a clearer test.

REFERENCES

"Byatt Rejects Access to Pricing Model," *Water Magazine,* no. 33, December 4, 1998.

Competition Commission (2000). Reports on Mid Kent Water plc and Sutton and East Surrey Water plc on the References under Sections 12 and 14 of the Water Industry Act 1991, London, September 13 (http://www.competition-commission.gov.uk).

Department of Trade and Industry (DTI) (1998). "A Fair Deal for Consumers—Modernising the Framework for Utility Regulation—The Response to Consultation," London, July.

Department of Trade and Industry (DTI) (1997). "Margaret Beckett Announces Review of Utility Regulation." Press Release, June 30, page 1.

Federal Power Commission v. Hope Natural Gas Company, 320 US 591 (1944).

Greenwald, Bruce (1994). "Rate Base Selection and the Structure of Regulation," *The Rand Journal of Economics,* Vol. 15, No. 1.

Houston, Graham (1999). "Regulatory Risk and the Cost of Capital." Paper presented at Utility Regulation Summit '99, London, May 26.

Hyman, Leonard, Andrew Hyman, Robert Hyman, Edward Meehan, James Hempstead, and Jose Kochen (1998). *Understanding the Water Supply and Wastewater Industry.* Vienna, VA: Public Utilities Reports, Inc.

Jones, Sion (1999). *NERA Topic 22,* "Comparatively poor? A Comment on the Ofwat and Ofgem Approaches to the Assessment of Relative efficiences," National Economic Research Associates, London, October.

Kydland, Finn, and Edward Prescott (1997). "Rules Rather than Discretion: The Time Inconsistency of Optimal Policy Plans," *Journal of Political Economy,* Vol. 85, No. 3, pp. 619–637.

Littlechild, S. C. (1986). *Economic Regulation of Privatised Water Authorities—A Report Submitted to the Department of the Environment.* Her Majesty's Stationary Office, London, January.

Maine, State of (1995). Public Utilities Commission, Docket No. 92-345 (II), January 10.

Phillips, Charles F., Jr. (1993). *The Regulation of Public Utilities: Theory and Practice.* Vienna, VA: Public Utilities Reports Inc.

Shleifer, Andrei (1985). "A Theory of Yardstick Competition," *Rand Journal of Economics,* Vol. 16, No. 3.

Vickers, John, and George Yarrow (1988). *Privatization—An Economic Perspective.* Cambridge, MA: The MIT Press.

Williamson, Brian (1997). *NERA Topic 20,* "Incentives and commitment in RPI–X Regulation," October.

Williamson, Brian (2001). "'UK Incentive Regulation'—International best practice?" in *CRI Regulatory Review—Millennium Edition 2000/2001,* ed. Peter Vass. Bath, England (RI Regulatory Review).

ADDITIONAL READINGS

American Water Works Association. *Manual of Water Supply Practices, Principles of Water Rates, Fees and Charges 2000,* Washington, DC.

Laffont, Jean-Jacques, and Jean Tirole. *A Theory of Incentives in Procurement and Regulation.* Cambridge, MA: The MIT Press, 1993, p. 76.

Ofwat. MD145, March 8, 1999.

Pennsylvania Public Utility Commission v. Bell Telephone Company, 43 PUR 3rd 241, 246 (Pa, 1962).

Permian Basin Area Rate Cases, 390 U.S. 747, 88 S. Ct. 1344, 20 L. Ed. 2d 312 (1968).

Chapter 3

Ten Years of Water Service Reform in Latin America

Toward an Anglo-French Model

Vivien Foster

During the 1990s, most countries in the Latin American region* undertook major reforms to their water supply industries. Chile was the first to attempt to modernize its water sector, with new legislation passed as early as 1988. By 1991, both Argentina and Mexico were beginning to conduct

This chapter is based on material originally prepared for a seminar given to the Water Cluster in the Finance, Infrastructure and Private Sector Network of the Latin American Region of the World Bank on July 1, 1999. The role of Abel Mejia (Sector Leader), Carlos Velez, and Yoko Kata-kura in commissioning and guiding the work is gratefully acknowledged, as are the very helpful comments received from the seminar participants. The following colleagues provided very helpful updates on the state of reform around the region: Oscar Alvarado (World Bank), Ventura Bengoechea (World Bank), Jaime Caller (SUNASS), Jonathon Halpern (World Bank), Jorge Rais (FENTOS), Marcela Restrepo (BBV), Lillian Saade (IHE), Felipe Sandoval (CORFO), Anna Wellenstein, and Francisco Wulff.

*For the current purposes, "the Latin American region" is defined as all mainland Spanish- and Portuguese-speaking countries south of the Rio Grande, which is to say, those countries listed in Figure 3.1.

Vivien Foster is an economist with the Information and Communications Technology Division of the World Bank. Formerly an economic consultant for Oxford Economic Research Associates in the United Kingdom, she acted as a consultant to many South American countries on water sector reform.

Reinventing Water and Wastewater Systems: Global Lessons for Improving Water Management, edited by Paul Seidenstat, David Haarmeyer, and Simon Hakim.

a series of experiments with private sector participation. In a second wave, Peru, Colombia, and Bolivia enacted ambitious new legislation in the mid-1990s. During the second half of the same decade, reform began to take root in Brazil and Central America. By the end of the nineties, there were barely any countries remaining that had not either completed reforms, had major reforms in process, or were actively considering reforms. (See Figure 3.1 for a country-by-country overview.)

However, while the breadth of the reform process in Latin America is impressive, the depth of reform varies substantially across countries and unquestionably falls far short of what has been achieved contemporaneously in other infrastructure sectors, notably electricity and telecommunications. More specifically, *regulatory reform in Latin America has gone significantly further than private sector participation* (PSP). Indeed, in some countries—for example, Panama, Peru, and until recently, Chile—regulation has been introduced without privatization. In others—such as Bolivia and Colombia—regulatory reform has been nationwide, yet privatization has been confined to metropolitan areas or a handful of major provincial centers. Overall, it is estimated that, although 41% of urban water consumers now enjoy regulatory protection, only 15% are serviced by private sector operators.

This widespread willingness to embrace reform was in large measure motivated by dissatisfaction with the traditional "clientilist" model of water service provision (Figure 3.2), according to which state-owned water companies were more often treated as part of the political apparatus than allowed to function as efficient service providers. Politicians exerted their control over the sector through the appointment (and dismissal) of water company directors, and by providing public subsidies to finance investments as well as prop up ailing enterprises. In return for this patronage, water companies were often obliged to supply political favors in the form of overemployment, artificially depressed tariffs, political targeting of new investments, and distribution of contracts based on political rather than economic criteria. The consequences of this regime have been spiraling costs, low quality of service, and precarious finances, while the scarcity of resources for investment has left substantial sections of the population unserved and therefore forced to rely on a range of expensive or inconvenient substitutes.

The reform process aims to break this pattern by addressing its underlying institutional causes. Hence, a key feature of the reform blueprint (Figure 3.3) is the separation of the functions of policy-maker, regulator,

	Regulation	PSP
Chile	100%	86%
Argentina	88%	62%
Bolivia	100%	28%
Colombia	100%	13%
Peru	100%	0%
Brazil	24%	1%
Uruguay	100%	17%
Mexico	19%	19%
Venezuela	3%	3%
Nicaragua	100%	0%
Panama	100%	0%
Paraguay	100%	0%
Ecuador	0%	0%
El Salvador	0%	0%
Honduras	0%	0%
Costa Rica	0%	0%
Guatemala	0%	0%

Figure 3.1   Regional overview of reform. The percentage numbers refer to the percentage of the urban population that enjoys regulatory protection and receives its service directly from a private sector operator [excluding build, operate, and transfer (BOT) projects]. PSP stands for private sector participation.
Legend: Dark grey = major progress has been made. Mid-grey = some progress has been made. Light grey = in process or actively being considered. White = nothing has happened yet.

and service provider, which tend to be blurred under "clientilism." In the new model, politicians are confined to supplying strategic guidance to the sector. A regulatory agency is introduced with a view to insulating the utility company from political interference and requiring that its business be conducted in line with sound operational and financial principles. The regulator sets tariffs at a level that allows the company to recover the efficient costs of operation, as well as a reasonable rate of return, while at the same time monitoring the achievement of quality and coverage targets. The actual operation of water services is delegated either to the private sector, or to a strongly corporatized public company.

The purpose of this chapter is to provide a synthesis and evaluation of the reform experience in Latin American water supply industry during the eventful decade of the nineties. In order to make the exercise tractable, the bulk of the discussion will focus on a "panel" of six countries—Argentina,

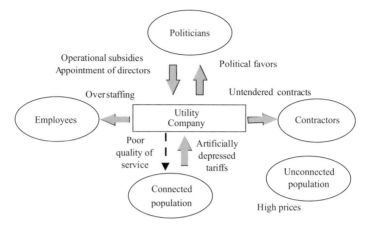

Figure 3.2   The "clientilist" model of water provision.

Bolivia, Chile, Colombia, Panama, and Peru—chosen to represent a variety of approaches to the reform process. A comparative analysis of the reform in these six countries will be complemented by use of more detailed examples and case material. The chapter will consider three major components of reform in Latin America: (1) the extent to which the water industry has been restructured and companies subject to institutional transformation; (2) the ways in which different countries have redefined

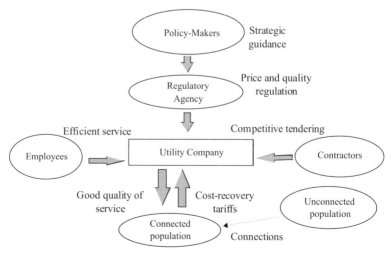

Figure 3.3   The reformed model of water provision.

the role of the state by separating out the functions of policy-maker and regulator; (3) the sorts of instruments that regulators in Latin America have developed to support the day-to-day functioning of the regulatory process.

1. SECTOR RESTRUCTURING, OR THE BALKANIZATION OF WATER

In order for reform to be successful, there must be an underlying coherence between the form of regulation and the institutional nature of the regulated entities. This need for coherence manifests itself at two distinct levels. First, the political and geographical jurisdiction of the regulator must be compatible with that of the service providers. Second, the choice of regulatory instruments must be suitably adapted to the managerial incentives of the water operators.

As this section will illustrate, a number of Latin American countries have embarked upon reforms that lack this kind of overall coherence. Thus, in some cases, there has been radical decentralization and municipalization of service provision, accompanied by simultaneous moves to impose regulation and PSP from the center, while, in a number of other cases, there have been attempts to apply British-style incentive regulation— premised on the existence of the profit motive—to state-owned enterprises that ostensibly lack this kind of motivation.

1.1 Sector Structure

It is helpful to distinguish between three models of water industry organization: the national monopoly, the regional monopoly (based on political boundaries such as states or provinces, or hydrographic boundaries such as the water basin), and the local (typically municipal) monopoly (Table 3.1).

Prior to 1990, many Latin American countries (e.g., Argentina, Chile, Colombia, Panama, and Peru) had chosen to organize their water industry as national monopolies under the direct control of the central government. Growing dissatisfaction with the performance of the national monopolies, combined with wider political pressure for devolution across all areas of government, created the conditions for a move toward decentralized control in the 1980s and 1990s. In countries such as Argentina, Colombia, and Peru, this entailed a sudden fragmentation of the industry into literally hundreds of small municipal providers (Table 3.1).

Table 3.1   Overview of Decentralization

Country	Model of Organization		
	National	*Regional*	*Municipal*
Argentina	OSN (1912)	1,497 providers[a]	
Bolivia			Municipal bodies
Chile	SENDOS (1977)	13 companies (1990)	
Colombia	INSFOPAL (1950)		1,380 providers (1989)
Panama	IDAAN (1961)		
Peru	SENAPA (1981)		136 municipal bodies

[a]Of which 14 are provincial, 462 municipal, 990 cooperative, and 22 private.

It is important to emphasize that, on the whole, decentralization was not a studied response to the specific problems of the water sector, but rather the by-product of a wider reform of the state. Indeed, in a number of cases, decentralization preceded subsequent water sector reform by a number of years. This lack of synchronicity between structural and regulatory reform was unfortunate, because it meant that regulatory reform had to be superimposed onto an industry structure that was often far from optimal in the economic sense. In many cases, it was difficult to contemplate subsequent restructuring because of the foregoing political commitment to municipalization, which in some countries (notably, Colombia and Brazil) was even enshrined at the constitutional level.

A number of problems have consequently arisen. The first of these is the loss of scale economies. Work by Yepes (1990) suggests that the minimum efficient scale for water companies in Latin America is 100,000 connections. Aside from technological economies, the scarcity of human resources may also make it undesirable to dilute technical capacity across a large number of service providers. Related to this is the questionable commercial viability of many of the small business units created (many of them serving low-income rural communities), which in turn leads to difficulties in attracting private sector investment except to the largest population centers.

Finally, municipal control of the sector has made it difficult to subsequently drive regulation and PSP from the center. For political reasons, municipalities may be unwilling to relinquish their recently regained control of service provision to the private sector or to accept tariff rulings from a national regulator. Hence, in Peru, the regulator's power is limited to designing tariff rules and proposing tariff levels, while ultimate approval of tariffs must come from the municipalities (who are typically also the service providers). In Bolivia, the 1999 water law had to be modified within a few months of its promulgation, owing to political pressure to give municipalities (who are once again, by and large, the service providers) a say in the process of tariff-setting. These kind of arrangements seriously risk undermining the basic principle of institutional separation between regulator and service provider. Even in the absence of opposition, the sheer fragmentation of a decentralized sector can paralyze the efforts of a central regulator. The most extreme example is Colombia, where the national regulator simply lacks the resources to monitor the operations of 1,380 municipal service providers.

One possible solution is to organize regulation at the provincial level, such as in Argentina, where 14 out of 23 provinces have created their own regulatory agency plus an additional one in the federal capital. While this is a valid approach, its cost effectiveness hinges on the size of the states or provinces involved, since many of the costs of regulation are fixed in nature, generating significant scale economies. This effect is visible in Argentina, where provincial regulation absorbs up to 6% of industry turnover, compared to around 2% for many national-level regulators. However, state-level regulation is probably ideal for a country such as Brazil, where individual states may be more populous than many Latin American countries (World Bank, 2001a).

Another approach to the problem is to try and promote ex post facto industry consolidation. For example, the 1999 Bolivian water law encourages "mancomunidad," or the creation of multimunicipal companies. Brazil offers a few examples of multimunicipal concession contracts, while a similar arrangement is currently under consideration in El Salvador. In Colombia, companies were legally required to prepare a study of financial viability and the regulator empowered to merge companies that were demonstrably not viable. However, in practice, the regulator did not take advantage of this opportunity for industry consolidation. Companies serving less than 8,000 customers (that is, those least likely to be viable) were exempted from the requirement to undertake the study of financial viability, and only those companies finding themselves nonviable were required to submit their conclusions to the regulator. It is hardly surprising, then, that only one company did so.

As noted above, municipalization is not the only possible response to dissatisfaction with monolithic national providers. In this respect, Chile provides an interesting and important exception to the pattern hitherto described. This is, first, because the restructuring of the water industry in that country was undertaken as an integral part of the water sector reform process rather than as part of a general thrust toward decentralization, but also because the decentralization of the sector was limited to the regional rather than the municipal level, with the creation of 13 companies, most of which grew out of the preexisting 11 regional directorates of the former national monopoly SENDOS. As a result, Chile has been able to avoid many of the difficulties described above.

Brazil is the other country where state-level water companies are important, servicing about 80% of the population. However, these companies did not arise from the fragmentation of any national monopoly. Rather, they were created as a result of voluntary agreements with municipalities, who temporarily ceded their constitutional right of service provision to the state authorities in return for an attractive investment financing package under the PLANASA program. (In this sense, there is a parallel with England and Wales, where 11 regional water companies were created from the amalgamation of hundreds of municipal service providers in 1974.) However, the recent expiry of the PLANASA agreements has left a great deal of legal ambiguity regarding state versus municipal responsibility for the provision of water services, particularly in metropolitan areas (World Bank, 2001a), and is already leading to industrial fragmentation.

1.2   Private Sector Participation

The institutional form of the operator is important because it affects the incentives faced by managers. In particular, public sector managers will tend to be influenced by political pressures, although this will depend on the degree of corporatization. Corporatization strengthens the political autonomy of a publicly owned enterprise by making it increasingly self-sufficient financially (depending directly on tariff revenues rather than state subsidies) and introducing rules that protect directors and senior managers from being removed on political whim.

On the other hand, private sector managers, motivated by profit, are more likely to be concerned with expanding sales revenues and reducing costs. Thus, in some respects, private operators are more easily regulated than public operators. The reason for this is that it is possible to design regulatory instruments that make it financially attractive for a company to act in the interests of the consumer. Two key examples are the "price cap," which provides an incentive for managers to drive costs down even in the absence of competition, and the use of fines to punish noncompliance with performance targets.

Water sector reform processes in Latin America have universally recognized the need to effect some kind of institutional transformation. Various models are shown in Table 3.2. However, private sector participation has proved difficult to implement. In larger urban centers, this is primarily for political reasons, while in smaller cities and rural areas, there is the additional problem of commercial viability. A common pattern is for PSP to take place in the metropolitan area, preceded and/or followed by a handful of provincial capitals (Table 3.3).

Where PSP has proved possible, the concession contract has proved to be the most popular vehicle (Table 3.4). However, service contracts and management contracts have sometimes been used as first steps toward a concession. Colombia has created a number of "mixed enterprises," and more recently, Chile has divested four of its regional water companies. A number of countries have also used build, operate, and transfer (BOT) instruments to finance the construction of water and wastewater treatment plants. Indeed to date, the lease contract ("affermage") is the only modality for PSP that remains completely untested in Latin America.

There have also been a considerable number of suspended or failed PSPs. For example, the attempt to award a concession for the city of Caracas in 1992 failed to attract a single financial bid, while, both in Lima (Peru) and in Panama City, bidding processes were suspended at the last moment due to political opposition. On two other occasions—in Tucuman (Argentina) and Cochabamba (Bolivia)—concession contracts had to be canceled after a relatively short period of operation. In both cases, the immediate cause was public opposition to the substantial tariff hikes following the award

Table 3.2   Models of Institutional Organization

Type of Organization	Ownership	Control
Direct provision	State	
Corporatization	State	Public corporation
PSP Contracts	State	Private corporation
Mixed enterprise	State and private investors	Private corporation
Private enterprise	Private corporation	
Cooperative	Customers	

Table 3.3   Overview of PSP by Country

Country	Metropolitan	Interior
Argentina	Greater Buenos Aires (1993)	Since 1991, in 11 out of 23 provinces
Bolivia	La Paz (1997)	Cochabamba (in 1999, rescinded in 2000)
Chile	Santiago (1999)	Since 1998, companies serving Regions V, VI, and X
Colombia	——	Since 1991, Barranquilla, Cartagena, and 20 small towns
Panama	Panama City (suspended in 1999)	——
Peru	Lima (indefinitely postponed in 1995)	——

of the concession, while the underlying cause was the large scale of the investment program required under the concession contract.

PSP transactions have attracted a total US $14.7 billion of private investment in the Latin American water sector over the period 1990–1999, equivalent to about half of total private sector investment in developing countries over the same period of time. However, these capital flows are highly concentrated in a small number of countries (Figure 3.4), with more

Table 3.4   Overview of PSP in Latin America by Modality

Country	Service Contracts	Management Contracts	Leases	Concessions	BOTs	Divestments
Argentina				√		
Bolivia				√		
Brasil		√		√	√	√
Chile	√				√	√
Colombia		√		√	√	√
México	√	√		√	√	
Panamá				(√)	√	
Perú	√					
Uruguay				√		
Venezuela		√		(√)		

Legend: √ = successfully undertaken; (√) = unsuccessfully attempted. BOT = build, operate, and transfer.

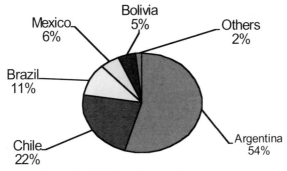

Figure 3.4   Capital flow by country.

than half going to Argentina. Nevertheless, it is noteworthy that a country as small as Bolivia, with a relatively high risk rating, has been able to attract almost as much private investment as Mexico.

The degree of competition for these transactions, as measured by the number of bidders, has been fairly limited. Figure 3.5 plots the number of bidders against the population of the city for a number of concession contracts bid during the 1990s. The record number of bidders for a Latin American water contract was six in the case of Mexico City. Otherwise, three appears to be the maximum number, while about half of the contracts considered attracted only one bidder (Barranquilla, Cochabamba, La Paz, Santa Marta, Tucuman).

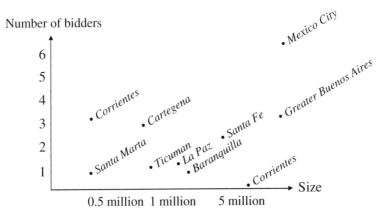

Figure 3.5   Degree of competition for PSP contracts.

Not only has the number of bidders for Latin American water contracts been comparatively small, but they have always tended to come from the same handful of predominantly French (and to a lesser extent Spanish and English) companies. Thus, the three most significant investors in the region are Suez Lyonnaise des Eaux, Aguas de Barcelona, and Vivendi (formerly Compagnie Generale des Eaux). For example, of the 20.2 million urban water consumers in Argentina served by private sector operators, some two-thirds receive their water from consortia headed by Suez Lyonnaise des Eaux.*

Despite the considerable amount of PSP in the Latin American region, it is important to recall that by the close of the 1990s still only 14.8% of the urban population received water services from an operator under private sector control.† If one excludes Brazil and Mexico—the two largest countries in the region where progress toward reform has been relatively modest—the proportion rises to 25.1%, and if one takes into account failed and attempted privatizations, the proportion rises further to 33.5%. Nonetheless, these percentages are much lower than those for the number of urban water consumers affected by regulatory reform, which was 40.9% regionwide, rising to 66.9% if Brazil and Mexico are excluded. The relatively limited scope of PSP has not been for lack of available urban centers. As illustrated by Table 3.5, the vast majority of Latin American cities across all size ranges remain under public sector service provision.

The difficulties associated with private sector participation have prompted many countries to embrace corporatization as a "second-best" solution, or (in the case of Chile) as an initial step in a slower paced privatization process. In Bolivia and Colombia, corporatization of water utilities is required by law, while Peruvian law stipulates that operators in the larger urban areas must transform themselves into "sociedades anónimas" (that is to say, public limited companies).

Whether the service is provided by private or public corporations, most countries require operators to hold some sort of concession or licence granted by the state. The authority that grants the concession differs across countries, and may either be the executive branch (Argentina and

---

*Specifically, consumers covered by the contracts for Greater Buenos Aires, Cordoba, and Santa Fe.

†Private sector control is here defined to be a management contract, a concession contract, or a privately owned company. Service contracts and BOTs are excluded on the grounds that they only entail private contracting for upstream services and do not give overall control of operations to the private sector.

Table 3.5   Overview of PSP in Latin America by Size of City

	Population Range (millions)			
	>5	*1–5*	*0.5–1*	*0.1–0.5*
Cities affected by PSP	3	4	12	71
Cities not affected by PSP	4	30	36	399
Total number of cities	7	34	48	470
Percentage of cities affected by PSP	42.9	11.8	24.0	15.1

Chile), the regulator (Bolivia), or the municipality (Peru). Such concessions invariably give the holder an exclusive right to operate in the designated area, and moreover, legally require consumers to connect to the network once it is made accessible to them. These measures have been justified in terms of the need to provide assurance to private investors that sunk investments in network expansion will yield the expected revenues; however, this comes at the cost of eliminating any competition.

An interesting exception is Colombia, where no concession is required to operate the service, and consequently, there is no exclusivity. Theoretically, a rival company would be free to enter at any time, although in practice this has yet to happen. In Panama, users or potential entrants can apply to the regulator for a licence to supply a subarea of the incumbent's jurisdiction. This will be granted if the current service level is inadequate or if the incumbent gives its approval. At the other end of the spectrum, in Buenos Aires (Argentina), exclusivity demands that new users actually close off previous alternative sources of supply (personal wells, etc.).

2. REDEFINING THE ROLE OF THE STATE, OR THE QUEST FOR REGULATORY INDEPENDENCE

The two central functions of the state under the reformed model of the water sector are to define policy and to regulate the service providers. The precise allocation of responsibilities between policy-making and regulatory bodies remains an area of some ambiguity in Latin America. However, in principle, the policy-maker should be responsible for the definition of long-term objectives for coverage and quality of the service, usually articulated in some kind of national plan. Furthermore, policy should define the broad strategy for meeting these objectives, in terms of investment financing, subsidy requirements, and the appropriate role for the private sector.

Regulation, on the other hand, should be primarily concerned with ensuring that the chosen strategy is implemented. In many cases, the overarching function of the regulatory body is to enforce the sectoral law "cumplir y hacer cumplir la ley." The more detailed functions typically ascribed to the regulatory body include monitoring compliance with the legal and contractual obligations placed upon operators, determining tariff levels, and resolving conflict between regulated companies and their customers.

2.1   Institutional Framework

Both policy-making and regulation are rendered more complex by the convergence of three different areas of strategic concern in the water sector. First, there is the economic perspective of the sector as a key public service, and component of urban infrastructure. Second, in a developing country context, provision of potable water and adequate sanitation has traditionally been viewed as a central public health concern. More recently, however, the environmental dimension of the water and sanitation sector has begun to be recognized, with the concomitant need to regulate abstraction of water and discharge of effluent.

This multifaceted character of water typically means that several different ministries will have an interest in the sector (Table 3.6). While the health and environmental aspects of the sector are typically assigned to the corresponding ministries, the agent responsible for the public service aspects of the sector varies considerably across countries. In some cases, the economic aspects are subsumed within the public health (Panama) or environmental (Venezuela) dimension. However, more commonly, these issues are the jurisdiction of a third ministry, be it public works (Argentina and Chile), economy (Colombia), or housing (Bolivia). Only in the case of Panama is a single institution (the Ministry of Health) responsible for all three aspects of policy, and even this is acknowledged in the relevant legislation as a stopgap measure while the necessary environmental institutions develop.

At the regulatory level, there have been greater attempts to try and integrate these three different dimensions of social concern (Table 3.7). One practical explanation for this is that the development of separate agencies charged with implementing the public health and environmental policies of government lags behind the creation of agencies for economic regulation. For example, it is not unusual for the economic regulator to have some re-

Table 3.6   Overview of Ministries Responsible for Policy-Making

Country	Economic	Public Health	Environment
Argentina	Public works	Health	Environment
Bolivia	Housing	Health	Environment
Chile	Public works	Health	Public works
Colombia	Economy	Health	Environment
Panama	Health	Health	Health
Peru	Presidency	Health	Agriculture

sponsibility for monitoring the quality of drinking water, although this responsibility is sometimes shared with the ministry of health. In some cases, the economic regulator also monitors the quality of effluent discharged from sewers. However, the issuing of licenses for the abstraction of water is almost always handled separately by the environmental authorities.

The existence of multiple political interests in the sector calls for some degree of coordination, a fact that has not yet been widely recognized in the design of regulatory frameworks around the region. In particular, policy decisions about suitable quality objectives for potable water and sewage

Table 3.7   Overview of Regulatory Bodies

	Regulatory Body[a]		
Country	Economic	Public Health	Environment
Argentina[b]		ETOSS	
Bolivia		SSB	SA
Chile		SSS	SSS (CONAM)
Colombia	CRA	SSPD	MINMA (CAR)
Panama	ERSP		ERSP (MINSA)
Peru	SUNASS	SUNASS (MINSA)	CONAM (MINAG)

[a]CAR, Corporacion Autonoma Regional; CONAM, Consejo Nacional del Medioambiente; CRA, Comision Reguladora de Agua; ERSP, Ente Regulador de Servicios Publicos; ETOSS, Ente Tripartito de Obras y Servicios Sanitarios; MINAG, Ministerio de Agricultura; MINMA, Ministerio de Medio Ambienta; MINSA, Ministerio de Salud; SA, Superintendencia de Agua; SSB, Superintendencia de Saneamiento Basico; SSPD, Superintendencia de Servicios Publicos Domiciliarios; SSS, Superintendencia de Servicios Sanitarios; SUNASS, Superintendencia Nacional de Servicios Sanitarios.

[b]Refers only to Greater Buenos Aires.

effluent can have major cost implications for water service providers and will need to be reflected in tariffs set by the economic regulator, while the need to obtain abstraction licenses from the environmental regulator may represent a barrier to entry or service expansion. There is some evidence that these factors are increasingly being taken into account. In Colombia, the minister of environment has recently been incorporated onto the commission responsible for economic regulation, where he sits alongside the ministers of health and economy. In Bolivia, the economic regulator is required to coordinate with the natural resources regulator over the granting of concessions for service provision and water abstraction.

The institutional framework for Peru (Figure 3.6) illustrates just how complex the water sector can become, as well as the problems that can arise with the allocation of roles between governing bodies. For example, the municipalities are both owners and regulators of the municipal water companies, having the final say over tariff levels and participating in quality of service regulation. While MIPRE (Ministerio de la Presidencia in Peru) has the power to appoint and remove both the Superintendent of the regulatory agency and the Director of SEDAPAL (which is by far the

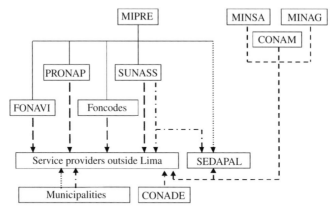

Figure 3.6   Institutional framework for the water sector in Peru. The interpretation of the acronyms is as follows (in alphabetical order): CONADE, Consejo Nacional de Desarrollo; CONAM, Consejo Nacional de Medioambiente; FONCODES, Fondo Nacional de Compensación y Desarrollo Social; FONAVI, Fondo Nacional de Vivienda; MINAG, Ministerio de Agricultura; MINSA, Ministerio de Salud; MIPRE, Ministerio de la Presidencia; PRONAP, Programa Nacional de Agua Potable; SEDAPAL, Servicio de Agua Potable de Lima; SUNASS, Superintendencia Nacional de Servicios Sanitarios.

largest of the regulated companies), in both cases, this introduces scope for political interference.

In addition, SUNASS has been charged with both regulating the sector and providing technical assistance to the regulated companies, two roles which are clearly incompatible. In practice, the highly fragmented and financially precarious circumstances of the service provider have meant that the role of technical assistance has, in many cases, proved more appropriate than the role of regulator.

2.2   Regulatory Structure

Perhaps the single most important consideration in designing a regulatory institution is that of maintaining an appropriate balance between independence and accountability. A certain degree of institutional independence is required if the regulator is to be able to function effectively as an arbitrator between the competing interests of the customers, the investors, and the state (Guasch and Spiller, 1999). On the other hand, as a public entity the regulatory agency must also be accountable for its decisions. All of the regulatory design issues to be considered below will contribute in one way or another either to the enhancement of regulatory independence or to accountability. For example, independence can be enhanced by means of financial autonomy and mechanisms to prevent day-to-day political interference in regulatory decisions, while the accountability of the regulator will depend on the establishment of an effective appeals process and the creation of vehicles for consumer participation.

Regulation tends to be organized at the national level, except in federal states such as Argentina or Brazil (Table 3.8). The sectoral jurisdiction of the regulator varies substantially between countries. Only Panama has introduced truly cross-sectoral regulation, with responsibility for water, electricity, and telecommunications vested in a single agency (ERSP). Although Bolivia has created a cross-sectoral regulatory institution (SIRE-SES), in practice it operates as a loose association of single-sector regulators. In Colombia, the regulatory process is divided into two tiers, namely the determination and enforcement of regulatory rules. First-tier activities are the responsibility of a regulatory commission (CRA), while second-tier activities are undertaken by a multisector superintendency (SSPD) that enforces the dictates of the water, energy, and telecommunications commissions, as well as providing a one-stop shop for customer complaints. In some provinces of Argentina, water and electricity regulation are jointly

Table 3.8 Overview of Design of Regulatory Agency

Country	*Institutional Locus*	*Geographical Jurisdiction*	*Sectoral Jurisdiction*
Argentina	Under Ministry	Provincial	Nine provinces: water only Six provinces: water plus electricity
Bolivia	Under Ministry	National	Water, energy, telecoms, and transport
Chile	Under Ministry	National	Water only
Colombia	Under Ministry	National	Commission: water plus garbage Superintendency: water, energy, telecoms
Panama	[?]	National	Water, electricity, telecoms
Peru	Under Ministry	National	Water only

undertaken. Elsewhere (Chile and Peru), water regulation is organized on a single-sector basis.

Regulatory agencies in Latin America typically report to a "parent ministry," rather than to the president or the legislature. The relationship with the parent ministry is probably at its closest in countries such as Chile and Peru, where a single-sector regulatory body reports directly to the ministry responsible for policy formulation in the sector. In countries with cross-sectoral agencies (such as Bolivia and Panama), the ministry to which the agency reports is not necessarily the ministry charged with policy formulation for the water sector, thereby creating greater distance between policy-making and regulation.

The Bolivian cross-sectoral regulatory agency, SIRESE, was originally designed to be a hybrid between single-sector and multisector systems of regulation. The benefits of single-sector regulation were to be achieved through the creation of five sectoral superintendencies—for telecommunications, transport, electricity, hydrocarbons, and water—each with his or her associated staff of technical specialists. The sectoral superintendents have full decision-making autonomy in their respective jurisdictions.

The benefits of cross-sectoral regulation were to be achieved by integrating the five sectoral superintendencies into a single system with a common legal framework, and placing a general superintendency at the center of the system. The superintendent general performs three main statutory functions:

- First, he or she acts as an appeals body for disputes that cannot be resolved by the individual sectoral superintendencies.
- Second, the superintendent general is required to monitor the performance of each of the sectoral regulators by issuing an annual statement on the efficiency and efficacy of regulation in each sector.
- Third, the superintendent general is responsible for approving the budgets of the sectoral superintendencies and presenting the insitution's consolidated budget to congress.

Over time, the SIRESE has begun to function increasingly as a collection of single-sector agencies rather than as a cross-sectoral body. This has been due in part to the desire of the sectoral superintendents to assert their own independence, and partly because the superintendent general's function as arbiter makes it difficult for him to act in a coordinating role.

2.3   Regulatory Leadership

Regulatory bodies are typically headed either by an individual director or by a commission of several directors. A commission arguably lends greater stability to the regulatory process, both by bringing to bear a variety of perspectives and by avoiding sudden changes of leadership (for example, by staggering terms). It is also sometimes argued that a group of individuals may be less susceptible to "regulatory capture." On the other hand, the search for consensus among a group of commissioners may blur and retard the decision-making process.

In Latin America, there appears to be a clear geographical divide on this question (Table 3.9). Countries west of the Andes (Bolivia, Chile, and Peru) have their regulatory agencies headed by superintendents, whereas countries east of the Andes (Argentina, Colombia, and Panama) have opted for commissions or directorates comprising between three and seven members. The commissioners typically elect a president from among themselves in rotation. The regulatory commission from Buenos Aires is an interesting case because the six members of the directorate represent the three levels of government with an interest in the metropolitan area—the federal government, the government of the province of Buenos Aires, and the municipal government of the federal district.

Most countries prescribe the duration of term for the leadership of the regulatory agency, which lies in the range three–six years, although in Chile and Peru there is no legally identified limit on the superintendent's

Table 3.9   Overview of Regulatory Leadership

Country	Leadership	Duration of Term	Appointed by	Criteria for Removal
Argentina[a]	Six directors	Six years (renewable twice)	Executive	Only with just cause
Bolivia	Superintendent	Five years (renewable once)	President (via senate)	Conflict of interest or legal conviction
Chile	Superintendent	Unspecified	President (via minister)	Unspecified
Colombia	Seven commissioners (three experts plus four ministers)	Three years	President	Unspecified
Panama	Three directors	Five years	Executive (via legislature)	Conflict of interest, legal conviction, or bankruptcy
Peru	Superintendent	Unspecified	Minister	Unspecified

[a]Refers only to Greater Buenos Aires.

term of office. In the case of Panama, the appointment of directors is staggered to ensure that the entire commission is not reappointed simultaneously. In some countries, the legal framework prevents directors from being reappointed to office more than once (Bolivia) or twice (Argentina). It is sometimes argued that these kinds of restrictions serve to enhance the independence of regulation because the regulator stands to gain nothing from humoring his or her political masters.

In order to be able to function as an effective arbitrator between the competing interests of the operator, the consumers, and the government, the regulator must enjoy a certain level of security of tenure. In Latin America, the appointment of regulators is almost always done through the executive branch of government, either by the responsible minister or directly by the president (Bolivia, Chile, Colombia). Only in Bolivia and Panama does the legislature participate in the appointment of regulators. In the case of Bolivia, the senate provides a shortlist of three candidates from whom the president must select the future superintendent. In the case of Panama, the legislature must give its approval to the candidate nominated by the executive.

In most cases, the law specifies technical criteria that must be met by the appointee. These vary in their breadth and rigor. However, they often include restrictions on nationality, age, years of professional experience, and nature of previous professional experience. Just as important as the criteria for appointment are the criteria for removal, since it is these that protect the regulator from arbitrary dismissal in areas where he or she may come into conflict with political masters. No such conditions are specified in the case of Chile, Colombia, and Peru, making the regulators particularly vulnerable. The regulatory framework for Buenos Aires includes the rather vague stipulation of a "just cause" for dismissal, while those for Bolivia and Panama are more explicit in citing conflicts of interest (such as commercial interests or blood ties), legal convictions, and bankruptcy. In Panama, the removal of a regulatory director requires the approval of the Supreme Court.

While Latin American legal frameworks are often exemplary in the degree of protection they provide to regulators, the reality can be very different. In particular, it is not possible to legislate against "voluntary resignations" by regulators, and in a public sector culture where political patronage is central, regulators have often been known to resign (more or less) "voluntarily" in the immediate aftermath of elections or at times of political upheaval. The typical length of time that a Latin American water regulator actually stays in office has been of the order of two–four years, compared with legal terms of five–six years. Indeed, only in the countries that have regulatory commissions have some of the regulators succeeded in serving their full legal term.

2.4   Regulatory Resources

The principle of financial autonomy for regulatory agencies is almost universal in Latin America (Table 3.10). Regulatory financing is by means of a percentage levy on the turnover of the industry. In the case of national agencies, this percentage tends to fall between 1% and 3%. An interesting exception to this pattern is the Chilean water regulator, which, due to its early establishment, predates the movement toward financial autonomy; hence, Chile continues to fund its regulatory agency through general tax. The levy for financing water regulation is often higher than that for other

Table 3.10   Overview of Regulatory Resources

Country	Staff	Pay Scale	Finance	Budget (US $ millions)	Approval Process
Argentina[a]	70	Public	Levy of 2.7%	7	Unspecified
Bolivia	20	Private	Levy of 3%	2	General Superintendent and Executive branch
Chile	80	Public	Tax revenues	2	Executive branch
Colombia	20	Public	Levy of 1%	3	Executive branch
Panama	XXX[b]	Public	Levy of 1%	XXX	Executive branch
Peru	100	Public	Levy of 2%	4	Superintendent

[a]Refers only to Greater Buenos Aires.
[b]XXX = data unavailable.

regulators (such as electricity and telecommunications). This simply reflects the fact that the water sector tends to have a lower turnover than these other industries, yet similar (or indeed more complex) regulatory requirements.

However, financial autonomy tends to be balanced by some degree of financial accountability. In most cases, the budgets of the regulatory agencies must be integrated into the general public sector budget and require government approval through the usual channels. In Bolivia, approval comes internally from the Superintendent General. A number of countries make the stipulation that surplus funds should count against the budget of the following year. However, Colombian law exceptionally allows surpluses to be transferred into central government funds.

In making comparisons between overall budget and staffing levels, it is important to bear in mind that the sectoral, geographical, and technical jurisdictions of these various agencies differ quite significantly. Budgets for the regulatory agencies tend to lie in the range $2–$4 million. The regulatory agency for the Buenos Aires concession, ETOSS, stands out as having a budget that, at $7 million is substantially larger than that of any of the other countries reported in Table 3.10. One explanation for this is that, in Argentina, regulatory agencies have sometimes been regarded as a vehicle for the reemployment of staff laid off from the utility as a consequence of privatization.

With regard to human resources, there is essentially a divide between countries such as Argentina, Chile, and Peru, which have relatively large regulatory bodies (over 50 employees), and those such as Bolivia, Colombia, and Panama, which have relatively small agencies (less than 50 employees) and rely on subcontracting services. In almost all cases, the staff are remunerated in line with civil-service pay scales. A notable exception is Bolivia, where salaries are legally required to keep pace with trends in the privatized companies subject to regulation. However, in practice, this has proved to be a bone of contention with the executive branch.

2.5   Regulatory Accountability

The opportunity to appeal regulatory decisions is an important counterbalance to regulatory independence. Given the shortcomings of the judiciary across Latin America, many countries have opted to complement judicial appeal with some form of administrative appeal to the executive branch of government. Administrative appeal processes tend to be more agile, and in some instances, better equipped to deal with the complex technical issues underlying regulatory disputes.

Administrative appeal takes a wide variety of different forms across countries (Table 3.11). At one extreme, in Peru, the regulator's decision is final with no administrative appeal at all. In Panama, appeals may be made to the regulatory body, but there is no subsequent channel for administrative appeals, while in Argentina, appeal is to the minister responsible for the sector. Chile provides an interesting exception in that administrative appeal is replaced by independent arbitration. Disputes arising in the tariff-setting process are settled by appeal to a commission of three experts: one selected by the regulated company, one by the regulator, and a third by mutual agreement between both parties.

With regard to judicial appeal, the picture is more uniform. Most countries allow direct judicial appeal through the standard court system. In some cases, appeal to the Supreme Court is also possible. For example, in Panama, the Supreme Court is the only route for judicial appeals.

The Bolivian regulatory framework incorporates a three-tier appeals procedure:

- In the first instance, those dissatisfied with regulatory decisions can appeal directly to the sectoral superintendent, who must respond within 30 days. In 1997, 86 out of 1,286 (or 7% of) regulatory decisions were appealed.

- If dissatisfied with the result of this appeal, the complainant may go on to appeal to the superintendent general, who must respond within 90 days. In 1997, 22 of 86 (or 26% of all) appeals were passed on to the superintendent general.
- If dissatisfied with the result of this appeal, the complainant may go on to appeal to the Supreme Court. In 1997, 3 out of 15 (or 20% of all) resolved appeals were passed on to the Supreme Court. As of early 1999, none of these cases had been resolved.

It is important to note that the first two layers of the appeals structure are administrative, while the final layer is judicial. The reasons for creating an administrative appeals procedure are to speed up the process and to provide a specialized appeals body.

Of all appeals to the superintendent general, 9% were initiated by consumers, and the remaining 91% by regulated companies. This suggests some imbalance in the use of the appeals mechanism, probably due to the absence of organized consumer bodies. In 80% of appeals made to date, the superintendent general has upheld the original decision of the sectoral superintendent.

A well-designed regulatory framework should give consumers as well as operators an opportunity to express their concerns. It is now commonplace, in Latin America, for regulatory agencies to have offices to deal with customer complaints (Table 3.12). The regulator effectively acts as an appeals body for customers whose complaints have not been adequately resolved by the regulated companies.

There has been a great deal less progress in the establishment of wider

Table 3.11   Overview of the Appeals Process

Country	*Administrative*	*Judicial*
Argentina	Executive branch	Direct judicial appeal
Bolivia	Superintendent, general superintendent	Direct judicial appeal Supreme Court
Chile	Independent arbitration	Direct judicial appeal
Colombia	XXX[a]	Direct judicial appeal
Panama	Regulatory agency	Supreme Court
Peru	None	Direct judicial appeal

[a]XXX = data unavailable.

Table 3.12   Overview of Consumer Involvement in Regulation

Country	Complaints Office	Public Hearings	Consultative Committees
Argentina	Yes	None	None
Bolivia	Yes	Optional	None
Chile	Yes	None	None
Colombia	Yes	None	Comites de Desarrollo y Control Social
Panama	Yes	Optional	None
Peru	Yes	None	Comites Consultivos Regionales

avenues for customer participation in regulatory decisions and debates. The mechanism of public hearings, for example, has received relatively little use in the Latin American water sector. In both Bolivia and Panama, the regulatory framework provides the option for public hearings. However, there have been no actual instances of its use to date. Nor have there been many attempts to set up consumer associations or consultative committees, either under the auspices of the regulatory agency or elsewhere.

An interesting exception is Peru, where the regulations require the creation of a consultative committee comprising representatives of the superintendency, the water companies, and the municipalities (who are supposed to represent the customer viewpoint). When the committee was set up, representatives of APIS (Asociación Peruana de Ingenieros Sanitarios) and ASPEC (Asociación Peruana de Consumidores y Usuarios) were also included. While the Peruvian case represents an important attempt to create a forum for regulatory consultation, the composition of the consultative committee seems to provide greater representation to the service providers than to the consumers. Given that the municipalities are invariably the owners of the water utilities, it is questionable whether they can really be relied upon to represent consumer interests. The implication is that, out of the 10 members of the committee, only 1, the ASPEC representative, provides an independent consumer perspective on regulation.

In Colombia, the creation of Comites de Desarrollo y Control Social at the local level is supported by the superintendency. The role of these committees is to collect information on customer satisfaction and lodge any resulting complaints to the companies and the regulators. However, to date very few such committees have been created.

3. DEVELOPING REGULATORY INSTRUMENTS, OR THE MISSING RULES OF THE GAME

Establishing a regulatory system is not only about creating the new institutional players, but also about defining the rules of the game. Such rules are typically laid down in a series of legal norms comprising instruments of public law (legislation, regulations, and decrees) and instruments of private law (concessions and licences). Some of these instruments (such as legislation) are much harder to modify than others, and hence their use entails a higher degree of regulatory commitment. This may be appropriate for establishing the fundamental principles of regulation, while areas requiring greater flexibility should be covered in instruments that are more readily amended (such as regulations or decrees). At the same time, public law is best suited to covering matters of universal application, while private law enables these to be applied to specific cases.

A key issue for potential private investors is the rate of return to be secured on investments. For countries with a high degree of investment risk, it may therefore be important to signal a high degree of commitment to providing an appropriate rate of return by specifying the level quite carefully in legislation, which is difficult to amend. Figure 3.7 illustrates three

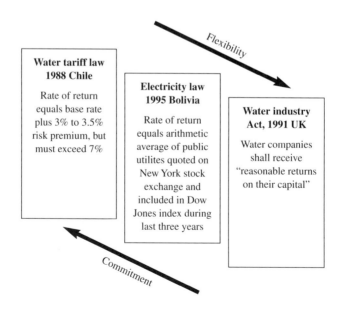

Figure 3.7   Comparison of rate of return rules across countries.

very different approaches to specifying the rate of return in regulatory legislation, exhibiting different balances between commitment and flexibility.

At one extreme, the Chilean water tariff law of 1988 states that the rate of return will be equal to a specific number, while at the other extreme, the UK Water Industry Act 1991 gives only a very vague indication of what returns the regulator will allow. The Bolivian Electricity Law 1995 illustrates an intermediate position, where the law is extremely explicit about the methodology to be used to calculate the rate of return, without fixing the number in advance.

3.1   Legal Framework

The most common pattern is that followed by Chile, Panama, and Peru, which each began with water sector legislation and went on to flesh out the details of the regulatory framework through a series of regulations leading on to the issuing of concession contracts for individual service providers (Table 3.13). This sequencing allows the regulatory framework to evolve in an orderly and logical manner. At the other extreme, the case of Buenos Aires (Argentina) is also of interest because of the absence of sectoral legislation of any kind, a situation that is also to be found in the telecommunications and transport sectors in Argentina. A regulatory body created by decree, rather than by legislation, is clearly more vulnerable to political change.

Only Bolivia and Colombia have enacted water reforms via cross-sectoral legislation affecting all of the utilities, although in the case of Bolivia, it has subsequently proved necessary to flesh out the cross-sectoral framework with sector-specific laws. This may owe something to the fact

Table 3.13   Overview of Legal Instruments

Country	Multisector Legislation	Sector Legislation	Regulations	Concessions
Argentina[a]				1993
Bolivia	1994	1999		1997
Chile		1988 →	1989 →	
Colombia	1994		1994 →	
Panama		1997		
Peru		1992 →	1994 →	

[a]Refers only to Greater Buenos Aires.

that the Bolivian Ley (law) SIRESE is much briefer and more general than the Colombian Ley de Servicios Públicos. Moreover, Bolivian experience indicated that the cross-sectoral law was not always an adequate basis for establishing regulatory authority in any particular sector. Another interesting feature of the Bolivian experience is the fact that the concession contract for water services in the metropolitan area was issued prior to the sectoral law and regulations, which are still in process. This is creating considerable regulatory uncertainty, due to a number of points on which the sector law contradicts the prior concession contract.

3.2   Tariff Regulation

One of the most fundamental aspects of the regulatory framework is the procedure for determining and reviewing price controls. Latin American water laws typically declare their allegiance to a number of basic tariff-setting principles that include allocative efficiency,* productive efficiency,† financial sustainability, social equity, and administrative simplicity (even if they differ in the degree to which they recognize and address the inevitable conflicts that arise between these objectives).

The central focus of tariff regulation in Latin America has been the definition of so-called "fórmulas tarifarias," which state the mathematical relationship between tariffs and underlying costs. These generally take the form of present-value formulas based on cost and demand projections discounted at the estimated cost of capital. The main objective of these formulas is to ensure that water tariffs are set at a level compatible with financial sustainability. In most countries, the formulas reflect average cost pricing principles, although, in the interests of allocative efficiency, Chile and Peru have espoused some degree of marginal cost pricing. The concept of financial sustainability includes a reasonable return on new investments. In Chile, the allowed rate of return has been set as 3.0–3.5% above the base rate, with a floor of 7%, while in Colombia, the regulator has used an indicative range of 9%–14% in tariff-setting.

Water sector tariffs are rarely cost-reflective from the outset of a reform process. Hence, in countries such as Colombia and Peru, a convergence period has been established. Peruvian legislation is particularly lucid in de-

---

*Allocative efficiency is achieved when prices provide consumers with accurate signals of the marginal cost of the service so that an efficient level of consumption takes place.

†Productive efficiency is achieved when the service is provided at the minimum feasible cost.

fining three stages of tariff convergence. In the first eighteen months following the legislation, described as the "etapa preparatoria," tariffs must cover operating expenditures while water companies work on defining their investment plans. There follows a five-year "etapa de mejoramiento," during which tariffs should rise progressively to the level of long-run marginal cost, where they remain in the "etapa definitiva." In Colombia, the regulator has come under political pressure to slow the rate of tariff increase, with the result that tariffs are currently lagging behind their legally envisaged convergence path.

While financial sustainability is an essential first step, tariff formulas that simply pass on current cost levels to customers provide no guarantee of productive efficiency. In many cases, the legal framework stipulates that tariff-setting should be based on efficient rather than actual costs. In reality, only Chilean law provides an explicit methodology for assessing efficiency. The Chilean approach to efficiency adjustment is extremely sophisticated, but perhaps in some respects unduly complex. One reason for the complexity is that, for legal reasons, it is necessary for the regulator to identify the exact nature of any inefficiency, and he or she is not at liberty to impose general efficiency targets. More modest steps toward efficiency analysis are being taken in other countries. For example, in Colombia, operating costs reported by companies are adjusted downwards to reflect a "benchmark" level of 30% leakage. However, other sources of inefficiency are not taken into account.

Regarding tariff review procedures, there has been a widespread acceptance of the British style "price-cap" approach across Latin America. The review cycle tends to be every fourth or fifth year, although in many cases, as in the United Kingdom, extraordinary reviews are permitted sooner. Where the regulated company is privately operated, the price-cap mechanism should provide adequate incentives for costs to be reduced over time, allowing the associated efficiency gains to eventually be passed back to customers. In this context at least, the absence of efficiency targets in price-setting merely slows the pace at which customers benefit from efficiency improvements. However, when the price cap is (more commonly) applied to a publicly managed company, no such incentive arises, so that the need for externally imposed efficiency targets becomes particularly critical.

The Chilean "empresa modelo" is particularly interesting. The Chilean regulator has developed a rather unique approach to regulatory tariff-setting, which is based on the concept of a model company (or empresa modelo). The model company is an extreme application of the concept of yardstick regulation, whereby the regulator creates an idealized benchmark specific to each regulated company and uses comparisons between the ideal and the reality to gauge the extent of any inefficiency.

Unlike more conventional forms of yardstick regulation, the model company combines two forms of efficiency:

- *Engineering Efficiency.* In the sense of undertaking an analysis of the optimality of the physical configuration of the infrastructure.
- *Economic Efficiency.* In the sense of applying least-cost functions to determine the cost of operating the optimized infrastructure.

Furthermore, the process of analyzing the model company takes place at a highly disaggregated level, with the 13 regional companies being broken down:

- Horizontally, into 320 water supply and 270 sewerage systems, which are grouped into 37 similar types, in order to facilitate the calculation of tariffs;
- Vertically into four distinct stages of production—water production, water distribution, sewerage collection, and sewerage treatment.

The model company is used to produce two parallel estimates of the tariff based on average and marginal cost. The latter is adjusted in the light of the former to ensure the financial viability of the company.

In order to ensure that tariff revisions are coherent with the operator's technical targets and the underlying investment program, many countries have developed quinquennial planning instruments. In the cases of Argentina, Bolivia, and Peru, although the plans are explicitly technical in content, the regulatory framework explicitly stipulates that they should be synchronized with the tariff-setting procedure, with tariff revisions usually following on once the technical planning process is concluded. However, in the case of Colombia, there is a notable absence of coordination between the Planes de Gestión—which contain efficiency and output targets and are approved by the ministry—and the tariff formulas—which are calculated as part of a separate process and are approved by the regulatory agency.

3.3   Regulation and Social Policy

While water is regarded as a socially sensitive sector throughout the Latin American region, few countries have attempted to provide a clear definition of their social policy objectives or to evaluate the efficacy of the instruments traditionally chosen to achieve them (Foster et al., 2000). The central concern of Latin American social policy in the water sector has been to ensure the affordability of the service to low-income households. This issue has typically been addressed through a complex array of cross-subsidies that include a rising block tariff structure, and the application of substantial surcharges to industrial tariffs. More recently, in the context of sector reform, attention has been turning to the inadequacy of service coverage among poor households. The standard approach to this problem has been to incorporate legally binding connection targets into concession contracts. In return for expanding the service into commercially unattractive areas, the operator receives the right to oblige local residents to connect to the network and to levy a substantial connection charge. The Buenos Aires (Argentina) and La Paz (Bolivia) concessions provide examples of this approach.

However, the accumulating evidence suggests that this standard blueprint for social policy carries a number of significant flaws. On the one hand, the existing cross-subsidies often do more to benefit the middle classes than the poor. For one thing, the poorest families tend to be those that remain unconnected to the network, and are hence unable to benefit from cross-subsidies. For another, the criteria used for allocating the cross-subsidies do not always correspond with the economic condition of the recipient. In the case of rising block tariffs, "subsistence consumption" thresholds are often set so high that they wind up benefiting the vast majority of domestic customers.

On the other hand, the charges levied for mandatory network connections have sometimes been set so high as to be well beyond the means of the poor households they were intended to benefit. The classic example is the Buenos Aires concession, where infrastructure charges of the order of $600–$800 led to widespread civil unrest and threatened the financial equilibrium of the concession contract until a compromise solution was found by spreading the costs of network expansion across all water consumers.

Chile and Colombia provide interesting exceptions to this broad pattern, in that both countries have developed more scientific approaches for identifying poor households, so as to improve the targeting performance of their subsidy schemes. In Chile, targeting is based on a socioeconomic score derived from a wide-ranging household interview, while in Colombia, targeting follows a nationwide socioeconomic classification of neighborhoods based on the physical quality of local housing and amenities.

3.2   Output Regulation

Quality of service regulation is a necessary complement to price regulation. Without it, regulated companies may have an incentive to compromise the quality of service as a cost-cutting measure.

Progress toward an operative system of quality of service regulation has tended to be much more significant in cases where private-sector participation has taken place. The general tendency has been to define quality of service parameters in lower level legal instruments, such as concession, rather than in the general sector law. Typical parameters include potability, pressure, and continuity. A common approach is to classify the various different types of potential noncompliance according to a hierarchy, and then to establish a range of fines for each level in the hierarchy. The fines are sometimes expressed in monetary terms and sometimes as percentage of revenues. Within these ranges, the regulator is given a certain amount of discretion to determine the exact value of the fine to be applied in any particular instance, depending on the gravity of the offence. However, while fines have proved to be an effective sanction for private sector operators, they have not had much of an impact in motivating public sector managers.

In Colombia, the four yearly Planes de Gestión are the primary vehicle for output regulation. The plans specify objectives for a range of key managerial and technical parameters over short-, medium-, and long-term planning horizons. The indicators include financial efficiency (billing and collection ratios, execution of investments), technical efficiency (leakage and labor productivity), and quality of service (service and metering coverage, continuity, potability). In practice, it has proved difficult to find effective sanctions for noncompliance with these targets where state-owned enterprises are concerned. Fines, which are supposed to be deducted from employees' salaries, have tended to become paralyzed in the judicial appeal process. Attempts to motivate management by means of publicizing good and bad performance using honors and awards have not been very successful. However, the government has proved unwilling to impose the ultimate sanction of restricting access to credit by poor performers.

3.5   Regulatory Information

The fundamental problem of regulation is one of asymmetric information between the regulated company and the regulatory agency. The regulated company will have a strong incentive to abuse this strategic advantage by undersupplying information or distorting the information supplied. It is therefore critical that the regulatory framework establishes the obligation of the regulated company to supply information to the regulator in the form required.

Where private sector participation has taken place, information requirements are usually specified in some detail in the concession contract. They typically include detailed periodic reporting of both financial and technical performance, as well as the maintenance of detailed registers of operating assets and customer records. All information must be audited before it is submitted to the regulator.

However, such information cannot be readily interpreted unless it is prepared in accordance with clearly specified regulatory accounting guidelines, which often go far beyond standard accounting requirements. A number of countries have been working on the development of such guidelines, for example, Colombia's Plan Unico de Cuentas. However, much remains to be done in this sphere.

4.   CONCLUSIONS

The scale of water service reform in Latin America has been truly impressive, and unparalleled in any other part of the world. By the close of the decade, just about every country in the region had undertaken or was actively considering sector reform measures. However, the real transition for most water consumers has not been from public to private operation but rather from unregulated centralized public provision to regulated decentralized public provision.

The reformed sector model that has emerged in the Latin American region might well be described as an "Anglo-French hybrid." On the one hand, it takes from the British model the creation of a centralized regulatory agency relying on incentive-based regulatory instruments, but rejects the equally British notion of regionally consolidated and fully privatized water operators. On the other hand, it takes from the French model the notion of a decentralized, municipally based industry relying on concession contracts as the primary vehicle for PSP, but rejects the equally French notion that regulation can be confined to municipal monitoring of contracts.

The combination of these two approaches has created tensions that tend to undermine the functioning of the new model. These include the problems entailed by imposing regulation and PSP from the center on a sector that is often legally under municipal control, as well as the difficulty of attracting private investment into a highly fragmented sector. While the attempt to regulate state-owned water utilities using incentive-based instruments—which are premised on the existence of a profit motive—has unsurprisingly not proved to be very effective.

In order to escape this impasse, the region faces one of two possible options. The first is to try and consolidate the structure of the industry by promoting the creation of larger regional companies, as is now the case in Chile and has been the case in Brazil. Depending on the country context, this might be achieved voluntarily (by "mancomunidad"), through financial incentives (such as PLANASA), or by fiat (as in England and Wales). The second option is to rethink the way in which regulation is done, adapting it to the requirements of a fragmented and still largely state-owned sector. Either way, it promises to be a challenging task.

REFERENCES

Foster, V., A. Gomez-Lobo, and J. Halpern (2000). *Designing Direct Subsidies for Water and Sanitation Services. Panama: A Case Study,* Policy Research Working Paper 2344, Finance, Private Sector and Infrastructure Sector Unit, Latin America and Caribbean Region. Washington, DC: The World Bank Group.

Levy, B., and P. T. Spiller (1996). *Regulations, Institutions, and Commitment: Comparative Studies of Telecommunications.* Cambridge: Cambridge University Press.

World Bank (2001a). *Brazil: Regulations for Better Water and Sewerage Services,* Report No. 19568-BR. Washington, DC: Brazil Country Management Unit, Finance, Private Sector and Infrastructure Unit, Latin America and the Caribbean Region, The World Bank Group.

Yepes, G. (1990). "Management and Operational Practices of Municipal and Regional Water and Sewerage Companies in Latin America and the Caribbean," Discussion Paper, Infrastructure and Urban Development Department, The World Bank Group, Washington, DC.

ADDITIONAL READINGS

Estache, A., V. Foster, and Q. Wodon. *Infrastructure Reform and the Poor: Learning from Latin America's Experience, Regional Study, Latin America and Caribbean Region and World Bank Institute.* Washington, DC: The World Bank Group, 2001.

Guasch, J. L., and Spiller, P. *Managing the Regulatory Process: Design, Concepts, Issues and the Latin America and Caribbean Story.* Washington, DC: World Bank Latin American and Caribbean Studies, Finance, Private Sector and Infrastructure Unit, Latin America and the Caribbean Region, The World Bank Group, 1999.

SIRESE. *La Regulacion Sectorial en Bolivia 1997.* Superintendencia General, SIRESE. La Paz, Bolivia, 1998.

World Bank. *Brazil: Private Participation in the Water Sector,* Report No. 19896-BR. Washington, DC: Brazil Country Management Unit, Finance, Private Sector and Infrastructure Unit, Latin America and the Caribbean Region, The World Bank Group, 2001b.

Chapter 4

The Pursuit of Efficiency Through Benchmarking

The Experience of the Water Industry in England and Wales

Stephen Ramsey

Following a long period of consolidation within the water industry in England and Wales, 10 multipurpose water authorities were created under the Water Act of 1973. These authorities were given responsibility for the planning and control of all uses of water in their river catchment area, including primarily water supply and sewerage. All 10 authorities were privatized in November 1989 by way of public share sale. The new companies both own the assets and hold operating licences.

A Department of Environment White Paper published in 1986 stated:

> Privatisation itself will encourage the water service plcs [public limited companies] to compete effectively in fields where they can do so. Where this is not practical the Government's aim is to introduce a system of regulation which will stimulate a competitive approach. Profit is a more effective incentive than government controls.

Stephen Ramsey is Director of WRc Utilities and Consulting, part of the WRc Group, London. He has a law degree and is a Chartered Accountant. An expert on benchmarking, he consults for private companies and international organizations.

Reinventing Water and Wastewater Systems: Global Lessons for Improving Water Management, edited by Paul Seidenstat, David Haarmeyer, and Simon Hakim.
0-471-06422-X   Copyright © 2002 by John Wiley & Sons, Inc.

Economic regulation of the water industry in England and Wales has been undertaken by the Director General of Water Services, acting through the Office of Water Services (Ofwat). The Director General also assumed regulatory responsibility for the remaining service providers, which continued as privately owned statutory water-only companies.

The water companies in England and Wales, both privatized and historically privately owned, face unique pressures, not only to provide a quality regulated service at determined tariffs, but also to invest to maintain and improve network and services, to meet ever growing and sophisticated customer expectations, and to meet shareholders' capital growth and dividend expectations.

The English and Welsh model of privatization is perhaps at one end of a spectrum of operational models for the water industry, ranging from fully privatized to municipally owned and operated, and is not one that is likely to be widely replicated. Nevertheless, the experience gained from the pursuit of quality and financial performance improvements should provide object lessons for the wider industry in other countries.

1. REGULATORY BENCHMARKING

The primary duties of the Director General relate to the efficient operational and financial performance of the companies. The position's secondary duties relate to the protection of customers and the promotion of economy and efficiency. Overall, the Director General seeks to promote real competition within the sector.

The Director General seeks to satisfy these objectives by the extensive collection and analysis of metric benchmarking data and explanatory factors, which are used both to determine comparative performance and as a basis for individual company price determination. A range of indicators (the Level of Service Indicators) has been established covering: (1) reliability of water supply (pressure, interruptions); (2) adequacy of water resource (availability, bans); (3) sewerage services (flooding); and (4) customer service. These indicators are consistently monitored between companies, against preset targets, and against other sectors. The regulator also requires reporting against a range of financial and nonfinancial measures. Capital expenditure assumptions are based on Strategic Business Plans prepared by companies, detailing investment proposals, and are reviewed against standard costs derived from the range of other companies. All of these data are used to assess the relative efficiency of the companies.

Based on these comparators, the Director General operates a price cap regime, where the average price rise to customers for a given period is defined as KPI + K, where KPI represents the retail price index and K is a company-specific combination of a real reduction in prices deriving from increased efficiency offset by an increase in prices to fund investment to meet new environmental standards.

In principle, companies are expected to provide existing standards of service (normal utility operations) at prices falling in real terms, and generally to fund improvements in standards of service through greater efficiency (and lower returns), rather than higher prices. Companies that are relatively less efficient are set tougher performance targets by the regulator.

Results reported by Ofwat show that industry performance has shown a steady improvement across the range of Level of Service Indicators. At the same time, operating expenditure has fallen in real terms.

The work undertaken by Ofwat is a particularly in-depth and sophisticated form of metric benchmarking. Nevertheless, it remains open to the criticisms that metrics cannot totally account for explanatory factors, particularly geographic variations, and that the underlying econometric models may be fundamentally too simplistic, and indeed, may run contrary on occasions to engineering common sense. Nor do metrics answer how to change or improve; that is for company management to resolve. From the companies' perspective, the data collected are inextricably linked to price-setting, leading to particular emphasis on presentational issues, and on occasions, to suboptimal decisions, such as the need to spend to capital limits.

Nevertheless, the regulatory regime has driven significant performance improvements and financial efficiencies. Some of these will have been achieved to climb the relative performance and efficiency tables, the desire to be upper quartile, and some on account of the regulatory stick. However, the key in the end is that fairly long regulatory delays under price-cap regulation maximize the incentives for productive efficiency in line with the length of time the company is allowed to enjoy abnormal profits secured through cost efficiency. As the 1986 White Paper declared, "Profit is a more effective incentive than government controls."

2. OBJECTIVES FOR COMPANY BENCHMARKING

While the specific objectives of companies in England and Wales will differ from those in other countries and regulatory jurisdictions, nevertheless, many of the objectives will be of wider relevance. The typical English water company has four main objectives, the first defensive and the other three proactive:

1. To beat the regulator—that is, to ensure that company performance moves consistently in line with the moving target Levels of Service Indicators set by the regulator, such that the company is not penalized for underperformance by way of reduction against the price-cap tariff increase. Benchmarking also is seen as a mechanism for reinforcing the message to the regulator that the company is actively pursuing performance improvement and cost-efficiency objectives and therefore carries no fat. These are the primary aspects of the defensive objective.

The company will, more importantly, have a number of key proactive objectives:

2. To be an upper quartile, if not the best, performer against the Ofwat Levels of Service measures.
3. To establish a center of excellence as a base for national competitive or global expansion, where reputation counts.
4. To maximize opportunities to increase profits by implementing operational and cost efficiencies. Under the regulatory regime as discussed above, companies retain the benefit of such efficiencies until the next price review, when the cycle restarts. Efficiencies become progressively more difficult with each cycle, reinforcing the benefits of learning from the experience of others. The regulatory system has also, in the past, encouraged companies to underspend their regulatory capital allocation on an item-by-item basis.

While the regulatory system evidently provides an impetus for performance improvement, it also clouds the issue in that any commercial company should seek to improve efficiency and reduce operating and capital costs, not as a regulatory objective, but to satisfy the normal commercial objectives of its owners or shareholders. Where the water sector differs is clearly the absence of a developed competitive marketplace.

In other instances, companies that are not profit-driven, such as a number of the Scandinavian operators, seek operational excellence even though this may be at the expense of cost efficiency. This does not preclude the benefits of benchmarking, where benefits to any company or operator focus on its stated objectives.

3. PROCESS BENCHMARKING

As companies have gradually exhausted the more obvious methods of costs cutting, primarily by staff reductions, and to avoid risking the companies' performance as measured by the Levels of Service Indicators, they have increasingly looked to formal benchmarking. A report commissioned by Ofwat in 1999 concluded that, from the data available, there appeared to be no uniform approach to benchmarking within the water industry (either in terms of use, focus area, depth, or quality), but that the evidence from other privatized industries was that cost reductions were achievable on a sustainable basis. This is a significant challenge to the water industry.

One English company commissioned a significant six-month process benchmarking exercise in 1998/1999, followed by an ongoing period of implementation of recommendations. The company's objectives were as discussed above. The company operates within the regulatory benchmarking framework, and has therefore already generated a significant body of metric data. It was therefore decided to minimize the collection of metric data to that essential for the process benchmarking exercise, and to focus efforts on identification of how things could be done better and/or more cost effectively and how far these lessons were applicable to the benchmarking company.

It is worth noting at the outset that the company, in line with the rest of the industry in England and Wales, has made significant improvements in its operating efficiency and capital procurement costs in the period since privatization. The company consistently ranked as average or above average in relation to the Ofwat sample, without any obvious areas of significant underperformance. The Ofwat sample could be perceived overall as relatively well performing by comparison with the sector worldwide. That said, Ofwat does not perceive the sector per se to be at the frontier of efficient practices. Our company could be a good performer in a relatively underperforming sector.

The exercise was developed on the basis of four fundamental principles:

1. Metric benchmarking is used to identify the performance gap between the company or within the company and against external comparators. The collection of metric benchmarking data is strictly limited, but is nevertheless essential for the calculation of key performance indicators (KPIs), measurement of interdivisional or other internal comparative performance, identification of key areas of opportunity within the company, identification of best prac-

tice performers as comparators, relative performance measurement of the company against external comparators, quantification of benefits and costs of recommended operational improvements, and measurement and monitoring of benefits and costs of change implementation. For these purposes, a clearly focused metric collection exercise is undertaken. Metrics are the means to an end rather than an end in themselves.

2. Process benchmarking is used to develop an understanding of the causes of the performance gap and how this can be closed or narrowed by implementation of performance improvements learned from benchmarking partners (whether within the company or externally).

3. The opportunity is taken to employ best practice from within the company across the board as far as is possible, establishing an internal level playing field. Experience shows that internal benchmarking can be a fertile source of improvement opportunities and ideas.

4. Best practice partners are identified both within and outside the water industry. In many cases, best practice is seen to reside outside the industry, where market forces and competitive pressure have proved effective drivers of improvement and efficiency.

4. METHODOLOGY FOR PROCESS BENCHMARKING

The methodology utilized falls into three distinct phases, namely, identification of what to benchmark, the actual exercise of benchmarking to identify appropriate best practice and implementation, and monitoring of improvement. Underpinning these is the organizational structure of the exercise, and a number of key principles emerge:

1. The exercise must have clear client ownership. This may be in the form of a benchmarking manager, a sponsoring technical manager, or ideally, and in this case, a high-level steering committee with responsibility for setting directional guidance, facilitating participation in and cooperation with the exercise team at both a managerial and operational level, adopting recommendations, and driving and monitoring implementation of recommendations.

2. It is often necessary to curtail the extent of ongoing benchmarking

and related activity, and it can be particularly important where a major process benchmarking exercise is being undertaken that benchmarking activity is consolidated and coordinated through one exercise.

3. There are many variations on the makeup of a project team, but an effective solution is a joint client and consultant team. This is a particularly hands-on exercise. It is essential that the client team members be familiar or very familiar with the area to be benchmarked. This will enable them to understand and evaluate opportunities from a technical or operational perspective, and will facilitate interface with client staff. It is important that client participating staff be of sufficient standing to be able to exercise some authority but be sufficiently hands on to be prepared to undertake some work and to be *au fait* with day-to-day operational issues. These staff will also need to win the confidence and participation of the general body of staff, who probably anticipate yet another head office data-gathering exercise. These staff will also need to agree and support recommendations (expert recommendations are rarely implemented) and present these to client management and win their support for implementation. Where possible, client project team members should be excused their normal duties. This is not a part-time exercise, and there is a clear and demonstrable linkage between effort and input and the quality and value of the results.

4. Consultant team members are expected to bring benchmarking expertise and an independence of mind to question established arrangements and methods, and to open doors to views that might not be apparent on an initial cut.

5. Overall, the client must believe in and support the exercise, and must be patient for results. While some benefits will fall out of the exercise on an ongoing basis, much of the benefit will only become apparent when the exercise has been properly completed, in this case over a period of six months to the point of recommendations for implementation.

6. Where possible, the project team should be free of political pressures such that the project can focus on the most important areas regardless of how politically sensitive these may be. Where there are political no-go areas, these should be clearly stated up front to avoid wasted time and effort.

7. Throughout the exercise, the project team should exercise a pragmatic approach. This is not an exact science; there are very few absolutes. The most important task is to identify workable improvements for the company, those that will generate real efficiency gains.

5. IDENTIFICATION OF KEY PROCESSES

The first major task for the project team was to identify what to benchmark. It is of key importance to ensure that what is to be benchmarked is of consequence to the company and is in line with company priorities. The company may have an overriding interest in benchmarking a particular area, and this should clearly take priority, although it is essential to establish that the area may or may not be an optimal area for benchmarking based on an objective view of the potential afforded by the particular area.

The joint project team will undertake and agree to an analysis of the company business or of the target area of the business by process rather than by function. It is preferable at this stage to leave the definition of key areas for benchmarking as open as possible, as it can be the case that the key areas turn out to be different to what was anticipated and are not always the obvious areas. Thus, it is preferable to analyze the whole company or business, although certain boundaries between water and wastewater, or between home and overseas activities, or indeed between core and other group business activities may be drawn.

Most companies are structured by function, for example, the finance department, the wastewater department, or the maintenance department. The structure will vary between companies even among those in the same sector. What the project team must do is to redefine the operations of the company by process, that is, for example, revenue collection, sewage collection, or maintenance of assets. Each process will probably combine activities from a number of functions. Revenue collection may include activities falling under the water, finance, call center, and legal departments. The process structure will also differ between companies even in the same sector. Thus, the project team must agree on an appropriate breakdown, probably at least to the level of processes and subprocesses, supported by a list of activities falling within each subprocess, and agree on definitions for each item. This is very important, and should be agreed on at an early stage. It is particularly important to ensure that no part of the company's operations is omitted from this review in error.

Once the process structure has been agreed upon, the processes and their constituent activities are costed using process and activity costing. This method defines which activities consume what resources, is independent of any overhead allocation system, and facilitates questioning of how activities are currently conducted.

Some of the data required for activity based costing will fall out of the accounts, where there is overlap, and some companies will produce activity-based costed management accounts. Nevertheless, the match may not be exact, and indeed accounting systems and timesheets are seldom accurate or sophisticated enough to capture this information directly, and some additional analysis will be required. It is possible to undertake this exercise as a first cut and, providing all costs within the business have been allocated to some process, this should be adequate for the purpose of focusing on key cost areas, which will be analyzed in considerably more detail at a later stage.

Some companies may identify the construction of an activity-based cost model as a beneficial exercise in its own right, and further analysis at this stage, along the lines indicated below for the more detailed analysis, would be required. Where necessary, financial analysis of the activity-based costs will be undertaken, including, as appropriate, rate of return and cash flow analyses. What is being sought is the identification of those processes or parts of processes which are or have:

A significant revenue item
A significant operating cost item
A significant capital cost item
A significant impact on revenue
A significant impact on operating costs
A significant impact on capital costs

It is important to think from all these perspectives, as a process that is not cost significant in itself, may have an enormous "knock-on effect" and merit benchmarking on that basis when it might not be seen to merit it on a first review. An example of this could include the capital planning process.

It is also important to bear in mind that cost may not be the only or main criterion. There will be areas where companies are prepared to overspend, perhaps to avoid environmental risk or to maintain a profile in an area or a reputation for excellence. The next step is to undertake a stakeholder analysis, starting with identification of the relevant stakeholders. These are likely to include the shareholders or owners, regulators, customers, employees, and the wider community, but may also include lenders and the media as examples. Where possible, use should be made of existing research, including customer surveys. It is also important to rank the stakeholders in order of importance from the project perspective.

Each stakeholder group will have its own objectives and requirements that need to be considered and ranked by group and between groups. To the list above should be added:

A significant impact on performance

Quite considerable effort can be expended at this stage, and indeed in the whole business of identification of key areas for benchmarking, which in some ways could be perceived as the starting point for the project. It can only be emphasized that all the time and effort input at this stage will be repaid as the exercise progresses. Fundamentally, time and effort at this stage will ensure that the full efforts of benchmarking will be brought to bear on the process or process area with optimal potential for benefit to the company. Secondly, much of the knowledge gained at this stage will be used directly at later stages of the exercise.

Some pragmatism should be brought to bear on the stakeholder analysis, and perhaps one of its key uses is to identify areas that should be ruled out on the basis of particular interests. However, caution should be exercised here, and good opportunities should not be eliminated lightly. There will be other areas where there are real obstacles, such as legal requirements or regulatory requirements, which might make cost-cutting difficult or impossible. There may also be "special cases," which are removed from the equation for whatever reason, and issues of practicality.

Armed with all of this information, the project team is well placed to make recommendations to company management on areas to be chosen as priority processes. These ought to be areas that are significant to stakeholders and/or that have financial significance for the business; in either case, they should be areas that are within the abilities of the company to change (see Figure 4.1). Anticipated future changes in policy and legisla-

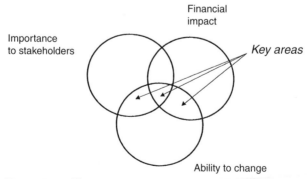

Figure 4.1 Choosing priority processes.

tion should also be borne in mind in terms of the company's ability to effect change.

As a conclusion to this initial analysis, the project team will identify key performance indicators and drivers of performance for each selected area for benchmarking. These will form the basis for measurement of company performance against external comparators.

6. IDENTIFICATION OF BEST PRACTICE

At this stage, one or more priority processes or subprocesses will have been identified as being of particular significance to the business. It is now essential to analyze these areas in much greater detail. The effort involved in this exercise will be repaid:

1. The joint project team will master the detail of the area being reviewed, and thus will be well placed to identify improvement opportunities in other businesses.
2. The team will be well placed to establish what improvement could be implemented from best practice in other businesses.
3. The exercise itself will involve self-examination by the team and will unearth a significant number of improvement opportunities from objective observation and from the collected views of employees involved in the detail of operations. It may be that the benchmarking visits simply serve to prove or develop some of the ideas identified internally.

It is therefore of particular importance that sufficient time be allowed for the team to gather the data required. This will entail an emphasis on direct observation and first-hand discussions with operatives, supervisors, and managers. From these will be gleaned a range of responses covering what ought to happen, or indeed what actually happens as opposed to what happens according to the book. The team should constantly question why things are done in a particular way.

The team will process map each chosen activity based on actual methodologies. Noting where these diverge from regulations or approved approaches may produce early suggestions for improvements. Similarly, where different divisions, groups, teams, or even shifts perform in different ways, these differences should also generate suggestions for improvement, at least as far as the establishment of a level playing field internally.

Some of the best performance improvements can originate at this stage, and can indeed be very obvious on an objective viewing. The team must remember to look for performance gains as well as cost efficiencies, in line with the project objectives.

The team will also develop the activity-based costing for each area by direct observation and questioning of time and cost allocations, to determine a more accurate baseline for external comparison. Understanding of this breakdown will also facilitate comparison on a like-for-like basis.

At all stages of the process, it is imperative not to overdo the level of detail. Too little will enable only a superficial analysis and benchmarking comparison, but too much can be equally damaging, leading to a project bogged down in detail. Always the team is looking for the areas of significant change, and the focus needs to be on the bigger cost items.

Armed with an accurate cost and map of each selected activity, the team is in a position to commence external comparison. Identification of appropriate benchmarking partners merits an appropriate degree of effort and is a time-consuming exercise. At least two months is a typical time lag for identification of suitable partners and setting up benchmarking visits at mutually convenient times.

There is a range of benchmarking clubs and forums offering introductions and facilitating benchmarking between willing partners. They tend not to identify the best partner for any particular situation, and every situation is different, but they do provide a good base of likely partners and contacts.

Nevertheless, the identification of best practice partners requires a degree of pragmatism. No database or club has all the answers. The best solution is to identify those sectors or types of business where one would expect best practice to lie in the target area. Thus, in the United Kingdom, it might be expected that utilities other than water would display superior performance on the grounds that they have faced significantly greater competitive pressures since privatization. They will, however, share many commercial similarities, such as dispersed asset bases, a wide variety of asset types, variable criticality of assets, large customer bases, and 24-hour service requirements. The precise area of similarity will clearly depend on the chosen area for benchmarking.

Beyond utilities, companies in highly competitive sectors should be considered, or those where the sort of activity is perceived to be critical to the performance of the industry. Traditionally, the Japanese car industry is perceived to display best production line practice, where downtime has a high cost implication. Other types of comparators could include the cement industry, which has per se a low value product like water but significant capital costs, and a highly competitive operating environment.

Other examples of best practice can be found in the oil industry, fast food industry, and banking and financial services, among others. No company is superior across the board. A superior company may not be superior in the required area of comparison, and best practice will often occur in unlikely places.

Once the sectors have been identified, they should be populated with high-ranking companies. These will be identified from official rankings, performance awards, industry recommendations, personal knowledge, and general recognition as an industry leader or other evidence of innovation or ongoing change programs. Obviously, partners must be willing to benchmark, and it is worth factoring in companies with established benchmarking programs or those known to be involved in benchmarking or as willing partners.

Companies identified at this stage are then ranked according to the criteria selected, where information is known. What is sought is best practice and the best match. A long list will be derived that can be further analyzed against the key performance indictors with data derived from publicly available sources. This should generate a shortlist covering a range of companies in a range of sectors. This is important because different companies will display best practice in different component parts of any process, and best practice for the company undertaking the benchmarking could come from one or more sources, and indeed an unlikely source.

The shortlisted companies are then contacted to confirm willingness to participate, and practically, their ability to benchmark within a given timeframe, and to define a performance gap, as defined by information derived from a questionnaire. It is also necessary to confirm the fundamental appropriateness of the partner and comparability and compatibility of the processes to be benchmarked. If necessary, a preliminary visit should be undertaken. If, at any stage, a partner is determined not to be an appropriate benchmarking partner, then no further effort should be expended in that direction.

Visits must be carefully structured and planned, with all sides having clear objectives, and a clear understanding of issues affecting confidentiality and reciprocity of visits and information share. A benchmarking code of conduct, such as the American Productivity and Quality Center–International Benchmarking Clearinghouse model, should be adopted and agreed upon with all participating partners. Planning, briefing of all participants, appropriate staffing, meeting the right people at the right levels in the partner companies, and the careful avoidance of "industrial tourism" all reap rewards in terms of the benefits to be derived from the visits.

The key to deriving benefits from these visits is to assess what the identified best practice partners are doing, and at what cost, and determining how or how much of these practices could realistically be implemented in the benchmarking company and at what cost. The benefit is the saving this would bring to the benchmarking company. There are no universal best practices, and what is right for one company in one situation may not be right for another in another situation.

Following a series of visits to designated partners, the benchmarking team will be in a position to draw up an agreed schedule of recommendations for performance improvement, together with cost benefit analysis, and a draft timetable for implementation. This must be presented to client management to achieve their support. This support will more readily be achieved when recommendations are made by client staff who are vouching that these are not simply consultant recommendations, but workable and supported realistic recommendations for performance improvement. Often, individuals making these recommendations will be responsible for their implementation. In any case, they will return to their department and will have to live with the implications of implementation.

Recommendations should be a combination of doing what is done now better and changing what is done now by the implementation of best practice. Recommendations will also need to be a combination of easily implementable recommendations and those that will take effect over the longer term. It is important for the client to see early changes following a process that has probably already lasted at least six months.

The cost benefit of all changes should be carefully analyzed. Clearly, changes should only be implemented where a clear case is made. Some changes may require significant capital investment or major changes in information technology or operational programming. These will not be undertaken lightly, and even when a clear case is made, the suggested change may still be subject to detailed scrutiny and delay. Some changes may never be implemented. Finally, the package of changes should fit together such that there is no overlap or contradiction in the recommended activities, all of which in their own right might be best practice.

7. IMPLEMENTATION AND MONITORING OF IMPROVEMENT

Implementation is clearly critical to the success of any benchmarking exercise, and should involve a planned, phased, costed, and monitored program with clearly defined targets, and with ongoing clear management ownership and direction. High-level ownership of the recommendations improves the likelihood of implementation, as does operational level input and support. But all such programs require vigilance as to the rate and effectiveness of implementation. Success should be measured against the timescale and against the key performance indicators established earlier in the process and the targeted improvement in those indicators (the performance gap).

The program should aim to achieve reasonable targets and a reasonable number of changes phased in over time. Improvement should be an ongoing process, and it cannot all happen at once. It is preferable to see effective implementation of changes in a phased and coordinated manner.

8. CONCLUSIONS

Benchmarking is a tool widely used worldwide and across a wide cross section of industry and government and noncommercial activities, and has established a track record over quite a number of years. But it has developed later in the water sector, which, in the United Kingdom, tends to a natural monopoly and inherent conservatism.

Ofwat has developed a regulatory benchmarking approach that has since been adopted or adapted for other jurisdictions. It has formed the basis for the regulatory stick and has served to incite companies to seek greater operational efficiencies and to improve standards of service to customers. It has also opened up opportunities for companies to make significant profits in the years between price reviews, although this issue has lately been tackled by Ofwat.

Ofwat has, in general, supported benchmarking without making it an operational requirement, although it is suggested that underperforming companies may be required to benchmark in addition to regulatory benchmarking. Ofwat has expressed the strong opinion that much is to be gained by the water industry by seeking best practice outside the sector where commercial and competitive pressures will have driven efficiency more effectively than any regulator.

A number of the water companies have undertaken significant amounts of benchmarking with varying results, and some are less than satisfied with the experience.

For a number of reasons, benchmarking exercises can generate little or no benefit. These can include overemphasis on metric data collection, lack of management ownership, lack of staff buy-in and commitment, lack of focus on key issues and areas of potential improvement, dissipation of knowledge, and nonimplementation of improvement recommendations. Identification of best practice in a partner is not necessarily the solution for the benchmarking company. The situation of every business is different. The benchmarking company must seek best practice and adapt it to its own requirements and situation, and even then, over time, this situation and these requirements will change.

What is clear is that benchmarking is not a perfect mechanism, nor an exact science, nor a panacea for performance shortcomings. Benchmarking does not have all the answers, nor does the exercise automatically lead to best practice performance or to immediate step changes. Key to any benchmarking exercise is the management of expectations.

Metric benchmarking can be used to put down markers on areas of potential underperformance, but a great deal of effort can go into endless metric benchmarking without any obvious benefit to the company. What can be of benefit is the use of metrics as a baseline for process benchmarking.

It can be seen that significant benefits can be derived from properly executed process benchmarking. These can be reflected in improved revenue performance, reduced operating costs, reduced or more effective capital expenditure or improved operational performance, or indeed, any combination of these.

Success is most likely to come from process benchmarking underpinned by focused metric benchmarking, with clear management support and ownership, integral staff participation and ownership of recommendations, prioritization of benchmarking exercises starting with those that have optimal potential impact on the business (this means getting beyond the obvious benchmarking of the finance function, customer services, or grounds maintenance), benchmarking against out of industry best practice comparators, adaptation of best practice to the company's situational requirements, and implementation. There is nothing to be gained unless the exercise is seen to its logical conclusion.

Where these core guidelines are followed, benchmarking can have a real and positive impact on the financial and operational performance of a company, even where the company is perceived to be a superior performer within its sector. Such benefits are evidenced by experience drawn from the water industry in England and Wales. Elsewhere, the business drivers, particularly the regulatory driver, will differ, but the potential for significant improvements in cost and operational performance must be at least as great, and should merit further investigation.

REFERENCES

Department of Environment (UK) (1986). "Privatization of the Water Authorities in England and Wales." London, February.

Office of Water Services (1999). "Final Report on Operational Benchmarking and Cost Reduction," II. London, February.

Chapter 5

The Development of Competition in the Water and Sewerage Industries of England and Wales

George Day

1. INTRODUCTION

The water and sewerage industries of England and Wales were privatized in 1989. Since then, they have remained as privately owned monopoly suppliers. Although contestable in some areas of their services, they have not been subject to direct market competition. However, competitive pressures are growing, and forms of market competition are likely to take center stage in the development of the industry over the next decade. The conditions are now in place for the market to play an increased role in delivering better services to customers, reflecting and adapting lessons from the liberalization of the United Kingdom's other utility sectors.

George Day is Head of Water Resource Economics of the United Kingdom's Office of Water Services (Ofwat). He was an economist for the British Natural Resources Institute and the Department for International Development.

Reinventing Water and Wastewater Systems: Global Lessons for Improving Water Management, edited by Paul Seidenstat, David Haarmeyer, and Simon Hakim.

2. FRAMEWORK AND STRUCTURE OF ENGLISH AND WELSH INDUSTRY

2.1 Industry Structure

Prior to privatization, most business and household water and sewerage services in England and Wales were provided by 10 state-owned water authorities structured according to "integrated river basin management" principles. In addition, 29 privately owned statutory "water only" companies provided water services to around 25% of the population. All were vertically integrated, controlling all parts of the supply chain from resource development to local distribution for water services, and from collection to treatment and discharge for sewerage services. As well as providing water and sewerage services on a commercial basis, the river basin authorities were also responsible for water regulatory functions, including drinking water quality, river quality control, abstraction licensing, and flood protection.

The structure of water and sewerage service provision was largely unchanged by privatization. The 10 regional water authorities became privately owned and vertically integrated regional monopolies. The water-only companies remained unchanged, but were brought within the new regulatory regime.* Each privatized company was licensed to provide water and sewerage services (or water only) within defined geographic "areas of appointment" covering all parts of England and Wales.† The only significant structural reform implemented as part of the overall privatization package was the separation of the water authorities' regulatory functions. River basin regulation was made the responsibility of the National Rivers Authority (NRA), while responsibility for drinking water quality regulation was given to a Drinking Water Inspectorate. In 1995, the Environment Agency was created and took over the NRA's functions, most notably in licensing water abstractions and controlling effluent discharges to the environment.

---

*Their number has now dropped to 15 due to merger activity.

†For sewerage services, the 10 former river basin authorities became the licensed suppliers for the whole of their respective areas. For water services, some areas were licensed to the former statutory water-only companies.

2.2   Regulatory Framework

The regulatory framework introduced at privatization reflected the "UK model" of privatized utility industries, heavily influenced by the thinking of industrial economists such as Littlechild and Beesley.* The central characteristic of this model is the use of "RPI–X" (retail price index plus an adjustment factor) price-cap regulation, under which price limits are set typically once every three–five years, reflecting future inflation less projected efficiency savings. Between periodic reviews, regulated companies can increase profits through achieving greater efficiency than assumed by the regulator in setting price caps. Such efficiencies may be shared or passed on to customers in later years through the price limits set for the subsequent period. Price caps under this model are applied to the overall level of charges, rather than individual charges, usually through a "tariff basket" mechanism. This form of regulation, it was argued, could be relatively light touch and provide for greater efficiency incentives than rate of return regulation.

The Water Act of 1989 put in place the statutory framework, creating an independent economic regulator (the Director General of Water Services, DGWS) and a system of price-cap regulation. However, the basic RPI–X model was adapted in a number of specific ways in recognition of the particular circumstances of the water and sewerage industries. In contrast to the statutes for other privatized utilities in the United Kingdom, the water regulator's duty to secure that regulated companies are able to finance their functions was strengthened by an additional reference to "securing reasonable returns on their capital." This wording reflected the cost of quality and environmental obligations placed upon the privatized water and sewerage (and water-only) companies (see Section 2.3).

In addition, the Act also set up a strict sector-specific merger control regime, reflecting the value placed on comparison between companies within

---

*Professors Stephen Littlechild and Michael Beesley were key architects of British regulatory policy. Both influenced the first British utility privatization (British Telecommunications) through their respective reports: *Regulation of British Telecommunications Profitability* (Littlechild, 1983) and *Liberalisation of the Use of the British Telecommunications Network* (Beesley, 1981). Littlefield subsequently influenced water privatization through his 1986 report, *Economic Regulation of Privatised Water Authorities,* and later served as the first electricity regulator for England and Wales from 1989 to the end of 1998. Beesley had influenced policy on nationalized industries through his involvement in both academia and government. He continued to advise and write influentially on the regulation of privatized utilities throughout the 1980s and 1990s.

the regulatory regime.* The consensus view at the time was that water authorities were a natural monopoly[†]; thus, it was anticipated that the ability of the regulator to make comparisons would form a key part of the framework. In this context, the DGWS's duty was only to "facilitate" rather than "promote" competition.

In implementing (company-specific) price caps, the Office of Water Services (Ofwat) has developed an approach reflecting the specific circumstances of the water and sewerage sectors. The most significant features have been:

- Allowance for prospective capital returns based on an efficient regulated asset base
- Provision for investment to meet rising quality and environmental obligations and, therefore, detailed output-related reporting of the capital investment program
- An emphasis on econometric modeling and other methods of assessing comparative efficiency
- The balancing of incentives with customer benefits (e.g., through allowing the gains from outperforming projected efficiency levels to be retained by shareholders for a full five years, regardless of the phasing of these gains).

2.3   Other Features of the Privatized Industry

A number of other features of the privatized industry are relevant to the development of competition. The assets of the water authorities were sold to investors at a large discount to their estimated replacement cost (the so-called "capital value discount"). On average, investors paid only around 5% of the estimated replacement cost.[‡] The concept of the "regulatory capital value" (RCV) is used by Ofwat to ensure price limits allow for returns

---

*S.34 of the Water Industry Act 1991 requires the Competition Commission to take account of the Director's ability to make comparisons, in considering mergers between licensed suppliers.

[†]Littlechild's 1986 report stated that "water authorities are for the most part natural monopolies. The work required to create their infrastructure makes it more efficient for the services to be provided by one than by several. . . ."

[‡]Market capitalization of the 10 water and sewerage companies (WASCs) at offer price in December 1989 was £5,239 million, while the cost of "modern equivalent assets" (i.e., the net MEA value) was estimated as £113, 969 in March 1990. Net MEA value is the cost of an asset of equivalent productive capability to satisfy the remaining service potential of the asset, less accumulated current cost depreciation.

on capital that reflect the flotation value of the assets* and net postprivatization capital investment. The RCV tool has ensured low prices for consumers, but raises a high hurdle for new entrants, who must earn a return
on the full value of their assets.

Rising standards in environmental and quality regulation have been a
defining feature of the privatized industry, and combined with past neglect
under state ownership, this has necessitated a massive capital investment
program. Total industry investment in the period to 1999 amounted to an
estimated £33 billion (in May 1999 prices), accounting for around £1 out
of very £40 invested in fixed capital in England and Wales. This represented a step change in the level of investment activity by comparison with
the preprivatization period, and meant that real prices rose in the years
following privatization.

Finally, the UK industry is noteworthy for the relatively low degree of metering. Only 17% of households are metered, with the remaining unmetered
households accounting for 59% of overall water consumption (excluding
distribution losses). From an economic perspective, the relevance is that,
for most customers, the marginal cost of extra water consumption is zero.

3. THE ROLE OF COMPETITION

3.1   Competitive Forces Within the Regulatory Regime

The regulatory framework put in place in 1989 opened a variety of channels for competitive forces to drive down costs and improve services for
customers:

- *Capital Market Competition.* Investors are able to compare the
  performance of companies and this is reflected in the share price.
  Where managers are perceived to be performing poorly, there is
  the risk of takeover.
- *Comparative Competition.* The existence of a number of comparators allows the regulator to set company-specific price limits based
  on econometric assessments of relative efficiency. This mechanism

---

*The RCV is the initial market value (measured by 200-day average) including debt plus sub-
sequent net (i.e., after allowing for current cost depreciation) new capital expenditure as assumed
at the time of initial (i.e., postprivatization) price setting and including new obligations imposed
since 1989.

creates incentives that mimic a competitive market by encouraging companies to compete on efficiency against each other.

- *Product Market Competition.* In input markets, the RPI–X regime has created incentives to find the lowest cost methods, resulting in an expansion of outsourcing and competitive tendering for a wider range of input goods and services.*

The price-cap regime has delivered major achievements through comparative competition in the 11 years since privatization. The result has been more efficient service provision, the delivery of a major program of quality improvement, and, following the 1999 Periodic Review, real price reductions for customers.

Table 5.1 quantifies some of the key achievements of price-cap regulation in England and Wales, in terms of the impact on average bills and the level of investment delivered.

3.2   The Role of Market Competition to Date

Opportunities for direct forms of market competition have so far been limited, but nonetheless significant. The regime has permitted various forms of direct competition including:

- Self-supply of water or wastewater services has always been available as a potential option. Private supplies of drinking water serve nearly 1% of households, and around 5% of households do not have a sewerage service. However, from a competitive perspective, it is the self-supply of large industrial users that has been most significant. Many industrial users have their own abstraction licences, or make use of specialized on-site effluent treatment. Trade effluent, in particular, has always been a contestable market, and many sewerage companies also provide on-site waste treatment and management services to customers.
- Inset appointments, whereby one company can replace another as the licensed supplier of water or sewerage services for a specified geographic area, have introduced an element of direct choice for

---

*This has also required increased vigilance by Ofwat in relation to transfer pricing in relation to contracts awarded by a licensed supplier to associated companies.

Table 5.1   The Outcome of Price-Cap Regulation (1999 prices)

Drivers of Bills	1990–1995	1995–2000	2000–2005	1990–2005
Quality improvements	£36	£34	£29	£99
Improved service and supply growth	£18	£3	£1	£22
Greater efficiency	−£5	−£18	−£60	−£83
Total change (to average 1999–2000 household bill of £248)	£49	£19	−£30	£38
Average Annual Capital Investment	1985–1990	1990–1995	1995–2000	2000–2005
Water and sewerage companies	£1.8 billion	£3.1 billion	£3.3 billion	£2.9 billion
Water-only companies	N/a	£0.2 billion	£0.3 billion	£0.2 billion
Industry	N/a	£3.3 billion	£3.5 billion	£3.1 billion

customers. This choice is available for greenfield sites, large industrial customers,* and where the incumbent consents. While there has been a relatively small number of inset appointments, the contestability that they have introduced has been reflected in the introduction of lower large-user tariffs across most of the industry.

- Choice of supplier is also open to customers in areas close to the border of two licensed suppliers, by means of a cross-border supply. Any customer can connect to another company's network and receive a supply, provided that they are prepared to pay the costs of making the connection.
- The DGWS also has the ability to determine the terms of new and existing bulk supply agreements between two licensed companies, including sewer connection agreements. This power is important in expanding the scope for inset appointments.

*Customers that use not less than 100,000 cubic meters of water per annum can choose an alternative supplier through the inset appointment mechanism. The consumption threshold was reduced from 250,000 cubic meters in August 2000, greatly increasing the number of customers who have this direct ability to choose.

3.3 The Potential for Market Competition: Developing Views

The UK system of regulation for privatized utilities has matured over time. In gas and electricity, the balance has shifted substantially in favor of market competition mediated over monopoly networks. Licensing arrangements implement the separation of natural monopoly networks from more competitive and service-oriented functions. The evidence from these industries consistently suggests a strong association between the benefits enjoyed by consumers and the intensity of competition. Real prices in gas and electricity have declined as a result of the growth of competition, as well as price-cap incentives. In electricity, for example, industrial electricity prices have continued to fall, despite the fact that large users had benefited from highly favorable deals prior to privatization. Competition in electricity supply for domestic customers was available across England and Wales by the end of May 1999 and was associated with price reductions ranging between 8 and 15%, while in gas, a recent study (Office of Gas Supply, 1999) estimates the value of price reductions following the introduction of competition to the domestic gas market in April 1996 at £1 billion.

In recent years, therefore, attention has turned to the potential for more market competition in water and sewerage services. The successful experience of introducing competition in the gas and electricity industries has led to more sceptical reassessment of the "natural monopoly" view of water and sewerage service provision. In particular, attention has focused on how needless barriers to competition, such as those created by incumbents to reduce contestability, can be removed, and how other more substantial barriers can be overcome.

The Competition Act 1998, which came into force in March 2000, has also been a catalyst in focusing attention on the possibilities for greater competition. The Act is a major strengthening of the general competition regime in the United Kingdom, and provides powers for the DGWS* to apply general competition law in the water and sewerage sectors. Extensive new powers allow the DGWS to investigate suspected infringements, to levy substantial financial penalties of up to 10% of UK turnover, and to issue directions to offending firms. Perhaps the most far-reaching implication of the Competition Act is that it opens up the possibility of competi-

---

*The other utility regulators gain the same powers in their respective sectors.

tion through common carriage (or the shared use of infrastructure assets). An obvious example of common carriage would be a competing supplier sharing the use of an incumbent supplier's pipe network. An unjustified refusal by an incumbent water (or sewerage) company to allow access to its network assets is likely, in many circumstances, to infringe the Act.* This was indicated prominently in guidelines issued by Ofwat in January 2000 on the application of the Competition Act.

Alongside these developments, the government has consulted on the scope for increased competition in the industry generally, and on how the market might play a greater role in the allocation of water abstraction licences (Department of the Environment, Transport and the Regions, 2000a, 2000b).

Thus, the last two years have witnessed a substantial reassessment of the potential for competition. In particular, industry opinions have changed from widespread scepticism to a more broadly positive view on the potential role of competition in general, and common carriage in particular. Companies have also begun to seek out competitive opportunities. Recent developments have included innovations by companies in developing differentiated tariffs and greater emphasis on customer-focused account management. Competitive activity through inset appointments and cross-border supplies has increased, and companies have also responded to Ofwat's call for them to develop access codes, with access codes published by most companies in August 2000.

Ofwat will have a significant role in facilitating the development of greater market competition, and will itself be obliged to adapt as competitive forces play an increasing role. The approach so far has been to focus on the creation of a level playing field for competition, most notably by signaling the intention to apply new competition powers robustly to prevent anticompetitive conduct. Wherever appropriate, Ofwat will seek to allow the market to drive change and innovation. In the future, reforms could be introduced to allow for the emergence of more competitive structures or to provide a more specific framework for common carriage.

---

*This reflects the doctrine of essential facilities, developed in European Commission (EC) competition case law, mainly involving ports. Under this doctrine, a facility can be considered essential if access to it is indispensable in order to compete in a related market; and duplication is impossible or extremely difficult owing to physical, geographic, or legal constraints. A refusal by the owner of an essential facility to provide access to a third party can be an abuse of market dominance.

3.4   Lower Barriers to Competition

At this stage, it is possible to identify a number of arenas where the barriers to direct market competition are being lowered, and where there will be an increased role for market competition.

3.4.1   Common Carriage

The Competition Act 1998 powers open up the scope for competitive access to incumbents' network assets. Ofwat has subsequently explored the implications of this in a series of public letters providing guidance to water and sewerage companies, and encouraging each company to develop access codes that set out the detailed terms and conditions for third-party access to their networks. Until recently, much had been made of the potential risks to water quality from common carriage, but it now appears clear that technical risks can be managed to safeguard water quality. The possibility of common carriage appears to be the most significant extension of contestability in water and sewerage service provision, with the potential to provide for greater customer choice and competitive discipline on regulated suppliers.

3.4.2   Water Abstraction

The system of abstraction licensing in England and Wales has been the subject of a government review, which has suggested an increased role for trading in abstraction rights. International experience suggests that greater flexibility in the trading of water rights could yield clear benefits.* This could take a variety of forms, ranging from outright trading in water rights to more temporary transfers. A greater degree of market flexibility and transparency may facilitate new entry and greater efficiency in resource allocation. Improved resource allocation could be significant, given the previously fragmented development of water resources in England and Wales, and the weight attached to environmental considerations (see Table 5.2).

3.4.3   New Infrastructure

There is significant interest from new entrants in competing to provide new infrastructure, in particular pipe networks, to serve new industrial

---

*The collection *Markets for Water: Potential and Performance* (Easter et al., 1998) contains varied examples of water markets and trading from the United States (Texas, California, and Colorado), Chile, Pakistan, and India.

Table 5.2   Water Abstraction in England and Wales

Purpose	Actual Abstraction (% of Total Actual Nontidal Abstraction)
Public water supply	49%
Spray irrigation and other agriculture	1.4%
Hydropower	25%
Electricity supply (net)	1%
Other industry	9%
Fish farms/cress/amenity ponds	12%
Other	3%

Source: Department of the Environment, Transport and the Regions/Welsh Office (1998).

and housing developments. Experience in gas (and electricity) suggests that many of the functions associated with network extension activity can be highly contestable. Competition Act powers may well be relevant in preventing licensed companies from restricting the ability of competing infrastructure providers to offer network connection and extension services to developers.

3.4.4   Outsourcing, Asset Management

Many companies already subject some of their operations to direct market competition via competitive tendering. More recently, companies have begun to explore the potential to extend these arrangements. Proposals so far have focused on the separation of asset ownership from operation, and appear to have been designed primarily to reduce the costs of capital for the asset-owning business, for example by using a debt-financed mutual ownership structure. While the earliest proposals have raised significant regulatory issues,* there appears to be no reason why market competition could not play a more extensive role in contracting out of operations (and other functions).

---

*Ofwat's response to one of the first of these proposals, involving the separation of asset ownership from operations for the provider of water and sewerage services in Yorkshire, stressed the need to demonstrate that customers would benefit from changes, that the asset owner should maintain sufficient control over operations to meet legal obligations, and that contracts for operating services should be let through a genuinely competitive process.

4. POTENTIAL BENEFITS FROM COMPETITION

Price-cap regulation can only be a second best substitute for vigorous market competition. Regulators rely on imperfect knowledge, and can only create relatively muted incentives when compared to genuine market competition. Markets create the incentives for the inherently unpredictable processes of innovation and, wherever possible, Ofwat's approach is to allow market mechanisms to reveal the most effective methods of meeting customer needs.

Regulators also have limited foresight, and cannot predict future innovation or the outcome of competition. Nonetheless, it is possible to anticipate some of the opportunities opened up by increasing the scope for market competition. These are set out in Table 5.3 under three broad headings, each of which is further discussed in what follows. Consideration of these provides good reasons to believe that significant benefits can be realized over the medium term.

4.1   Increased Efficiency

Market competition, through common carriage or other mechanisms, will provide sharper incentives: the most efficient and innovative companies will gain customers, grow their businesses, and generate higher returns. Conversely, companies that are less effective in meeting customer demands will face the ultimate sanction of lost customers. Price limits reinforce these incentives by limiting the returns that can be generated within the regulated business.

Opening up the possibility of common carriage is central to widening the scope for market competition. It will permit greater challenge to the implicit assumption that companies can excel in all parts of the supply chain, by opening up the possibility of companies specializing in resource development, treatment, network asset management, or in serving various customer segments as service suppliers. Similar specialization opportunities may also emerge through the development of competitive outsourcing markets for operational or asset management services. These ideas have been developed by companies, particularly in strategic proposals to separate asset ownership from operations. This might allow specialized asset owners to competitively outsource a larger proportion of operational and management services, on an impartial basis.

Table 5.3 Benefits from Competition

Source of Benefit	Possible Mechanisms
Increased Efficiency / Cost Reduction	
Opportunity for specialization	Emergence of specialized companies (e.g., resources, treatment, network managers, suppliers) Outsourcing of operational services
Opportunity for more efficient new entrants	Common-carriage entry by new entrant exploiting new resource/technology or network counterflows Common carriage by existing industry player Unrestricted access to heavily discounted assets
Increased efficiency by incumbents	Threat of new entry drives improved understanding of cost structure and drivers
Efficient tariffs and investments	Competition for customers leads to more innovative tariffs, incorporating incentives for customers to use resources in manner that minimizes cost burdens Cost-reflective tariffs drive innovation targeted at reducing costs
Efficient asset investments	Increased contestability may undermine any incentives to gold-plate new infrastructure (e.g., to meet quality program) Increased incentive to market-test asset design
Improved Customer Focus	
Improved choice	Multisite account management, or integrated services Wider range of tariff and repayment options
Value-added services	Competition based on value-added service, including on-site management, lowering customer search costs
Increased Flexibility	
Remove artificial boundaries and price differentials	Common-carriage competition by companies with cheap resources Increased cross-border movement of resources
Risk pooling	Common carriage allied with development of increased connectivity between networks Possible market mechanisms, e.g., dry year options Lower capacity margins for security of supply

Common carriage, in conjunction with reforms to the abstraction licensing regime, could provide much greater scope for new entry to the industry. Similarly, it will make it easier for existing players to challenge each other directly in their respective areas of appointment. The experience with the development of inset appointments suggests that the reality of a threat of new entry may be enough to induce major improvements in the terms offered to customers. In economic terms, the contestability of the market for supplying various customers will be greatly increased, and the geographic scope of this contestability will be greatly extended beyond that provided by the inset mechanism. A particular advantage provided by common carriage in the water and sewerage sectors is that it will allow new entrants to gain access on fair terms to the network assets that companies inherited at large discounts at privatization. This substantially reduces the scope of the in-built cost advantage enjoyed by incumbents over new entrants, who must earn returns on the full capital costs of their assets. In effect, new entrants can share in the benefit of the "capital value discountm," by gaining access to infrastructure facilities that were heavily discounted at privatization.

Much of the debate about the introduction of competition in water has emphasized the potential costs of competition. Variously, it has been suggested that competition will result in stranded assets, reduced economies of scope, increased transactions costs, politically unsustainable incidence effects, and inefficient entry based purely on regulatory arbitrage. The argument has been that these costs are large in comparison with the likely benefits of competition. In particular, it has been suggested that competition through common carriage will only impact on a relatively small proportion of the value chain, due to the relatively high proportion of the cost of water delivered accounted for by the monopoly network. This assertion does not appear to be supported by an indicative analysis of Ofwat's industry cost data, which suggests that the most monopolistic function (i.e., local distribution) accounts for only around 31% of total costs (see Table 5.4).

The fact that water is ubiquitous, but also expensive to transport,* gives rise to considerable incentives to innovate in optimizing the way the water is moved around, for example, through greater sophistication in network modeling and optimization. These incentives are sharpened by the possibility of common carriage. Competition might also be based on gains made

---

*Illustrative "order of magnitude" calculations suggest that the cost of transporting water in bulk for 100 kilometers would constitute around 20% of a typical volumetric retail price.

Table 5.4 Indicative Industry Costs by Function

Function	Average Costs ($£/m^3$)	Percent of Total Costs
Resources and treatment	25	33.3
Bulk transportation	17	22.7
Local distribution	23	30.7
Customer, scientific, and regulatory services	10	13.3
Total	75	100

Source: Based on Ofwat data for the 10 water and sewerage companies.

from reducing the need to transport water. New entrants could exploit "counterflow" effects, where they wish to transport water by common carriage against the existing network flow. Incremental costs caused by such counterbalancing flows within a network could be very low, or conceivably negative due to pumping cost and other savings. Grey water recycling might be another opportunity to cut down on the need to transport water.

Similarly, the network itself, particularly at the level of bulk transportation, can be considered to be at least partially contestable. For example, water drawn from river sources closer to centers of population with innovative treatment could substitute for the bulk transportation of water from more distant upland reservoir sources. Given the wide availability of water resources and high population density in England and Wales, it is perhaps only the local distribution network that remains genuinely monopolistic in almost all circumstances. Only experience will show how far bulk transportation or resource development prove to be contestable. Competition Act powers open up the opportunities, and can prevent powerful incumbents from restricting competitors. Price cap regulation remains in place until competition demonstrates how much it can achieve.

A real threat of competition will also sharpen the incentives for incumbents to improve understanding of their cost structures and cost drivers. This, in itself, may deliver efficiency benefits through improved cost management within companies. More cost-reflective tariffs will also encourage innovation to be directed most purposively at where costs are highest. It is also likely to contribute to improved effectiveness in the evolving regulatory regime. The response of incumbents to competition is likely to reveal important information to the regulator, partially correcting for the inbuilt information asymmetry between the industry and the regulator.

One example of this might be where a company sought to rebalance tariffs in order to respond to a competitive challenge. In doing so, the company would need to bear in mind that this revised attribution of costs could be used by Ofwat in assessing the terms on which it ought to supply water to competitors through bulk supplies, as well as to its own customers. Clearly, the introduction of commercial pressures expands the range of levers available to the regulator in assessing the information supplied by companies and combating information asymmetry, particularly in relation to estimates of long-run marginal costs (LRMC) submitted by companies for various components in the supply chain.*

Efficiency in the creation of new capital assets may also be an area where increased contestability may yield significant benefits. The prospect of competition is likely to encourage companies to seek the most efficient capital investment solutions to meeting demands, and associated quality obligations, rather than relying on in-house design and built in remuneration through the regulated capital base.† Inefficiently designed or located assets could be stranded, or partially stranded, if competitors can find more efficient ways of meeting demands. An example of this principle might be the large price increases imposed on some trade effluent customers arising from enhanced investments in sewerage treatment to meet quality obligations. Here companies' willingness to make investments without adequate consultation appears to have reflected their assumption that costs could be passed on to customers and recovered through the remuneration of regulated assets, regardless of whether customers would find on-site technology to be more cost-effective. With greater competition, and the possibility of asset stranding, companies will have incentives to adopt more efficient designs that embody customer demands to a greater degree.

4.2   Improved Customer Service and Focus

While price is likely to be the single most important motivation for customers, other factors are important to them too. Expanding the range of methods that competitors can use to poach customers, and increasing the range of choice open to customers will, almost self-evidently, promote im-

---

*Estimates of LRMC are a critical regulatory tool used to challenge the consistency of companies' approach to resource development, leakage control, and the terms offered for bulk supplies or supplies to large users.

†The tendency to overcapitalization was first noted by Averch and Johnson (1962) in the context of rate-of-return regulation.

proved customer service and focus. Up to now, there has been relatively little incentive for companies to explore customer needs. In particular, large users have varying requirements and problems, and many feel that their situations have not been adequately addressed by their existing suppliers.

Incentives will be sharpened by the real prospect of losing customers. New entrants competing via bulk supplies, inset appointments, or common carriage can work with customers on water efficiency and on-site services to reduce bills. Incumbent suppliers now have incentives to improve their customer service function, and to provide attractive value-added services such as on-site advice. They may also find new ways of matching customer demands and the capacity of their own infrastructure, in areas such as sewage treatment or the management of peak loads.

The beginnings of this process are already evident in new tariff developments, for example interruptible tariffs, or tariffs that are integrated with leak-detection services.* Among business customers, common carriage may offer the possibility of unified account management, allowing one supplier to provide the service to a large number of dispersed sites through one account. Two water and sewerage companies announced significant initiatives in this area during June 2000, providing account management and value-added services, respectively, to a major national food company and one of the United Kingdom's largest brewers.

4.3   Increased Flexibility

Common carriage may allow scope for a range of efficiencies by permitting increased flexibility in the use of infrastructure assets and water resources. The division of the industry into river basin based monopolies, and the historical pattern of asset development, appears to have given rise to a certain amount of rigidity in the approach to water resource use. The evidence suggests that companies have tended to prefer to develop their own resources rather than seek supplies from neighbors, despite what appear to be significant discontinuities in marginal costs at company boundaries. From a company perspective, it appears that the incentives have favored seeking allowance within price limits for capital investment for resource development within their own area over seeking resources via bulk supplies. New bulk supplies have been comparatively rare in the period since privatization. Competition by common carriage (or bulk trading

---

*One supplier now offers a tariff with a free leak-detection service.

in water) could smooth out price discontinuities, effectively by eliminating the ability of a relatively resource-scarce company to resist accepting more economical resources into its distribution system.

Competition also increases the incentive for companies to respond to changing demands in an innovative way. Climate change and demographic changes, and (perhaps) future changes in the value placed by society on water systems in the environment, are all leading to changes in the demands placed on existing infrastructure. Under the existing regulatory regime, the regulator provides the only challenge to companies' proposals to balance the changing demands on their systems. Companies have enjoyed an effective monopoly over the basket of resource development and management projects required to balance supply and demand, or equally the projects required to maintain serviceability. With the prospect of new entry, the challenge may increasingly come from innovative competitors, with innovative methods of bringing new resources into play.

Open access to infrastructure could also generate efficiencies by allowing pooling of the risk and variability inherent in the supply and demand of water resources. This has been a feature of the "smart" competitive pools that have been developed in electricity and gas supply systems, allowing competitive suppliers and trading mechanisms to play an important role in efficient balancing of variable supply and demand. Over time, these markets have taken on more of the characteristics of traditional commodity markets. The development of hedging, brokerage, and commodity exchange instruments in these sectors is illustrative of this, and the gains from market liquidity are significant.

Water transportation networks might also benefit from competitive pool concepts, perhaps through regional bulk water markets in areas where the price of water is high. Parts of England and Wales, particularly the southeast of England, may provide the basic conditions under which this type of concept could operate, although relatively high transport costs are likely to limit this to a local or regional basis. Some companies effectively already operate their own internal "pools" in the way they operate their networks. Future investments in new infrastructure and connectivity may take place if it is economic to create larger market areas.

Risk pooling is significant because it may allow for lower overall security margins, or the staggering or delay of sizable lumpy investments in resource development. Water's basic characteristics, with seasonal and weather-related variation in both supply and demand, give rise to risk. In particular, peak water demands tend to coincide with the periods of lowest resource availability. Thus, planning for the extreme events, when it may make economic sense to transport water over longer than normal distances, is an important feature of risk management in water supply. Individual companies could manage risk cooperatively by using dry year options or other instruments. Incentives to create the physical investments in risk management (increased connectivity and storage facilities) might be enhanced by clearly signaled financial penalties for environmentally damaging abstractions during dry periods.

5. STRUCTURAL ISSUES AND THE FUTURE

The focus on competition in the water and sewerage sectors has stimulated debate about the structure of the industry. At privatization, it was envisaged that economic regulation would be necessary in perpetuity, due to the inherently monopolistic character of the industry. In contrast to the electricity and gas industries in the United Kingdom, where separation of functions was imposed, vertical integration has until now remained unchallenged. Various models for restructuring are now being proposed, as summarized in Table 5.5.

Regulatory imposition of some form of vertical separation of the potentially competitive parts of the industry from the core monopoly elements is perhaps the most obvious format for restructuring. This would avoid discrimination in the terms offered by vertically integrated incumbents to competitors for access to network assets. The scope for delay through lengthy negotiation over the detailed terms of access in the absence of regulated third party access to infrastructure may be great. The experience in the gas industry, for example, suggests that regulated access to infrastructure, and ultimately vertical separation, may be required. So far, Ofwat has suggested that it will examine whether there should be revisions to regulatory accounting guidelines to ensure correct attribution of costs between monopoly and potentially competitive elements. Accounting separation is the mildest form of restructuring along a continuum to ownership separation. It is possible that future legislation may impose restructuring, or facilitate this by introducing separate licensing of various functions. Conversely, market forces may drive forms of vertical separation as companies seek to specialize, for example, in network asset management. A market-driven approach would have the advantage of allowing economies of scope to be internalized within decision-making. In the

Table 5.5 Models of Industry Restructuring

Dimension of Restructuring	Options	Drivers
Vertical separation	Separate competitive from monopoly elements	Secure nondiscrimination in terms of access to infrastructure
	Separate accounting, licensing, management, or ownership	
Horizontal restructuring	Mergers of vertically integrated businesses	Desire for increased financial weight in global markets
	Mergers/joint ventures in specialized functions	Synergies from functional specialization
	Power to enforce transfer of abstraction licences	Regulatory desire to provide for liquid market or prevent licence hoarding
Capital/functional	Separate asset ownership from operation	Lower cost of capital through increased debt financing or mutual ownership
		Scope for growth (outside regulatory regime) in markets for asset operations services

absence of legislative change, however, change may be limited by the fact that current licences are premised on vertical integration.

Horizontal restructuring could take place through mergers or joint ventures and could be combined with vertical restructuring. However, the scope for this will be limited by the need to maintain comparative competition. Vigorous development of market competition might increase the scope for merger activity in the industry, without prejudicing the interests of customers.

Functional or capital restructuring is also possible. This could be combined with increased market competition for the right to carry out operational service functions through competitive outsourcing by an asset owner. This would mirror the franchising model, such as that in France, where operations are run by private service suppliers, but assets are owned by municipalities.

The development of competition will clearly be a key driver of structural change, as will the ongoing search for greater efficiency. Ofwat's approach to structural issues is to avoid seeking to impose a structural blueprint. The regulator's role is to provide an overall framework within which companies are free to innovate and seek the most effective methods. Ofwat must ensure that proper incentives are provided for, that risks are properly apportioned, and that there are clear benefits to customers from any structural initiatives. Thus, Ofwat will continue to implement the regulatory regime in relation to issues such as corporate governance, appropriate ring fencing, and mergers. As competition develops, there are likely to be areas where regulation will need to evolve, for example, new forms of licence to allow for competitive entry. Similarly, changes to the detailed implementation of price-cap regulation may be needed to reflect market competition in parts of the business, and to maintain incentives.

REFERENCES

Averch, H., and L. Johnson (1962). "The Behavior of the Firm Under Regulatory Constraint," *American Economic Review,* 52: 1052–169.

Beesley, M. E. (1981). *Liberalisation of the Use of the British Telecommunications' Network.* London: Her Majesty's Stationery Office, April.

Department of the Environment, Transport and the Regions/Welsh Office (1998). *The Review of the Water Abstraction Licensing System in England and Wales,* A Consultation Paper, London, June.

Department of the Environment, Transport and the Regions (2000a). *Economic Instruments in Relation to Water Abstraction,* A Consultation Paper, London, April.

Department of the Environment, Transport and the Regions/National Assembly for Wales (2000b). *Competition in the Water Industry in England and Wales,* A Consultation Paper, London, April.

Easter, K. W., M. W. Rosegrant, and A. Dinar, Editors (1998). *Markets for Water: Potential and Performance.* Dordrecht, Netherlands: Kluever Academic Publishers, April.

Littlechild, S. C. (1983). *Regulation of British Telecommunications' Profitability—A Report to the Secretary of State for Industry,* London, February.

Littlechild, S. C. (1986). *Economic Regulation of Privatised Water Authorities,* A report submitted to the Department of Environment, London, January.

Office of Gas Supply (1999). *Giving Customers a Choice—The Introduction of Competition into the Domestic Gas Market,* National Audit Office, London, May.

ADDITIONAL READINGS

Kinnersley, D. *Coming Clean—The Politics of Water and the Environment.* London: Penguin Books, 1994.

Office of Fair Trading. *The Competition Act 1998—Assessment of Individual Agreements and Conduct,* September (www.oft.gov.uk).

Office of Water Services/Office of Fair Trading. *The Competition Act 1998— The Application in the Water and Sewerage Sectors,* February 2000 (www.oft. gov.uk).

Office of Water Services. *MD154: Development of Common Carriage,* Public letter to Managing Directors, November 1999 (www.ofwat.gov.uk).

Office of Water Services. *MD158: Common Carriage,* Public letter to Managing Directors, January 2000 (www.ofwat.gov.uk).

Office of Water Services. *MD159: LRMC and Regulatory Framework,* Public letter to Managing Directors, February 2000 (www.ofwat.gov.uk).

Office of Water Services. *MD162: Common Carriage—Statements of Principles,* Public letter to Managing Directors, April 2000 (www.ofwat.gov.uk).

Office of Water Services. *MD 163: Pricing Issues for Common Carriage,* Public letter to Managing Directors, June 2000 (www.ofwat.gov.uk).

Office of Water Services. *The Proposed Restructuring of the Kelda Group,* A preliminary assessment by the Director General of Water Services, July 2000 (www.ofwat.gov.uk).

Office of Water Services. *The Proposed Acquisition of Dŵr Cymru Cyfngedig,* A position paper by Owat, January 2001 (www.ofwat.gov.uk).

Part III

Financing the Water and Wastewater Infrastructure

Chapter 6

Global Water and Wastewater Project Criteria

A Rating Agency Perspective

Michael Wilkins and Jane Eddy

1. INTRODUCTION

Increased levels of capital needs for water and wastewater systems globally are resulting in many new entrants to the capital markets. As both publicly and privately owned systems need to increase resources and diversify capital structures, the need for credit information on market entrants is increasing. Standard & Poor's is a provider of objective financial information and ratings and risk analysis to the global financial community. Table 6.1 shows the outstanding Standard & Poor's ratings on global water projects.

Michael Wilkins is Director at Standard & Poors Rating Service. He heads up the analytical group covering utilities, project finance, and transportation in Europe, the Middle East, and Africa. With Standard & Poors since 1994, he has expertise in the European transportation, water, and electricity sectors. Previously, he worked at the Water Services Association of England and Wales, the trade body representing the UK water industry, and was a journalist for various UK newspapers. Jane Eddy is a Managing Director in the Corporate and Government Ratings Group at Standard & Poors Corporation. She heads the group responsible for sovereign and local and regional government ratings in Latin America. Her job also includes assisting in the development and implementation of criteria. She has had extensive experience in analyzing governments and infrastructure and project financings since 1982.

Reinventing Water and Wastewater Systems: Global Lessons for Improving Water Management, edited by Paul Seidenstat, David Haarmeyer, and Simon Hakim.
0-471-06422-X   Copyright © 2002 by John Wiley & Sons, Inc.

Table 6.1   Standard & Poor's International Water Credit Ratings[a]

Company	Corporate Credit Rating
Europe/Middle East/Africa	
Sociedad General de Aguas de Barcelona (Agbar)	AA−/Stable/A−1+
Severn Trent PLC	A+/Stable/A−1
Severn Trent Water Ltd.	A+/Stable/A−1
Thames Water PLC	A+/CW-Pos/A−1
Thames Water Utilities Ltd.	A+/CW-Pos/—
Kelda Group PLC	A+/CW-Neg/A−1
Yorkshire Water Services Ltd.	A+/CW-Neg/—
Northumbrian Water Group PLC	A/Stable/A−1
Northumbrian Water Ltd.	A/Stable/—
Suez Lyonnaise des Eaux S.A.	A−/Stable/A−2
Anglian Water PLC	A/Negative/—
Anglian Water Services Ltd.	A/Negative/—
Southern Water Services Ltd.	A/CW-Neg/—
North West Water Ltd.	A−/Stable/A−2
Enron Water PLC	A−/Stable/—
Vivendi Environment S.A.	BBB+/Positive/A−2
Wessex Water Services Ltd.	BBB+/Stable/—
Wessex Water Ltd.	BBB+/Stable/—
Rand Water	Local currency A−/Stable/—
	Foreign currency BBB-/Stable/—
Azurix Europe Ltd.	BB+/Stable/—
Latin America	
City West Water Ltd.	Local currency AAA/Stable/A−1+
	Foreign currency AA+/Stable/A−1+
Companhia de Saneamento Basico do Estado	Local currency BB/Positive/—
de Sao Paulo (SABESP)	Foreign currency BB-/Stable/—
Aguas Argentinas S.A.	BB-/Negative/—
Asia/Pacific	
Melbourne Water Corp.	Local currency AAA/Stable/A−1+
	Foreign currency AA+/Stable/A−1+
South East Water Ltd.	Local currency AAA/Stable/A−1+
	Foreign currency AA+/Stable/A−1+
Sydney Water Corp.	Local currency AAA/Stable/A−1+
	Foreign currency AA+/Stable/A−1+
Watercare Services Ltd.	A+/Stable/A−1

[a]Standard & Poor's ratings as of August 8, 2001. All ratings are subject to change. Ratings are statements of opinion, not statements of fact or recommendations to buy, hold, or sell any securities.

142

Water industry structures are evolving differently around the world; Standard & Poor's rating criteria address the diversity of credit risks inherent within each structure. This chapter provides a description of the rating methodology for non- or limited-recourse water and wastewater projects. Standard & Poor's rating criteria for the water industry are applicable across a wide range of projects. The criteria also have been designed to cover different ownership structures, regulatory regimes, levels of government involvement, and macroeconomic operating environments. The criteria include additional research and information gathered from project sponsors active in the field of build, operate, and transfer (BOT) projects worldwide.

2. REVENUE STRUCTURE

BOT projects in the water and wastewater sector typically rely on purchase contracts with a sole (usually municipal or sovereign) purchaser for offtake revenues. The offtake contract (equivalent to a power purchase agreement, or PPA, in the power sector) covers the principal elements determining the project's ultimate cash flows such as tariff levels, escalation and indexation, and fixed and variable cost reimbursements. These contracts are usually long term. However, such long contractual revenues streams are normally exposed to performance related conditionality, as well as to varying degrees of counterparty credit risk (see section 6, Purchase Credit Strength).

Revenue structure analysis in water projects must incorporate an assessment of the degree of flexibility afforded to BOT sponsors within the contract to raise the purchase price in a timely manner to mitigate the effect of a sudden rise in uncontrollable cost elements, such as inflation or foreign exchange movements. There also should be adequate protection provided in the case of enforced added capital or operating costs, force majeure (especially drought), or the interruption of bulk water supply or effluent source by a third party (see Section 5, Legal and Financial Structure).

This type of risk-mitigation process is typical for all project revenue contracts (PPAs, concession, and franchise agreements), although there are several important differences for water and wastewater projects:

- Input supply and effluent source (both quality and quantity) are often not controllable by BOT sponsors and, as such, this risk should not be assumed.

- The ability to deal with changes in environmental legislation (determining output compliance) is limited.
- Water or wastewater facilities are capital intensive and are usually dedicated to perform specific functions that may be catchment dependent. This reduces asset flexibility and heightens reliance on contractual specifications and protective contractual arrangements.

Take-or-pay (minimum guaranteed payment) or put-and-pay (minimum guaranteed volume) arrangements are preferable to lessen the project's exposure to demand risk. Where such arrangements do not exist, assumptions made about usage of plant and payment structure must be robust enough to withstand severe downside scenarios. A careful examination of contractual arrangements is required, especially where the level of take-or-pay reimbursements (where these exist) is below the expected usage of the plant during the period of the contract. The tariff mechanism, which compensates the plant for marginal usage, should be structured so that project creditors are not exposed to any potential mismatch between sudden and unexpected cost increases and payments. Usage payments also are sometimes subject to the project company meeting certain performance criteria. These may include stipulations requiring maintenance and efficiency investments, which, in essence, would transfer operating and obsolescence risk onto the project sponsors. These criteria and obligations under the contract should not be open to ambiguous legal interpretation, and mechanisms should be in place to prevent any arbitrary regulatory or political decisions adversely affecting the project's underlying economics. A robust revenue structure that minimizes operational, regulatory, environmental, and sociopolitical risks is, therefore, a prerequisite for an investment-grade rating of a BOT project.

3. TECHNOLOGY AND OPERATIONS

Technology risk in the water and wastewater sector is usually less acute than in some other infrastructure sectors. However, the rapidly evolving nature of water and wastewater processes, and an increasingly competitive environment in developed countries mean that sponsors are at times pushing the frontiers of new technology to win BOTs. The risks are also different in nature because of the acute consequences of malfunction or noncompliance. The supply of contaminated drinking water can lead to serious litigation, possible termination of contract, and, more critically, a serious public health incident that could damage the sponsor's reputation beyond repair. This being said, some international water project sponsors are well experienced in building water and wastewater treatment facilities and are often at the cutting edge of new processes and technology.

4. COMPARATIVE ECONOMICS

Often, BOTs are tendered in countries where the regulatory framework and monopolistic nature of the industry is such that exclusive contractual arrangements can be entered into for the provision of perceived needs in the sector. However, several factors need to be addressed, such as the offtaker's ability and willingness to access alternative suppliers and whether this could potentially undermine the economics of the BOT project in future years. While this is not usually considered to be a great risk to projects at present, technology improvements in 10 to 20 years' time could distort the underlying economics of a plant and force an offtaker to look elsewhere.

Often, project sponsors, in the absence of strict take-or-pay arrangements with the offtaker, will be required to take on a degree of volume risk. In considering this risk from a creditor's perspective, it is important that the baseline for debt service is set at a level that corresponds to conservative assumptions on demand and supply. Demand patterns for water services are unlike those found in the electricity sector. In mature economies, demand will, at best, remain flat or may even show slight declines over time as consumers respond to conservation initiatives or to pricing signals brought about by the introduction of metering. In developing countries, on the other hand, historical high levels of consumption may be misleading, as they may indicate high levels of leakage, theft, or the inelasticity of demand to artificially low tariffs. The introduction of private-sector operators and the need for significant capital expenditures can lead to higher prices at the consumer end. This, in turn, can have a sharp effect on consumption levels, which the offtaker utility must recognize in its tariff base if it is to avoid financial stress and subsequently expose the BOT sponsors to default risk. Sharp tariff adjustments can lead to sociopolitical tensions, which could ultimately undermine any reasonable economic rationale for undertaking the project. Therefore, it is crucial to consider output price and demand considerations in a wider context than just the framework of the BOT contract.

5. LEGAL AND FINANCIAL STRUCTURE

The legal structure of a BOT offtake contract should not expose project creditors to risks that are either uncontrollable, unsustainable on an economic basis, or subject to arbitrary decisions. As a general rule, Standard & Poor's will look to the allocation of risk that matches the ability of project participants to assume or mitigate the risks in question. The legal analysis of water projects incorporates a thorough examination of key contracts to ensure the rights of creditors are sufficiently addressed within these documents. For example, an analysis is undertaken of the potential for enforced termination without reasonable step-in rights, cure periods, or compensation to project creditors. The terms of the asset transfer at the end of the contract period should be fair to all parties. In some cases, the transfer payment mechanism leaves sponsors open to certain uncontrollable risks. The tenor of rated debt should preferably be shorter than the contract length to ensure full and timely amortization of principal. A financial structure that entails the use of a nonrecourse special-purpose company for the issuance debt should meet Standard & Poor's criteria for such entities.

6. PURCHASER CREDIT STRENGTH

In itself, exposure of project sponsors to counterparty credit risk is not a unique phenomenon of BOTs. However, for water and wastewater projects, this exposure is invariably to municipal or governmental entities, which are often not rated and which generally operate in fragmented, politically sensitive contexts. These purchasers also may have huge environmental catch-up obligations, inefficient networks, high levels of leakage, and noneconomic tariff structures. For water BOTs, offtaker credit risk also is heightened as a result of: (1) the essential nature of the product, which can politicize contractual arrangements, and (2) the uncertain legal and regulatory frameworks in emerging markets, which govern the award of BOT contracts and the water and wastewater sector in general.

Often, the critical requirement for a particular facility can drive the offtaker into entering into a long-term contract, which, at some later stage, it may live to regret. While the protection provided by "hard" contractual arrangements should not be ignored, neither should the political and social pressures that can force municipalities to renege on their obligations.

Experienced international water project sponsors typically assess these risks before entering into BOT contracts, especially in emerging markets. However, they also recognize that, where there are no existing credit ratings on the offtaker, this is a risk that is difficult to analyze with any degree of accuracy.

Water utilities in developing countries have put BOT contracts out to tender for a variety of reasons including: lack of funding sources, desire to improve efficiency; and the need to bring in outside expertise, technology, and project management skills. The BOT project itself may have a fair degree of local support, but in the grand scheme of the country's infrastructure requirements, may be relatively insignificant. As such, should the offtaker default on its obligations (financial or otherwise), creditors should not assume that the sovereign government will step in to the rescue. Often, a sovereign guarantee may not be available, although partial credit enhancement may be sought from multilateral banks and agencies. As water services are viewed as critical to the improvement of public health and environmental standards in both developed and developing countries, supporting BOTs has become a key priority for some development agencies.

7. FORECAST FINANCIAL RESULTS

Water projects typically display fairly standard operating cost profiles and predictable revenue streams. For this reason, financial profiles tend to display high levels of leverage (debt versus equity) and low coverage ratios. Nonetheless, operating margins should be sufficient to provide a cushion against an unexpected drop in demand or hike in operating or capital costs.

The sizing of the margin should be a function of the level of protection offered by the revenue and legal structures, as well as the nature of the BOT. Wastewater plants are inherently more risky, operationally, than water plants, hence, a higher margin would be required. For the better-rated projects, minimum annual debt service coverage ratio should not fall below a range of between 1.2 times ($\times$) to 1.5$\times$ over the life of the rated debt. Some projects can withstand lower coverage ratios depending on project structure and robustness to sensitivities. Again, for investment-grade projects, the proportion of equity should generally be at a minimum 15–25%, depending on other project features. This equity may be made up in part by deeply subordinated debt, usually provided by the sponsors, and should not be amortized in substantial preference to senior rated debt.

8. CONCLUSION

The growing magnitude of global water and wastewater projects that are necessary to update or replace current facilities or build new capacity requires a large influx of private funds. Standard & Poor's provides rating and risk analyses for these projects. Considered in the analysis are risk factors related to the project's revenue structure, technology, and competition, the legal structure of the contract, the credit strength of the contracting government agency, and the financial coverage of the debt.

Credit ratings provide important information that enables credit providers to make decisions on where to channel funds efficiently to support these projects throughout the world. The global community depends on financial information systems that are essential component of the world's credit allocation network.

Chapter 7

Water Concession Design and the Poor

Kristin Komives

Low-income urban households in developing countries have much to gain from private sector involvement in the water and sanitation sector. Public water and sewer authorities have been slow to serve the peri-urban areas where many low-income households live. Without access to water or sewer networks, the poor rely on a variety of alternative water sources and sanitation services, often at high prices and at great risk to their health. Private concessions offer a strategy for extending networks into these unserved areas: private companies receive the right to provide services for 25 to 30 years in exchange for meeting certain coverage and quality targets.

Much of the material in this chapter was previously published in *Water Policy,* Vol. 31, No. 1, 2001, pp. 61–79, Kristin Komives, "Designing Pro-Poor Water and Sewer Concessions: Early Lessons from Bolivia," and is reprinted here with permission from Elsevier Science. The case study also appeared in 1999 as World Bank Policy Research Working Paper No. 2243. This Working Paper provides detailed citations to specific concession contract clauses and Bolivian regulations, details that have been left out of this article in order to preserve space.

Research for this paper was conducted in June 1998 with the support of the Private Participation in Infrastructure Group of the World Bank and with the assistance and cooperation of many World Bank, UNDP-Water and Sanitation Program, Aguas del Illimani, and Bolivian government officials. Special thanks go to Penelope Brook, Alain Carbonel, Vincent Gouarné, Alain Mathys, and Luis Guillermo Uzin.

Kristin Komives is an economist with Sector and Utility Management Group, Institute of Social Studies of the Netherlands. Formerly she was on the faculty of University of North Carolina at Chapel Hill. She has written extensively on water issues affecting the poor in developing countries.

Although concessions have much potential, they are not guaranteed to help the poor. A private concessionaire will not necessarily expand service faster than its public predecessor. Under certain circumstances, the monopoly concession model could even hurt poor households. Much will depend on how the concession contract and supporting sector regulation is designed. Contracts typically try to guarantee service for the poor with expansion mandates and lifeline tariffs, but these measures each have flaws. Pro-poor concession contracts need to improve incentives to serve the poor, while avoiding policies that could hurt low-income households.

Bolivia is one of the many developing countries turning to the private sector to improve urban water and sewerage services. The country's first major contract in the sector, a concession for the neighboring cities of La Paz and El Alto, was awarded to Aguas del Illimani in August 1997. This concession was explicitly designed to expand service to the poor. Since the contract was signed, Aguas del Illimani and the regulator have shown active interest in finding ways to overcome some remaining barriers that could threaten the attainment of the contract's service expansion goals. For these reasons, the La Paz–El Alto concession is a good case for examining what to do and what to avoid in pro-poor concession design.

1. BACKGROUND ON THE CONCESSION

The La Paz–El Alto metropolitan area has a population of over 1.3 million, roughly divided between the two cities. The wealthiest residents of the area live in central and southern La Paz, located deep in a river valley. Lower-income families live on the steep slopes, or "laderas," surrounding central La Paz and in El Alto, which lies on the vast plain above. As in many Latin American cities, water and sewer services are more prevalent in the wealthier and older central areas than in outlying poorer, and often newer, parts of town. In 1998, somewhere between 83% and 93% of El Alto and La Paz residents had access to some form of piped water service: either an in-house water connection or a public tap near their home. An estimated 66% of La Paz homes and between 30% and 45% of El Alto homes had sewer service.

Until mid-1997, water and sewer services in La Paz and El Alto were the responsibility of Servicio Autónomo Municipal de Agua Potable y Alcantarillado (SAMAPA)—a municipally owned, semiautonomous water utility. The Bolivian government instituted several regulations in the 1990s to try to impose standards and increase national control over municipal water authorities like SAMAPA. The two most important regulations, which continue to affect Aguas del Illimani's operations today, are the 1992 National Regulations for Water and Sanitation Service in Urban Areas (1992 Regulations) and the 1997 Regulations for Institutional Organization and Concessions in the Water Sector (1997 Regulations).

The 1992 Regulations set national standards for urban water and sewer service delivery. They define in-house water and sewer service (as opposed to public standposts, tanker truck delivery, and latrines) as the only acceptable long-term water and sanitation solution for urban areas. The regulations direct utilities to offer private connections to neighborhoods prepared to cover the cost of secondary network expansion.

The 1997 Regulations created a new structure for the Bolivian water and sanitation sector. Until 1997, water and sanitation service providers had been controlled and supervised by municipalities, and the 1992 national regulations were not widely enforced. Now all entities (public or private) that provide water or sanitation services to the public need to obtain a concession for their operations from the new national regulatory body for the water sector, the Superintendencia de Aguas.

The first concession signed under this law was for the La Paz–El Alto metropolitan area. The Bolivian government awarded this concession to the Suez Lyonnaise des Eaux consortium, Aguas del Illimani, after a competitive bidding process. Aguas del Illimani assumed control of SAMAPA's operations in July 1997.

One of the government's principle objectives in privatizing water and sewer service in La Paz and El Alto was to extend services in poor areas.* The Request for Proposals for the La Paz–El Alto contract fixed the length of the concession period, the tariffs and connection fees, the required sewer expansion schedule, and a minimum acceptable water expansion schedule. Interested private companies were asked to submit bids of the number of in-house water connections they would commit to install in El Alto (the poorest subsystem of the metropolitan area) by December 31, 2001. The winning consortium agreed to install 71,752 new in-house water connections in El Alto by that date. No one knows exactly how many homes lack water connections in the El Alto subsystem, but by some estimates the 71,752 new connections will achieve 100% coverage in the area.† (See Table 7.1.)

---

*Expanding water service coverage in urban areas had been official government policy since at least the early 1990s. The 1992 Regulations were drafted in part to help achieve this goal.

†Over the life of the concession, Aguas del Illimani must keep pace with population growth and not drop below the water coverage level that it achieves by installing the 71,752 connections.

Table 7.1 Water and Sewer Coverage Targets in the El Alto Subsystem,[a] Fixed in the Request for Proposals and Bid by Aguas del Illimani

Year[b]	Private Water Connections	Sewer Connections
2001	100% (Bid)[c]	41% (Fixed)
2006	100% (Bid)	43% (Fixed)
2011	100% (Bid)	47% (Fixed)
2016	100% (Bid)	71% (Fixed)
2021	100% (Bid)	90% (Fixed)
2026	100% (Bid)	90% (Fixed)

[a]In the other two subsystems in the metropolitan area, Aguas del Illimani must achieve 100% water coverage by 2001 and 95% sewer coverage by 2021.

[b]Major expansion goals are set for five-year periods, but these goals are also broken into yearly connection requirements.

[c]This assumes that the 71,752 connections that Aguas del Illimani agreed to install will achieve 100% coverage. If the 71,752 connections do not achieve 100% coverage, the company must reach and maintain at least 90% coverage by 2011.

The contract that came out of this bidding process relies heavily on expansion mandates to obtain water and sewer services for low-income areas. The mandates are a combination of: (1) mandates set by the government and (2) the expansion plans bid by the concessionaire to win the concession contract.

Expansion mandates such as these demonstrate an interest on the part of the government to expand service to low-income areas. Expansion mandates alone, however, do not guarantee that concessionaires will actually meet those targets over the length of the contract. The remainder of this paper discusses the weaknesses of expansion mandates and how the design of other contract elements can either support or undermine the expansion goals.

2. EXPANSION MANDATES

Expansion mandates that are: (1) clear and reasonable; (2) enforced by a credible regulatory body; and (3) backed by incentives for compliance have the best chance of success. By this measure, the mandates in the La Paz–El Alto contract stand up fairly well.

The expansion mandates in the Aguas del Illimani contract take three forms: connection requirements, percent-coverage targets, and the requirement to connect households meeting certain criteria. Connection requirements, such as Aguas del Illimani's commitment to install 71,752 new water connections by December 31, 2001, are fairly straightforward to monitor, provided that what counts as a connection is well-defined. In this case, only private water and sewer connections count.

The percent-coverage targets (e.g., 90% sewerage coverage) pose more problems for the regulator because no one knows with certainty how many households live in the expansion area. The regulator and Aguas del Illimani deal with this problem by agreeing in advance how many connections will constitute compliance with the targets.

The third type of expansion mandate requires Aguas del Illimani to extend service to any area that meets specific population density criteria.* The company must program expansion to areas that meet these criteria, but the contract does not set enforceable deadlines for completing the expansion. Thus, this form of mandate creates a possible source of debate between the company and the regulator.

To strengthen these mandates, the contract establishes three penalties for noncompliance: increased connection requirements, fines, and cancellation of the concession contract. First, the number of required connections for a given year will increase by *one* connection for every *five* required connections the company fails to install by the end of the previous year. In addition, if Aguas del Illimani is more than 15% short of the connection goal, the company must pay a fine of US$500 per connection it fails to install. Finally, the Superintendent can cancel the concession contract and execute a $5 million guarantee if Aguas del Illimani falls more than 25% behind its expansion mandate.

These penalties would mean little without a regulator with the power and the will to take action against the concessionaire for not honoring expansion commitments. Bolivia's Superintendent of Waters has already demonstrated that he is willing to impose sanctions on Aguas del Illimani. In 1998, he fined the company for unscheduled interruptions in water service.

Expansion mandates have some inherent weaknesses that even well-designed mandates cannot overcome. When mandates call for anything less

---

*The concessionaire must serve areas that have more than 50 inhabitants or 15 buildings per *manzana,* or more than six buildings per block.

than 100% coverage, even concessionaires that meet their expansion mandates could avoid serving the poorest areas. Moreover, experience has
shown that contract provisions (such as tariff levels) are often renegotiated
in the first years of concession agreements. There is no reason to expect
that expansion mandates would be any less subject to adjustment than
other contract elements. It is important to remember that concessionaires
may have good reasons for being reluctant to serve poor areas. They may
seek to avoid large investments early in the concession if there is uncertainty about the sustainability of regulatory arrangements. Or, they may
consider certain low-income neighborhoods in particular too costly or risky
to invest in, because of location, density, neighborhood violence, or land
tenure problems. Expansion mandates do not by themselves resolve any
of these problems. Unless other measures are taken, concessionaires may
succeed in arguing that they cannot comply with the mandates due to the
magnitude of the obstacles they face in low-income neighborhoods.

As expansion mandates do not guarantee expansion to the poor, the financial incentive structure of the concession contract becomes important.
Financial incentives can either encourage or discourage a concessionaire
from serving poor areas. The remainder of the paper examines contract
provisions that create the financial incentive structure in concession contracts: input and output standards, which determine the concessionaire's
costs, and tariff structures, exclusivity agreements, and penalties for nonpayment, which affect the concessionaire's revenue stream.

3. COSTS: INPUT AND OUTPUT STANDARDS

Most concession contracts have something to say about what kind and
quality of water and sanitation services concessionaires must deliver. The
contracts typically specify outputs (type of service, minimum service quality, reliability standards) and inputs (materials and design standards, pro-
cedural requirements). Technical specifications that require high-quality
service in low-income neighborhoods seem appealing, but there is a potential conflict between these requirements and the utility's ability to
make efficient operating and investment decisions. Inefficient operators
will find it less profitable to expand or improve service. Moreover, the cost
reductions that governments hope to see after privatization will only ma-
terialize if private utilities make efficiency-enhancing improvements.

Technical specifications are best designed to tell concessionaires what to do, but not how to do it: output standards are preferable to input standards. But, as the Aguas del Illimani case demonstrates, loosening input standards is not enough. Output standards have their own problems.

The input standards in the Aguas del Illimani contract resemble those in most concessions. Aguas del Illimani must meet Bolivian water and sewer standards—high-quality equipment and traditional water and sewer system design. Aguas del Illimani can only deviate from these input standards in one case. The company can *recommend* exceptions to the materials' standards if expansion into the outer reaches of the concession area using existing standards would significantly raise total costs.

Both the company and the regulator (the Superintendent of Waters) recognize that the existing standards could make service expansion (even in central parts of the concession area) unnecessarily, and perhaps prohibitively, expensive. The Superintendent, therefore, approved a pilot project of *condominial* sewer and water connections to 10,000 households. Condominial systems provide private water and sewer connections at lower cost by using smaller diameter pipes and burying pipes in shallow trenches in yards or under sidewalks. Aguas del Illimani has been pleased with the condominial systems so far and is now using this technology in all expansion areas. This policy change could significantly reduce the dollar value of investment required to comply with the expansion mandates.

This loosening of input standards gives Aguas del Illimani some flexibility to reduce materials and design costs. But the company cannot choose to reduce costs further by offering different levels of service to its customers—the output standards in the contract are nonnegotiable. All connections the company installs must be metered private water and sewer connections. (Public standposts currently exist in El Alto and La Paz, but Aguas del Illimani must eliminate them over the first years of the contract.) On the surface, the private connection standard appears to be in the interest of low-income households. It ensures that rich and poor households will receive the same level of service. The problem is that the standard limits Aguas del Illimani's flexibility to offer various products, at different prices, to meet differing demand for water and sanitation service.

The Aguas del Illimani contract implicitly assumes that all households in the concession area will choose to pay for private water connections at the price set in the contract. This may well be true, but each household will make its own decision about whether or not to connect to the network system. Whether households currently using unmetered standposts will be willing to pay more for a household connection will depend on how much they value the benefits that a connection provides. Households may have other reasons, besides the monthly cost of service, for choosing not to connect to the water or sewer system. High up-front connection fees or the need to buy sinks, toilets, and other fixtures to take advantage of the service may be a barrier for some households. Other families may be wary of the uncertainty of variable monthly bills, and homeowners may not be willing to invest in connections for rental properties.

From a public policy perspective, it is appealing to use output and input standards to require concessionaires to provide a single uniform type and quality of service throughout the metropolitan area. But in reality, this could hinder service improvement in low-income areas in two ways. First, if households do not want to pay for the service Aguas del Illimani is authorized to provide, it may simply not be possible to achieve the Bolivian government's universal service goals (at least without increasing subsidies to households). Second, if some households in a neighborhood want connections and others do not, the density of connections will drop and the Aguas del Illimani's costs will rise, making cost recovery more difficult (and service expansion less financially feasible).

4. REVENUE: TARIFFS

Keeping down costs makes service expansion more viable, but cost is only half the story. Concessionaires must be able to recover costs through the tariffs, connection fees, and external subsidies set out in the concession contract. All else being equal, households would prefer to pay lower prices for water and sewer service. But setting prices below cost-recovery level can jeopardize service expansion. Concessionaires will demand reduced expansion obligations if they cannot cover their costs.

The tariff in the Aguas del Illimani contract was designed for full cost recovery. It was applied in La Paz and El Alto less than one month before Aguas del Illimani began operations. The contract calls for automatic tariff renegotiation once every five years, with adjustments calculated based the company's projected costs and the cost of the expansion plan for the next five years. Aguas del Illimani can also request extraordinary tariff revisions after the first two years if its costs go up more than 7% for any one of several reasons (change in tax code, etc.). Thus, in principle, the contract is designed to ensure that Aguas del Illimani always makes as acceptable return on its investment.

Two aspects of this tariff, however, could cause problems for Aguas del Illimani down the road. The first is that the tariff includes a strong industrial-domestic cross-subsidy: all households, except the tiny number using more than 150 cubic meters of water per month, pay less than the average per unit tariff; all industrial, commercial, and government customers pay more.* (See Table 7.2.) If industrial users are price sensitive, their usage will fall in response to these high rates. In La Paz and El Alto, many industrial customers have private wells. The high prices give these industries little incentive to connect to the Aguas del Illimani water network. Low demand by industrial users makes it hard for the concessionaire to collect sufficient revenue to offset the lower prices for residential customers.

A second difficulty is that La Paz and El Alto residents are strongly opposed to tariff increases. The current tariff was put in place just before the concession contract was signed. Residents associated Aguas del Illimani with the tariff hike, and the private operator met with much opposition early in its tenure. Recent protests over prices in the Cochabamba water concession in Bolivia add to the tension. This atmosphere makes raising prices a very sensitive issue, even if the concessionaire's costs rise enough to justify a tariff increase. Aguas del Illimani may not be able to fully take advantage of the contract's tariff revision procedures. Indeed, the general manager of Aguas del Illimani has publicly stated that he will not raise prices in the first five years of the contract, even if the company's costs rise.

When revenues do not cover costs, the alternative to raising prices is demanding reductions in expansion mandates. If this were to happen in La Paz–El Alto, poor households would be the most likely to lose out because the tariff structure gives Aguas del Illimani little incentive to serve the poor. All residential users pay less than average price for the water they consume, but the tariff for the first 30 cubic meters of water each month is especially low. Households in El Alto (where average monthly consumption is 7–11 cubic meters per household) do not even pay enough to cover operating and maintenance costs. The sale of water to these households is thus a loss-making proposition. This lifeline tariff structure creates financial incentives that undermine the contract's goal of expanding service to the poor.

---

*This tariff structure need not be a cross-subsidy if the per unit cost of serving industrial and commercial users is indeed higher than the cost of serving domestic customers. The information to make this determination, however, is not available.

Table 7.2 Tariff Structure for Aguas del Illimani (Varies by User Category and Monthly Water Consumption)

Tariff (US $/cubic meter)	Residential[a]	Commercial	Industrial
1.1862	301 m³ and above	21 m³ and above	1 m³ and above
0.6642	151–300 m³	1–20 m³	
0.4428	31–150 m³		
0.2214	1–30 m³		

[a]99% of all residential consumers use less than 150 cubic meters of water per month, so in practice very few consumers ever pay the same unit cost as the commercial and industrial users.

5. REVENUE: EXCLUSIVITY PROVISIONS

Raising prices for low-volume residential users would improve the concessionaire's incentive to serve the poor. But tariffs and fees that give the utility an incentive to expand service to the poor might also discourage some households from connecting or from using large volumes of water. Residential users, like the industries, are price sensitive and, as long as they have alternatives to the concessionaire's water network, Aguas del Illimani risks losing customers if prices are too high.* Concessionaires understandably would like to reduce this source of revenue risk.

There are a number of ways to handle the possibility of low demand when designing a concession contract. The most common approach is exclusivity, but this can be quite harmful to poor households. Most concessionaires are given the exclusive right to provide water and sanitation services within their concession areas. Exclusivity is often justified on the basis that it reduces revenue risk, and thus increases the attractiveness of projects with high capital costs and long amortization periods. It is also advocated as a way to protect the revenues of concessionaires that are required to cross-subsidize some groups of consumers; exclusivity prevents competitors from "cherry picking" customers who are charged high prices to support the cross-subsidy scheme.

Exclusivity is obviously to the disadvantage of industrial users, who are typically the cross-subsidizers. These users would benefit from lower

---

*Dissatisfaction with Aguas del Illimani's service could also lead households to choose another service option.

prices in a competitive service market. But exclusivity can also be quite harmful to the low-income households who are expected to benefit from a cross-subsidy tariff. Exclusivity illegalizes and suppresses the emergence of alternative service providers within the concession area, even in parts of the concessionaire area that the private utility does not plan to serve in the near future. This could delay improvements, or even reduce existing water and sanitation options, because only the concessionaire is authorized to provide service within the concession area.

In Bolivia, for example, sector regulations and the concession contract effectively eliminate competition from communal standpipes, wells, and septic tanks. Aguas del Illimani is required to meter and then eliminate all standposts in the first years of the contract. Private water sources (such as groundwater wells) require permission from the water utility, and the utility can charge well users for the volume of groundwater they extract. Similarly, households must obtain authorization to keep a septic tank open once sewer service is available. Should Aguas del Illimani fully exercise its power to restrict the availability of these alternative water and sanitation services, households within the concession area will have few or no legal service options other than network water and sewer service. This is an especially troubling prospect for neighborhoods that will not receive network service for some time: households could lose options even before a new service is available.

Exclusivity and other provisions aimed at reducing competition need to be treated with caution in the design of "pro-poor" concession arrangements because they have the potential to cause harm if strictly enforced by either the concessionaire or the regulator. In this case, however, Aguas del Illimani has permitted, and even expanded, some alternative water distribution systems within its service area. The Municipality of El Alto provides water delivery by tanker truck to some areas, and in July 1998 Aguas del Illimani initiated a similar service for households without access to the water network. Moreover, Aguas del Illimani requested and received authorization from the Superintendent of Waters not to meet the contract's schedule for metering and removing standposts. Nonetheless, Aguas del Illimani remains the monopoly decision-maker over all service options in its service area. We do not know whether poor and/or rich households would get a better deal in a more open service market, but exclusivity provisions in concession contracts clearly close down the possibility of employing competition in the market to reduce prices, improve service, or provide alternative service options.

There are several alternatives to exclusivity that could be used in concession contracts to address the possibility of low household demand for private water and sewer connections and that are less likely to harm poor households. First, governments could choose to boost demand for a particular service level by subsidizing the service. Subsidies of this type, however, often prove to be unsustainable. A second option is to allow concessionaires to offer a range of different services, at different prices, to satisfy differing levels of demand. Aguas del Illimani does not have this option; the output standards dictate that the company offer only private water and sewer connections.

Another safe alternative is to give concessionaires some flexibility in how they price their services. Aguas del Illimani's concession contract sets maximum tariffs and connection fees for water and sewer service, but it does not prevent the company from lowering prices or offering financing schemes to increase demand for the in-house water and sewer connections. Flexibility like this allows concessionaires to create payment packages tailored to particular demands, needs, or concerns while taking care that their offers make financial sense.

Aguas del Illimani has taken three measures to reduce the probability that lump-sum connection fees discourage households from soliciting connections.* First, Aguas del Illimani gives households the option of paying a reduced fee in exchange for supplying labor during the connection process. Households that help dig trenches pay only US $105 for a water connection and $130 for sewer (compared to the normal fees of $155 and $180, respectively). This option has been very popular. Only 20% of El Alto households that received water connections through August 1998 chose to pay the full fee.† Second, Aguas del Illimani offers "the population most in need" a three- to five-year financing plan for connection fees. Finally, in the poorest neighborhoods of El Alto, this financing plan includes a subsidized interest rate. Aguas del Illimani is choosing to use its pricing flexibility to reduce barriers for poor households, but concession contracts that *require* long financing periods or subsidized interest rates could be dangerous, because these measures may not be financially sustainable over the long-term.

---

*Aguas del Illimani committed to take the first two of these measures in an agreement with the Municipality of El Alto designed to address some of El Alto's preprivatization concerns.
†Data provided by Aguas del Illimani for connections through August 1998.

6. REVENUE: NONPAYMENT PENALTIES

Some concessionaires experience problems collecting bills from customers. Billing customers at cost recovery tariff levels does little good if customers do not pay their bills. Thus, concessionaires generally seek some leverage over nonpaying households during negotiations of the concession contract. In Bolivia, the 1992 Regulations permit utilities to cut off service to any households that do not pay their water and sewer bill within two months. Armed with this power, Aguas del Illimani has achieved high collection rates from domestic customers. This power is rare. Many concession contracts prevent disconnection for nonpayment or make cutting off service a very long and difficult process.

One could argue that low-income households *with* water and sewer connections would actually benefit in the short-run if governments prohibited disconnection; this move would effectively remove the obligation to pay water and sewer bills. Low-income areas *without* access to the water and sewer network, however, would lose out if a no-disconnection policy resulted in low collection rates. Low collection rates jeopardize cost recovery, and concessionaires that are unable to cover their costs are likely to reduce their service expansion plans.

7. CONCLUSION

The Aguas del Illimani concession contract, like most concession contracts in the water and sanitation sector in developing countries, relies heavily on expansion mandates and lifeline tariffs to achieve its objective of improving services for the poor. The expansion mandates seek to ensure that the concessionaire will extend service to low-income households, and the lifeline tariff is designed to make sure that service is affordable to poor households.

To date, Aguas del Illimani has met its service expansion obligations. But expansion mandates in the long run are no guarantee that concessionaires will reach the poor. Mandates are subject to renegotiation over the life of the concession contract. Where concessionaires have little incentive to serve the poor, poor households will likely lose out if expansion plans are reduced or delayed. Lifeline tariffs make low-volume consumers unattractive to private operators. In El Alto, the low-volume users are also the low-income households, so the lifeline tariff creates financial incentives that contradict the expansion goals in the concession contract.

The weaknesses of the La Paz–El Alto concession suggest some general guidelines for the design of future "pro-poor" concession contracts. First, government officials can increase the probability that a concessionaire will improve service in low-income areas by making the contract's financial incentive structure consistent with overall objectives and mandates. Second, contract drafters need to be extremely cautious about any provision that restricts the emergence of new service providers or service alternatives. Exclusivity provisions could cause more harm than good for poor households.

Finally, low-income households, like all utility customers, are best served by a utility that can offer them service that they want. Demand for service improvements usually varies by household and neighborhood and will change over time. Thus, poor households would benefit from private operators that have sufficient flexibility in operations and product offerings to meet this varied and changing demand.

Chapter 8

Development of the Mexican Water Sector

Edward J. Sondey and Neel A. Shah

Despite delays in the privatization of Mexico's electric power industry, Petroleos Mexicanos (Pemex), the state-owned oil company of Mexico and one of the largest oil companies in the world, has made significant strides in outsourcing to the private sector the construction and operation of wastewater treatment facilities associated with its refining operations. Deeming the provision of such services to be outside of its core competency, Pemex sought assistance from water industry experts and successfully implemented these projects. In contrast, the municipal water market has struggled. Although several municipalities in Mexico have attempted to attract private investment in their water treatment operations in recent years, many of the projects have languished due to financing obstacles.

Pemex Refinacion (Refinacion), the refining subsidiary of Pemex, is the offtaker for the wastewater treatment projects. These projects generally have been viewed as the best investment opportunities available in the

This chapter was previously published in the *Journal of Project Finance,* Fall 1999, Vol. J, No. 3, pp. 53–61. Used with permission.

Edward J. Sondey is Director of the Development Group of PSEG Power Development Corporation. He was formerly founder and Senior Vice President of Poseidon Resources Corporation. Neel A. Shah is President and General Manager for Six Open Systems, Inc., the U.S. division of a German software firm. He was formerly a Senior Associate with Poseidon Resources that managed the Atlantis Water Fund.

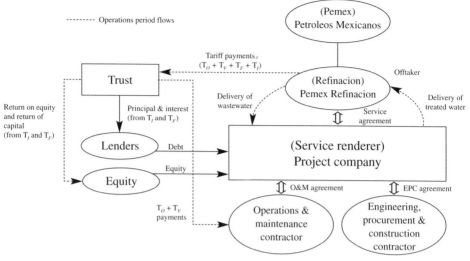

Figure 8.1   Refinacion project structure.

Mexican water sector. This is largely due to Pemex's fluency with risk allocation in project financings and the lending community's comfort with the credit of Pemex. Following the success of Pemex's competitively bid refinery and off-shore oil production projects, Refinacion structured its wastewater treatment outsourcing program to provide an equitable allocation of risks between itself and the service providers, resulting in a winning situation for all parties.

A typical project structure is illustrated in Figure 8.1. By harnessing private sector innovation through a series of competitive tenders, Refinacion has benefited from lower treatment tariffs and enhanced technical solutions for pollution and conservation concerns. At the same time, investors and lenders have been able to achieve attractive returns relative to the level of risk they assumed.

1. BACKGROUND ON REFINACION PROJECTS

In the mid-1990s, Mexico's federal government enacted major environmental legislation designed to curb industrial wastewater discharge and thereby reduce the pollution of urban systems and conserve water resources. This step was taken in order to address pressures from Mexico's

NAFTA partners, and also to force companies such as Refinacion to avoid competing with local populations for potable water supplies.

These laws and associated regulations required Refinacion to reconsider its usage of process water in its refining operations. In order to comply with the strict federal regulations, Refinacion initiated a wastewater treatment program in 1994, under which it would treat and recycle effluent from each of its refining operations. Refinacion owns and operates six refining facilities throughout Mexico: Cadereyta, Madero, Tula, Minatitlan, Salina Cruz, and Salamanca. Refinacion bid out five of the required treatment facilities as build, operate, and transfer (BOT) schemes through competitive tenders (the Salamanca facility was built and is operated by Refinacion). The original bids for the Refinacion projects were led by Mexican construction companies. Of the five projects, two were awarded to Cydsa, one was awarded to a Bufete-Degremont consortium, and the remaining two were awarded to local engineering companies.

Unfortunately, Refinacion and the winning bidders were caught in a difficult situation shortly after the awards were made, as the peso devaluation in later 1994 created an economic shock in the country and roiled debt markets. The contractors' own financial resources were strained as their dollar-denominated debt burdens were exacerbated, and new sources of debt became essentially unavailable. Throughout 1995, all the winners sought to restructure the Refinacion projects, even while dealing with other crises caused by the devaluation. As part of the restructuring effort, the contractors began looking to international water companies to participate as partners in order to access international sources of equity and debt.

With the addition of these new participations, the Cadereyta, Madero, and Salina Cruz facilities were ultimately restructured to reduce exposure to currency risks, which helped these three projects attract debt financing and achieve financial closing by late Fall 1997. However, the sponsors of the Tula and Minatitlan projects were unable to achieve financial close on either project, and once again each was bid out to the private sector in 1998 and 1999.

2. BIDDING STRUCTURE

Respondents to the Refinacion bid requests were required to perform engineering design, equipment procurement, construction, operation, and maintenance for each facility, as well as arrange debt financing and contribute owners' equity. Each concession was designed to have a 14-year life, from concession award through the end of the 12-year operating period.

Refinacion's evaluation of bids was based on: (1) the technical merits of facility design; and (2) the cost to Refinacion of the tariffs bid by the service provider, which was evaluated on a net present value basis, using a 12% discount rate—presumably Refinacion's cost of capital. The tariff structure was designed to allow bidding consortia to analyze and efficiently allocate various responsibilities and risks among consortium members. This goal is evident in the water treatment tariff, which is divided into four parts:

T_O: Fixed operating expenses tariff

T_V: Variable operating expenses tariff

T_F: Financing tariff, covering:
- Mexican corporate taxes
- Interest payments on debt
- Return on equity invested (investor's profit)

T_I: Investment tariff, covering:
- Principal payments on debt
- Return of equity capital (equity "principal")

Because of this rationalized bidding structure, the winning consortium was able, in each case, to separately analyze and structure engineering, procurement, construction (EPC); operations and maintenance (O&M); equity; and debt arrangements. The sponsor in each project was able to pay the T_O and T_V tariffs directly to the operator, which simplified the financial evaluation and structuring to an isolated analysis of the T_F and T_I tariffs. Refinacion also benefited from this structure, as bidders were required to be transparent in submitting their actual financing costs, including both the interest rates charges by lenders and the specific returns required by equity investors.

Refinacion ultimately conducted thorough, well-defined, and transparent bidding processes in requesting, analyzing, and awarding its wastewater concessions. In hindsight, it is apparent that this successful track record is unique. Refinacion overcame challenges that have hampered privatization initiatives in the power sector, in addition to efficiently dealing with multiple debt and currency market difficulties that occurred between 1994 and 1998.

3. PROJECT RISKS FACED BY DEVELOPERS

The Refinacion wastewater program has proven attractive to equity investors and salable to lenders, largely due to the rational allocation of risks among project participants. With distinct tariff payments and a high degree of support from Refinacion, as demonstrated through the terms of the projects' service agreement, project sponsors were able to significantly mitigate the following project risks, by assigning them to third parties or by obtaining protection in the service agreement:

Construction risk
Operator default or bankruptcy
Financing risk (creditworthiness of offtaker)
Floating-rate debt
Currency devaluation (repayment of debt and equity)

The standard form of service agreement between Refinacion and the service renderer was included in bidding documentation for each project. This agreement provided a straightforward means for project developers to negotiate credit terms, and was structured so that the primary construction and operating responsibilities could be subcontracted to capable, creditworthy third parties. As a result, debt and equity investors were able to focus on risks appropriately associated with debt and equity, not those associated with building and operating the facility.

Of additional comfort to investors was the fact that, in worst-case scenarios, Refinacion bears a substantial portion of both operating and construction risk for each project. Refinacion's refineries are critically dependent on the wastewater treatment facilities, as the treated water is used in the refinery operation process. For this reason, Refinacion included in the service agreement a provision whereby, in the event of a material construction failure, Refinacion has a right to terminate the service agreement and is obligated to purchase the facilities.

With respect to operations, the service renderer is required to meet certain standard performance measures. However, in the event that the service renderer does not deliver 100% of the water flow required under the terms of the service agreement, or delivers water that does not meet the predetermined effluent standards, Refinacion still must pay the service renderer on a timely basis. Refinacion is able to reduce payment of the financial and investment tariffs only in the highly unlikely event that the facility is totally shut down for a period of more than eight consecutive hours.

To ensure that the EPC and O&M contactors do not take advantage of these protective terms, the service agreement includes stiff penalties that encourage the contractors to perform as promised. Considering that the construction and operation processes are straightforward and low risk, the termination procedures and the measure of operating risk assumed by Refinacion provides lenders with sufficient comfort to provide construction and longer-term financing commitments. These terms also provide the developers the comfort with which to pursue more aggressive capital structures in their bids and reduce their overall cost of capital. While these measures are not guarantees of the debt and equity, they do provide a strong measure of asset protection for the lenders and the equity investors. Refinacion directly benefited from these attractive financing terms through a reduction of their final tariff payments.

The lenders in each of the projects also derived comfort from the service renderer's subcontracting of its operating obligations, and its ability to replace the operator in the event of default, and thus make timely repayments of debt. The tariff structure allows each project's service renderer to easily subcontract operations, with a defined revenue stream (T_O and T_V) that can be used to pay the operator. Since the operator is not necessarily the same entity as the service renderer, the service renderer's ability to perform is isolated from operator failure. It is not uncommon to find project structures where the operating company is a partial owner of the service renderer, in which case operator bankruptcy can cause the remaining equity invertors severe restructuring headaches.

Lenders were ultimately comfortable with Refinacion's credit as an offtaker because of Pemex's "trophy" commodity base and the familiarity of the capital markets with its credit (rated BB by Standard & Poor's and Ba2 by Moody's,* equivalent to the sovereign ceiling at the time). Although the revenue base of Refinacion is somewhat weaker than that of its parent, Pemex, the priority nature of payments made by Refinacion for wastewater treatment was an important element in convincing lenders of Refinacion's credit quality.

With respect to floating-rate debt, Refinacion's bidding conditions spe-

---

*Standard & Poor's rating as of September 23, 1999, was BB (with positive outlook); Moody's rating was Ba1.

cifically stated that, during operations, it would reimburse the service renderer for floating-rate interest expenses. This was significant in allowing the developers flexibility in selecting low-cost sources of debt.

In order to effectively mitigate Mexican inflation risk, Refinacion's bidding criteria indicated that it would index tariff components that were bid in Mexican pesos to domestic inflation, beginning on the date of bid submission. However, Refinacion also allowed respondents to bid any of the tariff components in foreign currencies, with Refinacion paying these tariffs in pesos indexed to the selected currency. As Refinacion is a worldwide exporter of crude, and has an ability to raise capital in a variety of foreign currencies, it was able to avoid passing currency risk to bidders, who were attempting to raise debt and equity in the international capital markets. This allowed developers to seek out the cheapest debt financing available in the markets, and then bid tariffs in the appropriate currencies for reimbursement of debt costs as well as equity returns, without necessitating the use of long-term foreign currency hedges.

Technical aspects of four Refinacion water projects are summarized in Table 8.1 and financial aspects are summarized in Table 8.2. A more in-depth discussion of each project's development and structuring follows.

Table 8.1   Technical Aspects of Four Refinacion Water Projects

Project	Cadereyta	Madero	Tula	Minatitlan
Location (state):	Nuevo Leon	Tamaulipas	Hidalgo	Veracruz
Type:	Wastewater treatment	Wastewater treatment	Wastewater treatment	Wastewater treatment
Treatment process:	Biological, chemical, reverse osmosis (RO)	Dissolved Air Flotation (DAF), biological, chemical, RO	Biological, chemical, multimedia filtration	DAF, biological, ultra filtration, RO
Treated flow (MGD):	13.6	4.2	5.0	15.7
Reused flow (MGD)	12.6	3.2	5.0	15.7
% recycled water	92.6%	76%	100%	100%
Municipal treatment (population served):	Yes (65,000)	No	No	Yes (73,000)
Construction startup	Nov. 1996	Nov. 1997	Apr. 1999	Oct. 1999
Commercial operation	Nov. 1998	Nov. 1999	Oct. 2000	May 2001
Status:	Operating	Construction	Construction	Financing

Table 8.2   Financial Aspects of Four Refinacion Water Projects

Project:	Cadereyta	Madero	Tula	Minatitlan
Concession life (years):	14	14	14	14
EPC contractor:	Atlatec	Atlatec	Atlatec	Atlatec
O&M contractor:	Atlatec	Atlatec	Atlatec	Atlatec
RO membrane supplier:	Hydranautics	Fluid systems	N/A	TBD
Total project cost (US $)	$35 m	$35 m	$18 m	$39 m
Debt as % of capitalization:	70%	70%	75%	82%
Initial lender(s):	Chase Manhattan Bank	Chase Manhattan Bank	Hypo Bayerische Vereinsbank	TBD
Loan maturity (years)	10	10	10	25
Political risk insurance	No	No	No	Yes
Currency denomination of loan:	US dollar	US dollar	US dollar	Japanese yen

4. CADEREYTA AND MADERO: SUCCESSFUL TEMPLATE AFTER THE MEXICAN CRISIS

The need for a wastewater project was especially acute in the case of Cadereyta, where the refinery was using scarce groundwater supplies shared with the local community, and then discharging the untreated wastewater into the nearby river. This contamination created a health threat for the nearby community and a danger for the environment. The BOT scheme for Cadereyta, which is representative of the other Refinacion projects, called for the collection of wastewater from the refinery and the local municipality, treatment to the standards required by the refinery operator, and recycling of the treated water for reuse by the refinery.

Poseidon Resources Corporation (based in Stamford, Connecticut), a developer of and investor in water infrastructure projects, partnered with Grupo Cydsa, a well-respected Mexican firm, to invest in the Cadereyta and Madero treatment facilities. Cydsa, a Monterrey-based conglomerate that manufactures chemicals and textiles, had won concessions for both of these facilities in June 1994. In each project, Cydsa secured construction and operations contracts for Atlatec, its environmental services division, which were renegotiated in 1996.

During and after the Mexican peso devaluation in December 1994, debt financing became very expensive, as many lenders suddenly became bearish on Mexico and the creditworthiness of local companies. As a result, many infrastructure projects in Mexico languished, including several wastewater projects in the municipal sector. Despite the tight credit markets, the Poseidon-Cydsa team successfully negotiated loans with Chase Manhattan Bank for both projects through project-financed debt facilities covering both the construction and term periods. With the participation of Poseidon Resources and the projects' equipment supplier as equity investors in both transactions, Cydsa was relieved of the obligation to invest all of the equity required in the projects. Cadereyta reached financial close in October 1996, and Madero was closed in October 1997.

5. TULA AND MINATITLAN: WEATHERING THE ENVIRONMENT DURING THE POST-ASIAN/BRAZILIAN CRISES

As mentioned previously, the concession for the Tula wastewater treatment project was rebid in 1998, after the initial sponsor had difficulty securing debt financing. The consortium made up of Cydsa and Poseidon's affiliate, the Atlantis Water Fund (an international private equity fund comanaged by Poseidon Resources), was awarded the project in the fall of 1998. Once again, the management team from Poseidon and Cydsa put together a successful financing package with a loan from Hypo Bayerische Vereinsbank, despite difficulties in the emerging markets following the Asian crisis and the Brazilian devaluation.

The capital markets' acceptance of Refinacion's long-term credit obligations, as well as the risk-mitigation features of the service agreement, enabled the sponsors of the Tula transaction to achieve excellent debt pricing. As an example of Pemex's strong reputation in the financial markets, over $1 billion was syndicated in the bank market in late 1998 for the expansions of Refinacion's refineries, despite the fact that the financial markets were largely unavailable to Latin borrowers during the latter half of 1998. Mexico's economy since then has been performing extremely well, despite the global turmoil in late 1998 and the fear of Latin American contagion in early 1999 following the Brazilian devaluation. Despite the initial negative outlook regarding Mexican risk, the sponsors of Tula were able to shop the deal to a number of project lenders and close a debt financing with tight risk spreads.

Tula is a good demonstration of how governments and industrial companies might structure a project. Refinacion developed a transparent bidding structure and well-defined project specifications. The service agreement terms adequately addressed the risks inherent in an emerging market project financing, and this, in turn, encouraged qualified project developers to invest the time, resources, and effort to formulate technically feasible and economically competitive tenders. In the final analysis, the Refinacion project structure allowed the developers of Tula to perform best at their core ability, which was to obtain long-term construction, operating, and financing commitments with competitive pricing, despite the difficult economic environment. By leveraging private sector expertise in shopping out the construction, operating, and financing obligations, Refinacion ended up maximizing its benefit through a reduction of the net present value of tariff payments.

Minatitlan was the most recent Refinacion project to be bid out, and was awarded to the Poseidon-Atlantis-Cydsa consortium earlier this spring. As described earlier, Refinacion's request for proposals afforded developers the opportunity to choose the currencies of their bids as well as to specify the interest rate benchmarks. The sponsors of Minatitlan, therefore, were able access the lowest-cost financing available by taking advantage of the low interest rate associated with a yen-based floating-rate bank loan.

While the bidding structure enabled the sponsors to recover the cost of their loan in yen, thereby mitigating operating period currency risk, the sponsors were still confronted with development and construction period currency risk, because the EPC contract was denominated in pesos and US dollars. As draws from the yen facility would be made by the EPC contractor throughout the 18-month construction period, in order to fund peso and dollar costs, the issue was to manage the exchange rate, from the bid date through the end of construction, at which time the yen debt proceeds would be converted. As exchange rates almost certainly would differ from those existing at the time of bid submission, there were three distinct time periods during which this risk needed to be hedged:

Bid submission to bid award
Bid award to financial close
Financial close through completion of construction

The sponsors met the first challenge by purchasing a US dollar call option on the date of bid submission, which afforded the option of swapping the yen facility into dollars six months after the bid date (the approximate time estimated for selection of the final bidder and financial closing) at a predictable exchange rate. After the bid was awarded, and the sponsors were more certain that they would reach financial close, they were able to replace the initial hedge with a series of forward contracts. These forwards allow the sponsors to swap the yen funds into US dollars, again at predictable exchange rates, in a timing sequence that matches the required payments to the EPC contractor throughout construction. Ultimately, these hedging mechanisms allow the sponsors to avoid exchange rate volatility during the prefinancial closing and construction time periods.

The Tula and Minatitlan projects highlight Refinacion's success in reducing its water treatment costs, through: (1) the rational assumption of project risks, which simplified lenders' credit analysis and encouraged investors to bid aggressively; and (2) the ability of project developers to engineer creative, lower-cost financial solutions, if they are free to seek the most efficient sources of available capital.

6. LESSONS LEARNED

The key to success for the Refinacion wastewater projects was the appropriate allocation of risks among project parties. As discussed above, one of the most important elements of this process was allowing developers the flexibility to determine the appropriate financial structure of each deal, on terms that they found to be acceptable. Additionally, Refinacion was logical in assuming a significant amount of construction and operating risk for each project, as its refinery operations are highly dependent on the water supplied by the water treatment facilities, and the operations of the facilities have a low level of technological risk. This sends a strong message to capital providers of Refinacion's commitment to the projects.

A final point with respect to the bidding process is that Refinacion designed the bids to encourage participation. In each of its tenders, Refinacion has requested that bidders put up security packages during the bidding and construction phases of the projects to ensure that they would carry out their commitments. The size of the required security instruments was at an appropriate level to provide Refinacion protection, but not so onerous as to discourage bidders from participating. In contrast, some municipal bids have required that developers guarantee as much as 100% of project costs to secure their bids. Such requirements impede competition by discouraging developers from submitting bids.

7. THE MUNICIPAL MARKET: POTENTIAL AND CHALLENGES

Just as the government enacted environmental legislation to regulate industrial discharge, it also legislated a requirement that all municipalities with more than 2,500 persons meet certain wastewater discharge standards. Along the United States–Mexico border alone, this new regulation has resulted in over $400 million in wastewater investment opportunities. However, due to a shortage of financial resources and delays at the municipal level, federal environmental authorities have been less stringent with municipalities than with industrial entities in enforcing compliance of wastewater treatment regulations.

To date, only a few states and some of the larger municipalities have begun developing integrated water concessions, with the encouragement of federal environmental regulations drafted by the National Water Commission (CNA). These concessions allow operators to supply all of the water supply and water treatment service needs of an entire territory, as well as to perform metering and bill collection services, effectively acting as a regional utility. However, some local government officials have been unwilling to entrust the control of municipal water supplies to private enterprises, as they are sensitive to public outcries if private parties cannot deliver on their promises to achieve greater efficiency and quality of service. This obstacle is not insurmountable, however, as demonstrated by the Refinacion water treatment initiative. In fact, addressing this concern with the inclusion of appropriate buyout provisions is consistent with the type of risk mitigation that appeals to capital providers.

As private sector participation in the water sector gains greater public acceptance, the municipal market in Mexico is expected to yield significant investment opportunity. The key challenge for sponsoring municipalities and the federal government will be to follow Refinacion's lead in structuring bids that draw on the private sector's innovative practices. As in all infrastructure finance, the key to accomplishing this objective is rational risk allocation.

If Mexican municipalities sponsoring water treatment projects can mitigate the risks perceived by providers of capital and seize upon the momentum created by successful closings of industrial projects, there is a positive outlook for municipal BOT schemes. As municipalities begin to bid out their projects, developers will look for bidding conditions that set guidelines for addressing: (1) rational risk allocation among project participants; (2) currency protection for debt and equity investors that allow developers to raise the cheapest available capital; and (3) the credit quality of the offtaker (ability to pay tariffs as promised).

In the municipal BOTs tendered to date, the inability of the governing authorities to assume currency risk has necessitated the use of local funds to finance projects. Unfortunately, local capital costs remain high throughout much of Latin America, and the resulting tariff prices are frequently prohibitive. Municipalities are more accustomed to indexing water usage and sewage tariffs to domestic inflation (as required to obtain local financing), which is politically acceptable, than they are to indexing tariffs to foreign exchange rates (as required to obtain international financing). Despite this state of affairs, there are a number of examples, including projects in Argentina, Chile, and China, where government-owned utilities, municipalities, or end users are actually bearing currency risk.

In many emerging markets, resistance by government-sponsored offtakers to indexing tariffs to foreign currencies is an obstacle for developers. Developers usually predicate their willingness to commit long-term equity on the ability to raise project debt on the type of attractive terms found in highly liquid capital markets, and on the ability to earn equity returns in a hard currency. The indexing of tariff payments to local inflation can create a situation whereby a sponsor raising debt in US dollars, but generating project income in local currency, is subject to a mismatch of revenues and expenses. Confronted with this issue, developers are forced to either borrow local capital at higher costs or to reduce the project leverage (substituting low-cost debt with high-cost equity), both of which increase the price for municipalities and individual consumers.

Another major concern of investors is the creditworthiness of Mexican municipalities. Tax collection in Mexico is centralized, and municipalities generate relatively modest local tax revenues. As a result, municipalities are highly dependent upon allocations of federal tax receipts. Creating further concern about the financial viability of municipal entities is the fact that municipal water companies often subsidize the cost to end users, rather than charging households the actual production cost of water. Therefore, most of these water companies run significant operating deficits; without federal subsidies, many of them would be bankrupt. In order to finance water projects, municipal credits are thus heavily reliant on the payment guarantee program for water projects provided by the National

Public Works Bank of Mexico (BANOBRAS). BANOBRAS provides infrastructure loans and grants, as well as payment support mechanisms, in order to encourage further investment from the private sector.

8. GOVERNMENT-SPONSORED SUPPORT PROGRAMS

BANOBRAS' payment guarantee program allows municipalities tendering BOT schemes to access a BANOBRAS line of credit that provides up to six months of tariff payments to pay private concessionaires in the event of a municipal default, thereby enhancing the municipality's credit. If BANOBRAS is required to pay under the credit line, it turns to Hacienda (the Ministry of Finance), in order to recollect its money from the tax revenues allocated to the state in which the municipality is situated. The BANOBRAS payment guarantee program must be renewed every six months, due to the limitations BANOBRAS has in assuming long-term infrastructure obligations. This renewal risk typically has been a source of discomfort for foreign lenders, and has not yet sufficiently mitigated their concerns regarding long-term debt repayment.

FINFRA, the infrastructure investment fund within BANOBRAS, has been a vital source of financing for municipal projects, providing up to 40% of a project's total capital costs in the form of a direct subsidy. This program has been instrumental in helping local developers to salvage troubled water concessions and restart construction projects delayed by a lack of available debt financing.

While the support mechanisms described above are somewhat effective, none of the programs currently in place addresses the issue of currency risk. Although BANOBRAS currently does not help mitigate this risk, progress on this issue will certainly help developers to get over their current hurdles. The Refinacion projects continue to provide the best examples in Mexico for addressing all of the issues required to finance. While several of the right elements exist for completing municipal projects, they need to be presented with more emphasis on investors' concerns.

9. FUTURE OF THE MUNICIPAL MARKET

While BOTs often present investors with a lower risk profile than integrated concessions, given their contractual nature, the latter model is expected to become increasingly prevalent in Mexico. As upcoming megaprojects that follow the concession model come on line, for example, in Mexico

City and Guadalajara, the hope is that the introduction of more international world-class service providers will encourage operating efficiencies, benefit end users, and offer expertise in bringing projects to rapid operational status.

Nonetheless, the concession model will not solve all problems confronting municipalities in Mexico. Investors in concessions still continue to face many of the BOT-type risks discussed above. Furthermore, many municipalities in Mexico undoubtedly will find the BOT structure more compelling than the complete outsourcing of services through integrated concessions. In either case, further development of the water sector requires supporters at the federal and local levels to continue rationalizing the available financial mechanisms in order to encourage further private investment and the probability of successful project closings.

Part IV

Reinventing Public Sector Operations

Chapter 9

Mexico's Federal District Water Reform

Lessons and Experience

Lillian Saade Hazin

This chapter analyzes the involvement of the private sector in the water sector in Mexico's Federal District. In October 1993, after an international bidding process, four private firms were awarded service contracts to implement universal water metering, rehabilitate the distribution system, and carry out a loss-detection program. The decision to involve the private sector was motivated by the urgent need to reform and improve water services in one of the largest cities in the world. Given the lack of information regarding the customer base, water consumption levels, and network conditions, a phased approach to private sector participation (PSP) was considered most appropriate.

The first section of this chapter provides a description of the situation of the water sector in Mexico. The next part focuses on the main characteristics of water supply and sanitation, particularly in Mexico City. In

The author is grateful to Mr. Jean-Denis Hatt for his invaluable inputs.

Lillian Saade Hazin is an independent consultant who previously was an economist with the Sector and Utility Management Group at the International Institute for Infrastructure, Hydraulic and Environmental Engineering. She was formerly Advisor to the Deputy Minister of Planning at the Mexican Ministry of Environment, Natural Resources and Fisheries.

Section 3, the water reform strategy adopted in the Federal District is presented, assessing the public–private sector arrangement. This section focuses on the factors that motivated and limited the reform, as well as the reasons for the selection of service contracts, the scope of the contracts, the consortia selection process, and the progress so far. The chapter reviews the relative merits and risks of the adoption of the use of a phased approach to PSP, as well as the decision to use four different consortia for the service contracts, seeking to introduce certain competition for improved performance. Section 4 highlights the parties involved and institutional arrangements. Section 5 provides a discussion on tariff issues. Finally, lessons learned are presented, and factors to be considered in the implementation of new schemes are contemplated.

The transition to PSP has achieved several goals. The change to metered consumption is one of the most important achievements. Substantial improvements have been made with the customer database, metering, and billing. Yet there is significant potential for effective demand management of water use. The implementation of the contracts has been hampered by a number of setbacks, including the 1995 financial crisis of the country and the lack of coordination among the institutions involved in the sector. The political changes in the Federal District have had an impact on the implementation of the project; however, the need for this type of project in Mexico's Federal District clearly goes beyond political considerations, and PSP may assist in ensuring that this is so.

1. BACKGROUND

1.1 Water Availability

The provision of water and sanitation services is a real and growing problem in Mexico. The mean annual rainfall is around 772 millimeters (715 in the case of the Federal District), equivalent to a volume of 1,528 cubic kilometers (CNA, 2001). Despite the high rainfall, water distribution is difficult and costly in the country, since the geographical distribution of water does not reflect the distribution of the population and its needs. Around 77% of Mexicans live in the northern and upland part of the country (where the Federal District is also located), where approximately 28% of rainfall occurs. In addition, this region contains 92% of irrigated land,

and generates 84% of the country's gross domestic product (GDP) (23% being generated in the Federal District) (CNA, 2000). The per capita GDP in the country is of the order of US $5,000.* Water supply is also unevenly distributed seasonally, since annual rainfall is mainly concentrated during the summer period.

Approximately 100 of the 650 Mexican aquifers identified are severely overexploited. (CNA, 2001), with the area of the Valley of Mexico being one of the most affected. While the aquifers in that region are recharged with approximately 700 million cubic meters annually, nearly double that amount is extracted. Groundwater exploitation first began last century. The falling groundwater levels have resulted in an average subsidence of 7.5 meters in downtown Mexico City. Subsidence has exacerbated the natural propensity of the city to flood and has damaged the city's infrastructure (Noll et al., 2000). The city has been obliged to install pumps to push the water to what is called the Great Canal, and to move water through urban drainage networks. This situation has left the Great Canal operating at 10% of its initial capacity (Adelson, 1999).

The basic funds concerning the district are as follows, with data drawn from the Comisión Nacional del Agua (National Water Commission, CNA, 2001) and the Comisión de Aguas del Distrito Federal (Federal District Water Commission, CADF):

- Population: 9 million people
- Surface area: almost 1,504 square kilometers
- Population with access to potable water: 98%
- Population with access to sewerage services: 94%
- Average annual rainfall: 715 millimeters
- Total water supply: 35 cubic meters per second

1.2 Water Supply Networks

While Mexico has invested significant funds in the construction of public infrastructure, numerous problems remain. The administrative and pricing policies in the country in general have not been effective in ensuring that the increasing demands for water are met. In particular, revenues

*Estimate is based on information from the National Institute for Statistics, Geography and Informatics.

from water supply have contributed a small part of the total cost of financing the system. Furthermore, there are important cross-subsidies among sectors. Private sector participation was sought as an option to improve water distribution in Mexico City.

2. WATER SUPPLY AND SANITATION IN MEXICO CITY

The Mexico City Metropolitan Area (MCMA) comprises two political-administrative entities—the Federal District and the peripheral urban municipalities of the State of Mexico. The total area of the Federal District is approximately 1,504 square kilometers. The metropolitan area extends to the east, north, and west of the Federal District into 17 municipalities of the State of Mexico, having a total area of 2,269 square kilometers. Approximately one-fifth of the Mexican population live in the MCMA, with around 9 million people in the Federal District and slightly more in adjoining parts of the MCMA in the State of Mexico (National Research Council, 1995). In total, the area served by the common water distribution and wastewater disposal systems equals 1,287 square kilometers. Management of water and wastewater services within the MCMA is shared by the Federal District and the State of Mexico, each of which is responsible for providing water and sanitation services within their jurisdictional boundaries. The CNA is in charge of providing bulk water to the service areas. Of the 62 cubic meters per second of water received by the MCMA, the Federal District consumes 56% and the peripheral municipalities of the State of Mexico 44% (Casasús, 1994).

2.1   Water Supply

Approximately 32% of water supplied to homes is lost through leakage.* This leakage is mainly due to the poor maintenance of the infrastructure and inadequate installations in homes. In addition, the average cost of providing water services to the fast-growing population is increasing.

Water consumption in the past five years has corresponded on average to 360 liters per day. While this is less than in the period 1984–1990, when 400 liters were consumed, it is still high relative to other cities in Mexico

---

*According to the CADF (personal communication), this figure has dropped from 37% in 1997 to 32% in 1999.

and the world. This is partly attributable to the high leakage rate, but also due to the fact that, prior to the introduction of the new system, the tariff structure gave users no incentive to save water. The daily per capita water usage in the State of Mexico is of the order of 230 liters. Authorities attribute the larger per capita water use in the Federal District to the fact that the Federal District is more developed and more commercial and industrial activities are carried out there. In addition, there are many private wells in the State of Mexico that are not included in the estimates (National Research Council, 1995).

Two-thirds of the water supply comes from the aquifer underlying the city and the rest comes from external sources, mainly from the Cutzamala system. The Cutzamala basin incurs high pumping costs, since it is located at a considerable distance from Mexico City (some 130 kilometers) and lies at a lower altitude (1000 meters below the city).* Mexico City has been rapidly drawing down the aquifer under the city. Because of the nature of the soil, the depletion of the aquifer has led the City center to sink by several meters. Overexploitation is due to the great size of the city's population and to the consumption patterns prevalent in the area.

According to CADF, 98% of the residents of the Federal District have access to piped water, either through an in-house connection or through a common distribution faucet in the neighborhood. The remaining 2% rely on water trucks, from which they fill containers for home use. It is estimated that about 74% of the residents of the Federal District have an in-house water source. In the metropolitan service area of the State of Mexico, roughly 52% of the homes relied on an in-house water source (National Research Council, 1995). Figures from the CNA indicate that, in the State of Mexico as a whole, 90% of the population has some access to piped water. 90% of the water supplied to the Federal District is for domestic use, and the remaining 10% is for nondomestic users (i.e., industry and commercial use) (CADF, personal communication).

2.2   Wastewater Treatment

Around 60% of wastewater discharges are generated in the Federal District and in 7 out of the 31 Mexican states. Studies show that the Valley of

*For a description of the water distribution system and interconnection between the Federal District and the State of Mexico service areas in the MCMA, refer to National Research Council (1995).

Mexico and the Lagunera region are among the areas with the heaviest groundwater pollution (CNA, 1994). Although 94% of households in the Federal District had access to sewerage services, only 10% of the municipal wastewater from the MCMA is properly treated. The remaining 90% of wastewater generated is diverted out of the Basin of Mexico through the general drainage system. The untreated wastewater is then used to irrigate over 85,000 hectares of farmland in the Mezquital Valley in the neighboring State of Hidalgo (National Research Council, 1995).

As in the rest of the country, the provision of water services in Mexico's Federal District has been highly subsidized. Since 1996, the service has required an annual subsidy of over 2 billion current pesos (over US $200 million) from other Federal District Government resources. The evolution of the financial situation of water services is presented in Table 9.1. Expenses include investments in large projects that are also financed through external loans. With the recent changeover to metered consumption, more efficient physical and commercial operation, and an increase in tariffs, it is foreseen that the situation will improve as revenues increase. (See below.)

Table 9.1   Financial Situation of Water Services in the Federal District Government

Year	Revenues[a]	Expenses[b]	Revenues	Expenses	Budget Deficit
	Millions of current pesos[c]		Millions of US dollars		
1992	471	1,694	137	493	−356
1993	596	1,599	158	424	−266
1994	766	1,761	187	431	−244
1995	809	2,182	143	387	−244
1996	1,140	3,147	155	428	−273
1997	1,590	4,493	183	518	−335
1998	2,053	4,100	208	416	−208

Source: based on Diaz, 1997, and information from CADF.

[a]Revenues include water charges, wastewater discharge fees, and water operation services.

[b]Expenses include investments in large projects [sanitation for the Valley of Mexico, deep drainage, Great Channel (Gran Canal), water transmission line (acauferilco), etc.].

[c]In 1998, the exchange rates was 9.865 pesos to the US dollar.

3. PRIVATE SECTOR PARTICIPATION IN WATER SUPPLY AND SANITATION SERVICES IN THE FEDERAL DISTRICT

Private sector participation was partly encouraged with the passing of the 1992 National Water Law, resulting in a considerable amount of activity. This law provides a modern regulatory framework for water management. The main focus of this law is the importance given to concessions, permitting greater private sector involvement in the water sector. However, it should be emphasized that this law only refers to federal water infrastructure.

In Aguascalientes and Cancún (both medium-sized cities), integrated concessions to private companies have been in operation since 1993 and 1994, respectively. In addition, service contracts were signed in Navojoa in 1997 and in Puebla in 1998.

Other experience with PSP in the country includes the build-operate-transfer (BOT) projects. Out of the 43 BOT municipal wastewater treatment projects existing in 1997, only 12 have been constructed and are in operation, representing investments for $170 million (Barocio Ramírez, 1999). 13 have been canceled, and 18 are in a renegotiating process. Of those in operation, most face financial problems and their treatment capacity is underutilized. The main reasons for these shortcomings are: the poor operating and financial performance of municipal water companies in general and their resulting lack of creditworthiness, the inadequate structuring of contractual and concession arrangements, and the general procedures of the bidding process. Associated with these factors are gaps in the legal, regulatory, and institutional framework at the state and municipal levels.

The primary concern of this chapter is the use of service contracts in Mexico's Federal District, which is examined in detail.

3.1 The Selection Process

In October 1992, the Federal District government launched an international bidding process to select qualified private companies to conduct a census of users, install meters, and rehabilitate the distribution system. The consortia were required by law to maintain majority Mexican ownership. In February 1993, seven firms submitted bids, out of which four were selected by the Federal District government (Casasús, 1994). The starting date of operation was delayed because one of the losing consortia contested the awards, but the four successful consortia signed general contracts with the government in October 1993 (see Table 9.2). The contracts were designed to overcome a number of deficiencies in the city's water distribution system, including information gaps in the customer database, an inadequate tariff structure, and deteriorating water distribution and drainage infrastructures.

Each consortium was responsible for one zone (see Figure 9.1), with approximately the same number of water connections. In order to allow for certain economies of scale, zones containing adjacent municipal districts were awarded. Moreover, it was thought that having four different companies would yield a certain degree of competition, and would ensure that the contractors used the best technologies available.

The objectives stated in the general contracts were very ambitious. They included*:

- Rationalization in a more effective way of water consumption in the Federal District
- Reduction in the aquifer abstraction
- Expansion of the secondary network
- Improvement in the operation and maintenance of the secondary network
- Use of the means that the Government of the Federal District had to provide water services
- Increase of the collection in line with the financial requirements of providing an adequate service

3.2   Multistage Process

Given the lack of information regarding customer base, water consumption levels, and the actual conditions of the network, a phased approach to private sector participation was considered most appropriate. It can be said that the contracts consisted of a "menu" of tasks, including the operation and maintenance of the secondary water distribution system, as well as other activities that at the time were not separately priced, such as opening customer service centers. Each task had a specific price, and the

---

*The contracts remain confidential. The information provided is based on a draft and López Roldán (1999).

Table 9.2   Consortia Involved in the Federal District's Water Distribution System

Zone	Delegations	Company	Consortium Partners[a]	Number of Connections[b]	Contract Value[c]
Northwest (Zone A)	G. A. Madero Azcapotzalco Cuauhtémoc	Servicios de Auga Potable, S.A. de C.V. (SAPSA)	Constructora ICA (Mexico) Vivendi (France)	350,064	969
Northeast (Zone B)	V. Carranza Iztacalco B. Juárez Coyoacán	Industrias del Agua S.A. de C.V. (IASA)	Socios Ambientales de México (Mexico) Azurix (USA)	290,000	981
Southeast (Zone C)	Iztapalapa Milpa Alta Tlahuac Xochimilco	Tecnología y Servicios del Agua S.A. de C.V. (TECSA)	Peñoles (Mexico) Suez Lyonnaise des Eaux (France)	304,059	979
Southwest (Zone D)	M. Contreras Cuajimalpa M. Hidalgo A. Obregón Tlalpan	Agua de México S.A de C.V. (AMSA)	Grupo Gutsa (Mexico) United Utilities (UK)	291,990	880

Sources: Compiled from Secretaría de Obras y Servicios (1998), Haarmeyer and Mody (1998), and personal communication with CADF staff.

[a]The composition of some of the consortia has changed. In September 1999, Azurix acquired 49% of the capital stock of IASA from Severn Trent. In January 2000, Peñoles acquired 51% of TECSA's shares from Bufete Industrial.

[b]The number of connections represent general intakes as of the date the general contract was signed.

[c]Value is in millions of pesos; pesos are in constant figures as of March 1993, when the exchange rate was 3.11 pesos to the US dollar.

Figure 9.1   Allocation of areas within the Federal District to private companies. *Source:* Adapted from Fundación para la Consevación del Agua en México-CADF (1994).

options were the same for each zone. In many aspects, these general contracts were "unique." Prior to one of the contractor's performing any task, CADF had to issue a specific execution order. Each contractor had the right of first refusal to execute orders in its zone. This permitted the Federal District authorities to sole-source a number of relatively small tasks over time that otherwise would have been required by law to go to competitive tender. The private firms were expected to perform all of the tasks on the menu by the time the contract period ended. The preferred tender would be the one offering the lowest present value for the menu of activities. Rates agreed were backed by a cost breakdown showing expected direct costs, indirect costs, and profit for each task. The specific tasks of the consortia were to be accomplished in the following three phases (Saade Hazin, 1998) (not being successive).

3.2.1 Phase I

During this phase, which started in May 1994, the companies installed meters, created a customer database, and drew up network plans. The Federal District government signed contracts with each of the consortia in which the latter agreed to install a specified number of meters or carry out a census in a particular area. The Federal District government pays the contractors per task accomplished on the basis of a unit price.

By December 31, 1996, with the exception of meter installation, the first phase had been completed—1.7 million users had been registered—and there had been maps produced covering an urban area of approximately 680 square kilometers (Secretaría de Obras y Servicios, 1998). Two types of meters are being installed—conventional and electronic. By December 1998, more than 1.1 million meters had been installed (see Table 9.3 for data through August 1998). The total potential number of meters to be installed is around 1.4 million, although it will not possible to achieve 100% coverage since certain areas of the city have problems with water quality and continuity of service and will therefore require attention prior to the installation of meters. More than 20% of the consumers that still do not have meters pay a flat fee for water, regardless of consumption (Urban Age Magazine, 1999). However, this is a significant improvement relative to the situation prior to Phase I, when 50% of households did not have meters.

Table 9.3 Progress Up to August 1998

| Task | Company | | | | |
	SAPSA	*IASA*	*TECSA*	*AMSA*	*Total*
Bills issued	407,289	410,804	442,229	349,654	1,609,976
Meters installed	282,489	273,653	240,844	252,677	1,049,663

Source: CADF, personal communication.

Note: Billing: figures refer to bills issued in each bimester; meter installation figures indicate total installations. This does not include the large users, which are under the control of the CADF.

3.2.2 Phase II

Phase II followed a contractual mechanism similar to that of the first phase, that is, the Federal District government has paid the consortia for each task accomplished. Specific contracts under Phase II include meter reading; meter maintenance; bill issuing and distribution; bill collection; customers' service; and new connections. Prior to December 31, 1996, CADF (acting through the four companies) had been processing 72.4% of all water bills, issued every two months, with the Treasury Department handling the remainder. By September 1997, the entire task of water services billing had been transferred from the Treasury to CADF, with over 1.6 million bills being issued every two months in 1998.

It has been observed that bills issued have increased around 50,000 per year, totaling around 1.7 million at the moment. The total number of meters installed at present is of the order of 1.2 million (CADF, personal communication).

The delay in the implementation of these two phases was partly due to the economic and political situation that had prevailed in Mexico since December 1994, which affected the financial situation of the City Treasury. Many contracts depended on imported items, such as meters and other equipment. The effect of the 1994 devaluation was to more than double the price in pesos of these articles. In addition, there were budget cuts at all levels within the government. These circumstances made it necessary for CADF to delay several contracts.

In the case of Phase II, in the general contracts, no specific rate was requested to cover the cost of the activities related to customer care and bill collection. The tasks included the setting up and operation of a customer care center for each 50,000 general intakes, a telephone information center, and field inspection teams. The contractors thus included the corresponding costs in the rates for other activities of Phase II for which a price was requested. Since no specific guideline had been set for this cost allocation, the four consortia came up with different solutions, some of them increasing their rate for bill issuing, other increasing their rate for meter maintenance, and one company distributing those costs as a constant percentage of all its rates for Phase II (Agua de México staff, personal communication).

This led to extreme differences between the contractors with respect to rates for Phase II for the same activities. Furthermore, the partial award of Phase II has caused important losses to the above-mentioned contractor, which distributed its customer care and collection costs as a constant percentage of all Phase II activities, recovering only 30% of those monthly costs.

3.2.3   Phase III

Phase III consists of the operation, maintenance, and rehabilitation of the water distribution and drainage networks. It began, to a limited extent, in 1997. Its implementation was severely delayed as a result of internal political problems relating to the institutional framework (discussed below), and eventually split into two separate substages. The first substage, which initiated with pilot projects, was suspended. It was based upon specific contracts between the CADF and the four firms. During the pilot program, firms were responsible for subdividing their zones into District Meter Areas (DMAs). Flow meters were installed at all entry- and exit-points of the DMAs, making it possible to calculate the amounts of water entering and leaving the area. Since the quantity of water consumed within a DMA can be established from readings of the micrometers installed during the first phase, the idea was that firms should be able to determine the quantity of water unaccounted for within the DMA. Unaccounted water was expected to be mostly due to physical leakage or clandestine connections. In principle, by comparing the situation in various DMAs, companies should be able to determine where priority needs exist for extensive leak detection and, in some cases, rehabilitation of the network.

The second substage of Phase III (started in July 1997) comprises the operation and maintenance of the distribution network within the area of the DMAs that form part of the first substage. Activities include the detection and repair of leaks, operation and maintenance of the water distribution network and its accessories (manhole covers and frames, valves, etc.), and the replacement of valves and service pipes. Some of these activities were not included in the general contract, but it allowed for certain flexibility for this type of activities to be carried out.

The 1997–2000 Federal District administration gave priority to activities such as leaks repair and upgrading or repairing the secondary network on a fee-per-action basis. By August 2000, the four companies had rehabilitated around 542 kilometers of network, substituted more than 68,000 service connections (ramales), and identified and repaired leaks (CADF, personal communication). See Table 9.4 for some indicators of their efficiency, and Figure 9.2 for the revised schedule for the implementation of the three phases.

For certain routine activities, such as billing and meter reading, the subscription of specific contracts has not been a problem. However, for other activities, such as meter installation, rehabilitation, leak detection and repair, and so on, the approval is contingent on the budget available and with

Table 9.4 Some Efficiency Indicators for 1995 and 1999

Indicator	Value 1995	Value 1999
Water system coverage	98%	98%
Drainage coverage	94%	94%
Daily per capita water supplied (liters)	360	360
Employees/1,000 water connections[a]	12	Na
Employees/1,000 users[a]	8	Na
Water produced (million cubic meters)	1,135	1,104
Revenues from water supply fees		
(current million pesos)	771	2,507
Efficiency (revenue/billed)	69%	82%
Efficiency in metering (water metered/		
water distributed)	68%	75%

Sources: Based on information from Haarmeyer and Mody (1998), and the CADF, personal communication.

Note: The exchange rate in 1997 was around 8 pesos per US $1.

[a] 1996 estimates.

the agreement of the CADF. General contracts clearly state that the CADF reserves the right not to subscribe specific contracts for the tasks considered in the following cases: when the necessary budget is not approved by the Federal District Legislative Assembly (yearly process); when the responsibility for those tasks is not transferred from other departments of the Federal District government; or when the performance of those tasks is not deemed in the interest of the city.

Finally, it is worth noting that currently no less than 40 specific contracts are signed every year with each one of the four contractors, which requires an enormous amount of administrative time (according to the contractors).

Due to the nature of the service contracts, there are two primary social and environmental consequences arising from the involvement of the private sector in the Federal District's water supply system: conservation of scarce water resources and greater access for poorer households. Through a household census of users, metering of connections, more effective billing, and rehabilitation of the distribution system, it is expected to manage water demand more effectively. It is also hoped that these arrangements can ensure that poorer households do not face an excessive financial burden as costs of provision rise.

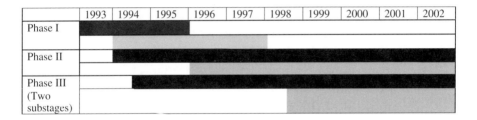

Figure 9.2   Revised schedule for the implementation of the three phases. *Source:* Author's analysis.

4. PARTIES AND INSTITUTIONAL ARRANGEMENTS

In July 1992, as a step toward solving the City's water problems, a special commission was created: the Comisión de Aguas del Distrito Federal (Federal District Water Commission, CADF). CADF works as an autonomous body (órgano administrativo desconcentrado) under the supervision of the Works and Services Secretariat (Secretaría de Obras y Servicios) of the Federal District Government.

Since its inception, the CADF has been entrusted with the responsibility of engaging the private sector. The CADF is in charge of supervising the four private consortia that undertake these tasks, and is responsible for the review and analysis of their financial statements. It also maintains the database of users and is responsible for customers' relations, establishing and maintaining plans of water distribution and drainage networks, and issuing water bills.

One of CADF's most important duties is water distribution to major users (grandes usuarios), who in 1992 totaled approximately 10,000. The number of these accounts has increased to around 17,000. It should be noted that major consumers comprise only 1% of all users in terms of numbers, but contribute approximately 50% of water fees collected.* CADF has retained control over these accounts.

In addition, a number of public authorities continue to play an important role in the sector.

---

*The four companies do not participate in issuing bills to major users.

- The Comisión Nacional del Agua (National Water Commission, CNA). The CNA is the main authority responsible for water management at the federal level. It is a financially autonomous body (organismo desconcentrado) under the Secretaría de Medio Ambiente y Recursos Naturales (Ministry of Environment and Natural Resources, SEMARNAT). The CNA is responsible for providing water to the city.

Under the Federal District Government, tasks are allocated to several public agencies:

- The Dirección General de Construcción y Operación Hidráulica (General Directorate for Waterworks Construction and Operation, DGCOH) is attached to the Works and Services Secretariat of the Federal District Government. It is the administrative unit responsible for receiving water and distributing it throughout the city. The DGCOH is also responsible for water purification and treatment plants throughout the city; supply wells and the arterial drainage system (including a network of deep tunnels that take surface and wastewater out of the city); and operation, maintenance, and new works.
- The Treasury Department of the Federal District retains overall responsibility for the management of income generated from water, as well as the allocation of departmental budgets.
- The 16 political municipal districts (delegaciones)* are in practice in charge of the operation and maintenance of the secondary water distribution and drainage networks, under the supervision of the DGCOH.
- Finally, the Presidential Decree of October 21, 1997 (Mexico, Government of, 1997b) gives the Secretaría de Obras y Servicios (Works and Services Secretariat) under the Federal District Government, responsibility for operation and maintenance of the secondary water distribution and drainage networks. The Decree does not

*The Federal District is subdivided into political and administrative units, delegaciones. At the end of 1997, the heads of the delegaciones were elected by the Legislative Assembly after being nominated by the Chief of Government and in July 2000, they were elected by the capital's citizens by a universal and direct vote.

specify if this function will actually be undertaken by CADF or by DGCOH. However, it is likely that CADF will be given the responsibility to accomplish the task with the work undertaken by the four consortia.

Having so many entities involved in the running of water operations causes certain problems. Although, in theory, each agency has a distinct role to play, in practice, there is a lack of coordination, and duplication and overlap of functions often occur. In particular, the four consortia have difficulties in determining the commercial efficiency in their respective areas (bills issued/ bills paid), and have no incentives to pursue nonpayers since, although they know which customers have paid in their offices, they do not have ready access to complete information on payments made to banks or the various Treasury offices.*

The length of the general contract continues to be an issue for discussion. A 10-year period was considered to be the minimum acceptable period to achieve the goals of the project and allow the companies to recover their investments. However, subsequent delays have reduced the effective period of operation.

In order to try and overcome some of the problems associated with lack of coordination, the CNA will continue to provide bulk water to the City. However, at the Federal District level, it is intended to have only two agencies involved in water distribution in all stages—the DGCOH and the CADF. (See Table 9.5.) Both of them are attached to the Works and Services Secretariat of the Federal District Government.

The role of the CADF as sectoral regulator is rather more straightforward for service contracts than it would be for more comprehensive forms of PSP. The companies' earnings have been derived solely from the completion of specific contracts with CADF. The CADF also carries out physical surveys of the work undertaken by the companies. The General Contract contains a clause specifying that, if any company performs poorly, either financially or technically, its contract can be revoked. Initially, there was the intention to include a penalty/payment, depending on the improvement/deterioration in the level of physical leakage as calculated through the analysis of information obtained from the DMA It will be in-

---

*This has been changing, as companies do not have to wait as many months to get the information.

Table 9.5   Activities and Responsibilities for the Water Distribution System in the Federal District

Activities	Currently	In the Future
Transportation[a]	National Water Commission (CNA)	CNA
Distribution[b]	Works and Services Secretariat (SOS) through the DGCOH	SOS through the DGCOH
Water connections[c]	SOS through the DGCOH	SOS through the DGCOH
Operations and maintenance[d]	Municipal Districts (delegaciones)	SOS (most likely through the CADF)
Updating[e]	SOS through CADF (tasks undertaken by the four companies	SOS through CADF
Billing[f]	Treasury	SOS through CADF
Sewerage System[g]	SOS through DGCOH	SOS through DGCOH
Sewerage System[h]	SOS through DGCOH	SOS through DGCOH

Source: Based on information from CADF.

[a]Bulk water to the city and water purification.

[b]Bulk water to the primary network. Water disinfection through underground wells. Operation of treatment plants.

[c]Greater than 15 millimeters in diameter (CADF is responsible for water connections of 15 millimeters or less).

[d]Secondary network (in transition).

[e]The customers' register, installing meters, and billing.

[f]Revenues are destined for the Treasury.

[g]Overall responsibility.

[h]Responsibility for the deep sewerage system.

teresting to see if this is actually implemented and can be effectively regulated.

The project was also affected by the entry of the first elected and opposition-led government in the Federal District. When the new party (PRD) took office in December 1997, a careful review of the contracts was made. As a result, some amendments were made to the parameters of Phase III, which led to the revision of the prices applicable to the execution orders in that stage. It is hoped that the change of administration in December 2000 will not severely affect the future of the contracts.

5. TARIFF ISSUES

In the Federal District, the general tariff level for water is specified in the Federal District Financial Code. Tariffs are updated every year, based on the recommendations of CADF experts, the General Directorate for Waterworks Construction and Operation (DGCOH), municipal districts, and the Treasury Department and are approved by the Legislative Assembly (Mexico, Government of, 1997a). Prior to the implementation of the present water management scheme, tariffs in most places were fixed and highly subsidized (Casasús, 1992). Despite some attempts to motivate the population of the Federal District to use water more carefully through ecological information programs, users perceived water to be a cheap and unlimited resource.

It is worth noting that, in 1997, the Legislative Assembly approved the application of a new tariff system, charging for every additional cubic meter consumed rather than a flat rate for consumption falling within a given range. In the Federal District's new water management scheme, responsibility for tariff setting remains with the Federal District authorities, although the private companies are often consulted. Moreover, the application of metered consumption to the majority of households has only been made possible through the census and metering tasks set out in the service contracts.

After setting up billing and metered consumption, domestic water consumption in several zones has decreased by an average of 20%.* However, it is important to note that, even with these changes, the current water tariffs still do not cover even 50% of the cost of pumping water up into the City, let alone the scarcity value of water.

Wasteful water use was also exacerbated by a lack of payment culture. While 82% of bills are paid, this is more than a 30% increase in the number of paying customers, due to more up-to-date information. According to the Federal District Financial Code, if domestic users fail to pay their bills, users can have their water services rationed to a "vital level." In addition, as mentioned before, CADF estimates that only 68% of the water supplied to the network is received by customers, some 32% being lost through leakage. The extensive rehabilitation of the network has contributed a reduction in water losses. However, significant problems remain. In particular, groundwater continues to be exploited unsustainably, particularly in the

---

*According to estimates made by the contractors.

eastern part of the city. This is mainly due to the fact that additional water sources from outside of the city flow in from the west. Since demands in the western and central parts of the city are filled first, the external source is often exhausted before reaching the eastern part. As a consequence, four municipal districts in the eastern part of the City have to rely on local wells, which are overexploited, and which contain iron and manganese that give the water a poor appearance.

The Federal District government has tried to alleviate this problem by distributing water to affected areas using tanker trucks.* However, since this solution appeared inadequate, the Federal District government is now looking for a more permanent and socially acceptable solution through the construction of a water transmission line (acuaférico) that will transport water directly from west to east. Construction of the transmission line initially was expected to be completed by 2005, although there have been some delays. In reality, one should not discount the fact that this project will be irrelevant if the fourth stage of the Cutzamala system (which means bringing 5 additional cubic meters per second to the city) is not carried out, as there will be no water to be distributed. The main reasons for the delay in the Cutzamala project are associated with social issues. People living in the areas that would be affected by the construction of the fourth stage have opposed the project, as they perceive it would supply water only to the people of Mexico City; they feel they should not have to suffer because water is needed in another part of the country (Tortajada, 1999).

The service contracts may play a role in alleviating some of the financial burden for poorer households. Depending upon the tariff structure, metering and use-based fees may allow for the application of cross-subsidies from richer households to poorer households. While the increasing block structure of the tariff schedule prevalent in the Federal District allows for this type of cross-subsidies, the real implications are not clear. Between 1996 and 1999, tariffs had been slightly increased (relatively) in nominal terms for residential users who consume less than 30 cubic meters every two months, and they have been considerably increased for users who consume more than 120 cubic meters every two months. For those who consume between 30 and 90 cubic meters every two months, there have not been important changes in the tariffs they pay; they have,

---

*In addition, efforts have been made to restrict urbanization in the southeastern portion of the Federal District because of the difficulties in providing basic services and because it represents an important groundwater recharge zone.

in fact, even enjoyed certain tariffs' reductions to encourage customers to pay their bills. It is worth noting that, according to the CADF, around two-thirds of residential users consume less than 60 cubic meters every two months. In 2000, residential customers consuming 30 cubic meters of water pay around US $0.13 per cubic meter. Payments for water services are made bimonthly. Finally, there have been slight increases for users who consume between 90 and 120 cubic meters (see Figure 9.3).

In real terms, more than half of all residential consumers (using between 30 and 220 cubic meters per bimester) saw 1998 tariffs substantially lower than in 1996. This tariff reduction for residential consumers was accompanied by an increase in the cross-subsidy from nonresidential to residential consumers.

At the same time, nonresidential consumers have seen rising tariffs in nominal and real terms, leading to a growing price differential. Since the consumption of nonresidential consumers accounts for around one-quarter of total consumption, the total effect has been that average real metered prices have decreased. The fall in water tariffs in real terms can be explained by: (1) the devaluation of the peso in December 1994 and (2) political fear of raising tariffs (Haggarty et al., 1999).

Moreover, such cross-subsidies only benefit households who are connected. According to the Mexican Human Rights Commission, the very poor currently pay five times more per unit of water, compared to the average domestic water tariff, since they are less likely to have access to water in

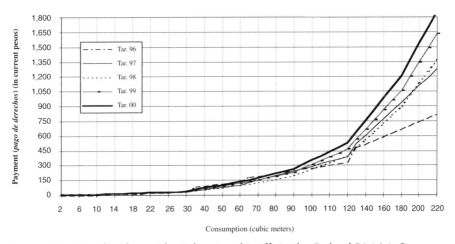

Figure 9.3   Trend in the residential metered tariffs in the Federal District. *Source:* Federal District Financial Code for the corresponding years.

their homes and therefore have to rely on more expensive water vendors. In general, data are unavailable regarding the provision of water services to the poorer households in the Federal District who do not receive piped water to their home. Research conducted in some irregular settlements showed that some residents were served by government tank trucks free of charge, other residents paid for an installed and metered network, and others paid a flat fee (National Research Council, 1995). In addition, rising block tariffs may be biased against households who rely upon multiple-household connections. Yet important efforts are being made to install meters per household and issue separate bills.

In the Federal District, a 50% discount has been granted to retired citizens or residents over 60 years old who prove their condition. In addition, cancellation of charges for late payments and fines' exemptions from debts generated between 1995 and 1998 have been granted to residential customers who regularize their situation in one installment (Mexico, Government of, 1999). It is worth mentioning that the Government of the Federal District has had a tradition of making late payments' exemptions or reductions for the different services provided with the aim of regularizing debtors. An example of this is the program of late payment charges exemptions was in place from June 15 until September 15, 2000 (see Table 9.6). These types of programs have been continuously repeated, and if they continue, they will provide less incentives for consumers to pay on time.

Finally, the Federal District Water Commission has faced social pressure from certain groups of the population that have complained about water services. In particular, several housing complexes do not allow for the installation of meters and are displeased with the new billing and collection systems. Others complain about the fact that they do not perceive that tariffs increases are accompanied by significant improvements in service (Secretaría de Obras y Servicios, 1998).

Table 9.6   Program of Late Payments' Exemptions

Deadline	Exemption of Late Payment Charges
June 15–July 15, 2000	100%
July 16–August 15, 2000	80%
August 16–September 15, 2000	60%

Source: CADF, personal communication.

6. CONCLUSIONS

Private sector participation in the Federal District has been implemented since 1993 in the form of service contracts. The implementation of the service contracts was slowed down as a result of the 1995 financial crisis and by weaknesses in the regulatory environment, particularly in sectoral coordination, leading to renegotiations. However, a number of important lessons have been learned from private sector involvement in water in Mexico.

Performance under the contract has so far proved quite satisfactory. Despite several ambiguities, the contract foresaw and covered some changes, such as the transfer of responsibility for the secondary network and billing to the CADF.

The focus in the Federal District has been mainly on improvement of performance, basically upgrading the service. This was probably due to the fact that coverage was not considered as the major issue. It is worth noting that water authorities in the Federal District administration did not have an important pressure to face the full costs of developing new sources of supply (due to Federal funding for projects such as the Cutzamala system). Therefore, the project did not require high levels of investment by the private sector.

Important advantages were gained from the use of service contracts rather than concessions. A concession contract would have been far more complex and risky to implement than service contracts, and would have required a more complex legal and regulatory framework. Despite the recent modifications to the legal and regulatory framework, there is still scope for improvement in the regulatory environment.

Furthermore, the phased approach has allowed for the generation of information necessary for designing better contracts and for improving the regulatory capacity of the sector. The staged approach allowed sufficient flexibility for mistakes to be corrected and adjustments made to cater for unexpected situations. A concession contract would not have enabled the same degree of flexibility.

Splitting the contract into four should have assisted the public authorities in regulating the sector, since it allows for benchmarking of performance. In practice, however, the regulator has faced difficulties in comparing the performance of the firms. Since the project was conceived in three stages, some firms have charged more in the first phase, subsidizing the second and third, and vice versa. Unit prices presented by the firms are not comparable with respect to real costs. In fact, different municipal districts have different prices in the same zone.

The change of administration in 1997, and again in 2000 (although the same political party remained) may have had an impact on the project. However, the need for this type of projects in Mexico should go beyond political considerations. The new administration is aware of the importance of continuing the project. There is little question that metering, billing, and rehabilitation have laid the groundwork for better management of scarce water resources. Important changes to the institutional framework will be necessary to achieve a more efficient water management. As the contractors expand the scope of their responsibilities and generate new information about consumers and the condition of assets, further progress will depend, among other things, on the quality and response of the regulatory bodies. It would be convenient to analyze the role of CADF as a contractor and regulator. It would be necessary to evaluate the convenience of separating these two functions.

Finally, the potential benefits of PSP are also circumscribed by the scope of the contract. Given the strong interconnection and combined water and wastewater systems serving both the Federal District and the State of Mexico service areas in the MCMA, some considerations have been made to create a Metropolitan Water Commission to coordinate efforts in the area. This is certainly a subject for further discussion, particularly since water is drawn from the same sources and deprivation is more pronounced in parts of the MCMA outside the Federal District.

ACRONYMS

- CADF—Comisión de Aguas del Distrito Federal (Federal District Water Commission).
- CNA—Comisión Nacional del Agua (National Water Commission).
- DF—Distrito Federal (Federal District).
- DGCOH—Dirección General de Construcción y Operación Hidráulica (General Directorate for Waterworks Construction and Operation).
- PRD—Partido de la Revolución Democrática (Party of the Democratic Revolution).
- SEMARNAT—Secretaría de Medio Ambiente y Recursos Naturales (Ministry of Environment and Natural Resources).

- SOS—Secretaría de Obras y Servicios (Works and Services Secretariat).

REFERENCES

Adelson, Naomi (1999). "Water Woes: Private Investment Plugs Leaks in Water Sector," *Business Mexico Magazine.*

Barocio Ramírez, Rubén (1999). "Participación del Sector Privado en el Sector del Agua Potable y Alcantarrillado en México." Paper presented at the Seminar on Tools of Private Participation in Water and Wastewater, Monterrey, Nuevo León, Mexico, March 10–12. Jointly organized by the World Bank and the Comisión Nacional del Agua.

Casasús, Carlos (1992). "Una Nueva Estrategia de Agua Para la Ciudad de México," in *Ensayos sobre la Economía de la Ciudad de México,* edited by Ricardo Samaniego. Mexico City: Librería y Editora Ciudad de México.

Casasús, Carlos (1994). "Private Participation in Water Utilities: The Case of Mexico's Federal District," Reinventing Government: New Opportunities for Public/Private Enterprise. Federal District Water Commission, paper presented in Toronto, Ontario, November 15.

Comisión Nacional del Agua (CNA) (1994). "Informe 1989–1994." Mexico City, November.

Comisión Nacional del Agua (CNA) (2000). "El agua en México: Retos y Avances." Mexico City, October.

Comisión Nacional del Agua (CNA) (2001). "Compendio Básico del Agua en México 2001." Subdirección General de Programación. Gerencia de Planeación Hidráulica. Mexico City, January.

Díaz, Luis F. (1997). "Una Nueva Estrategia de Agua para el Distrito Federal," paper presented at the Conference on Sanitation Projects in Mexico, Mexico City, September. Organized by the Center for Business Intelligence.

Fundación para la Conversación del Agua en México—Comisión de Aguas del Distrito Federal (CADF) (1994). "Agua: Una Nueva Estrategia para el Distrito Federal," Mexico City.

Haarmeyer, D., and A. Mody (1998). *World Water Privatization: Managing Risks in Water and Sanitation.* London: Financial Times Energy Asia Pacific Publishing.

Haggarty, L., P. Brook, and A. M. Zuluaga (1999). "Thirst for Reform? Private Sector Participation in Urban Water Supply: The Case of Mexico City's Water Sector Service Contracts." Mimeo, Draft 8. The World Bank, Washington, DC, October.

López Roldan, Raúl (1999). "La Perspectiva en México del Socio Privado: La Concesión en Aguascalientes y el Contrato de Servicios en el Distrito Federal." Paper presented at the Seminar on Tools of Private Participation in Water and Wastewater, Monterrey, Nuevo León, Mexico, March 10–12. Jointly organized by the World Bank and the Comisión Nacional del Agua.

Mexico, Government of (1997a). *Código Financiero del Distrito Federal, Leyes y Códigos de México.* Mexico City: Editorial Porrúa.

Mexico, Government of (1997b). *Diario Oficial de la Federación, Organo del Gobierno Constitucional de los Estados Unidos Mexicanos. México, D. F. Martes 21 de Octubre de 1997.* Mexico City: Government of Mexico.

Mexico, Government of (1999). *Código Financiero del Distrito Federal.* Mexico City: Editorial Sista.

National Research Council (1995). *Mexico City's Water Supply: Improving the Outlook for Sustainability,* Academia de la Investigación Científica, A. C. & Academia Nacional de Ingeniería. Washington, DC: A.C National Academy Press.

Noll, R., M. Shirley, and S. Cowan (2000). "Reforming Urban Water Systems in Developing Countries, "Stanford Institute for Economic Policy Research, Palo Alto, CA, Discussion Paper No. 99-32, June.

Saade Hazin, Lillian (1998). "New Strategy in Urban Water Management in Mexico: The Case of Mexico's Federal District," *Natural Resources Forum, A United Nations Journal.* Vol. 22, No. 3, August.

Secretaría de Obras y Servicios (1998). *Comisión de Aguas del Distrito Federal, Memoria de Gestión 1994–1997.* Mexico City.

Tortajada, Cecilia (1999). *Environmental Sustainability of Water Management in Mexico.* Mexico City: Third World Centre for Water Management.

Urban Age Magazine (1999). "Delivering Water to Mexico City." Vol. 6, No. 3, Winter.

ADDITIONAL READING

Mexico, Government of. *Código Financiero del Distrito Federal.* Mexico City: Editorial Sista, 1998.

Mexico, Government of. *Código Financiero del Distrito Federal.* Mexico City: Editorial Sista, 2000.

Chapter 10

The Largest Water System Privatization

The Manila Concession

Mark Dumol and Paul Seidenstat

By 1994, Philippine government officials were increasingly critical of the operation of the government-owned and -operated water and wastewater system in the Manila area. Since privatization of government operations was beginning to be considered as a viable alternative, a policy change began to evolve that would lead to the privatization of the water system. The process of privatization was completed on August 1, 1997.

The stakes were high for the Philippines as water and wastewater are key services that can be crucial in promoting economic growth. The recitation of this privatization effort, the largest of any water system to date, is instructive as to the pitfalls, the required steps, and the policy path that is likely to be successful for a developing country.

Based upon Dumol (2000).

Mark Dumol is Consultant and former Advisor to the Government of Philippines on Privatization. Paul Seidenstat is an associate professor of economics at Temple University. A specialist in the fields of public finance, public management, and water resources, he has written or coedited eight books and several articles on these subjects, including Contracting Out Government Services *(Praeger). Seidenstat and Simon Hakim are coeditors of* America's Water and Wastewater Industries *(Public Utility Reports),* Privatizing the United States Justice System *(McFarland),* Privatizing Correctional Institutions *(Rutgers),* Privatizing Education and Educational Choice *(Praeger), and* Privatizing Transportation Systems *(Praeger). He has conducted funded research projects for the federal government in the fields of manpower and water resources. His current research is in applying economic analysis to the management of public agencies.*

1. BACKGROUND

1.1 Historical Background

The Philippine government established the Metropolitan Waterworks and Sewerage System (MWSS) in 1878 to supply water to the Manila area. It was the oldest water system in Asia. By the last part of the twentieth century, it had grown to service the entire Manila metropolitan area, which includes 15 cities and municipalities with a population of 11 million, stretching over 2,000 square kilometers. About 10 million of its customers were in urban areas, with the remaining one million in rural areas.

The MWSS was set up as an independent, government corporation. It operated as an enterprise fund, able to maintain budget autonomy. However, it often depended upon government subsidies to remain solvent. Moreover, the corporation had to follow government procurement, personnel, and other rules.

1.2 Poor Performance of MWSS

It was generally conceded that the MWSS failed in its mission to provide adequate water and sewerage services. By 1994, the utility was supplying water to only two-thirds of its coverage population for an average of only 16 hours per day. Further, it was providing sewerage services to only 8% of its coverage population. More than half of the water from its major water source was unaccounted for and non-revenue-producing.

2. THE THRUST TOWARD PRIVATIZATION

The concern with the unsatisfactory performance of MWSS led President Ramos to shake up the agency. In June of 1993 the President appointed an experienced project engineer, Gregorio R. Vigilar, as Secretary of the Department of Public Works and Highways (DPWH), which position automatically made him Chairman of the MWSS Board of Trustees.

Although the new management had great concern about the agency's performance, a major event in 1994 began a process of accelerating action. A Malaysian company joining with Biwater, a large British water company, offered to purchase MWSS on a negotiated, government-to-government basis. The proposal offered to expand distribution services, provide new water supply sources, and to reduce leaks.

2.1   Enabling Legislation to Privatize

Although this purchase proposal was rejected on both political and legal grounds, since the shares of MWSS could not be sold, especially to a foreign-owned corporation, there were laws in place that could accommodate some form of privatization. In 1986, the Aquino government had passed legislation that encouraged widespread privatization. A Committee on Privatization (COP) was created with a mandate to privatize government operations. From 1986 to 1992, COP privatized 122 government-owned companies, generating $2 billion in revenue.

The Aquino government had also passed a build-operate-transfer (BOT) law that permitted BOT contracts. Using that law, the government entered into its first privately financed BOT contract for a gas turbine plant.

Building on its predecessor's initiatives, the Ramos government continued the privatization process. From 1992 to 1998, 132 government companies were sold, generating $4 billion in revenue. Demonstrating the effectiveness of BOT arrangements with infrastructure projects, the Ramos administration was able to expand electric power capacity rapidly.

Prior to December of 1994, an expanded BOT law had been passed. Therefore, in December, Biwater, along with a local real estate firm and multinational partners, initiated a new effort to purchase MWSS. Since a direct government negotiation was proposed, the effort failed, as government officials felt that only a bidding process would be feasible politically.

2.2   Recognition of the Disadvantages of Government Operations

The leadership of MWSS, Secretary Vigilar, and Mark Dumol, his chief of staff, researched the feasibility of privatization and became convinced that it was a viable option. It was felt that, not only would substantial net benefits likely be achieved, but also that many operational problems connected with a government-operated utility could be avoided. Chief among these operational issues were procurement, personnel, and financing.

Government procurement often involved red tape and delay. Protests from losing bidders were commonplace and often obstructed the process. Government personnel policies limited the optimal utilization of personnel and elevated labor costs. MWSS found that civil service regulations strained personnel relations and led to staffing levels much higher than in comparable water utilities in Asia. MWSS had 13 employees per connection, which was two to five times higher than similar utilities. Common with government-operated utilities worldwide, salaries of lower classification workers were higher than in the private sector, but management salaries were not comparable. It was difficult to generate high morale when there were few incentives for employees.

The difficulty in financing large-scale infrastructure investments is a major shortcoming of many government-operated utilities. There are always pressing claims against government funds, and infrastructure improvements are often viewed as postponable. In the MWSS case, since the government guaranteed the utility's debt, there was always pressure to go slowly with new commitments. Additionally, since MWSS was expected to earn profits, the corporation often skimped on operations and maintenance, especially preventive maintenance.

3. STARTING THE PRIVATIZATION PROCESS

To enhance the prospects for a successful privatization, two major steps were undertaken in the spring of 1995. It was felt necessary to enhance the financial viability of the utility and to explicitly permit the privatization of MWSS. The national legislature incorporated these steps in the passage of the Water Crisis Act (WCA).

This legislation made it a crime to steal water from the water system. More sweeping was the requirement to reorganize and downsize MWSS. In addition, the law granted explicit power to the President to privatize MWSS. He had six months to commence the privatization process.

3.1   Hiring a Consultant and Establishing Clear Legal Authority

Since the privatization process of a massive water/sewerage system could be very complex and involved very large sums of money, the top management of MWSS decided to hire a consulting firm with experience in privatizing water systems. With the financial help of the Philippine and French governments, and with the cost to be reimbursed by assessing a fee on the winning bidder, MWSS sought the services of a major international consulting firm that understood both the technical and financial issues. MWSS chose the International Finance Corporation (IFC) and a contract was signed in November of 1995.

To ensure a firm legal basis for privatization, an executive order of the

President was issued in March of 1996. The executive order made clear the desire for privatization and required presidential approval of any privatization arrangement.

3.2   The Concession Plan

The privatization plan that evolved was in the form of a 25-year concession. It was felt that this arrangement would give considerable flexibility and a powerful financial incentive. The government would operate in a regulatory role, but by limiting the time for the concession, it could reassume operational control if it were necessary. The plan had to address a number of specific issues.

3.2.1   Ownership

The Philippine constitution mandates that all public utilities must be owned and controlled by Filipinos. Consequently, the concessionaire must be 60% domestically owned and all its officers had to be Filipinos. Since no local firms had the size or the expertise in running water utilities, all bidders would be composed of international water companies with local partners.

3.2.2   Reduction of the Labor Force

In order to interest bidders, the bloated labor force of MWSS had to be reduced. Using provisions of the Water Crisis Act, MWSS could legally reduce the workforce. Working with the unions, MWSS negotiated an early retirement package, and 30% of the labor force accepted it. Those employees not retiring would be terminated prior to the private takeover, and their severance pay would not be taxed. The winning bidder would be required to rehire all interested employees on a probationary basis, and those not rehired would receive the full early retirement package. Also, those rehired by the new enterprise would be offered a stock option plan.

3.2.3   Costs and Water Tariffs

From a public relations standpoint, it was essential that the new company reduce water rates. To insure that outcome, the MWSS Board set the preprivatization tariffs as the cap in setting the bidding rules. In addition, five months before the bid submission, water tariffs would be increased about 38%, which was the level consistent with the actual costs of the current operation. This tariff increase reassured the bidders that the government had the will to raise rates if justified by economic conditions.

Other cost considerations were introduced to improve the chances of aggressive bids. The required expansions of water and sewerage service were designed such that major capital expenditures would not required until five years into the contract. The early focus was on less expensive expenditures, such as replacing meters and reducing water theft. Further, an automatic tariff adjustment was set for the fifth year to cover the funding for the required sewerage investments. Tax concessions and tax breaks were also provided. Finally, the Department of Finance guaranteed the contractual obligations of the MWSS.

3.2.4   Two-Zone Concessions

Following the French precedent of splitting the Paris concessions into two separate units with the Seine River as the boundary, the Manila contract was split into two concessions divided by the Pasig River. The thinking behind this provision considered both perceived advantages and disadvantages.

The Paris model would allow a performance comparison of the two concessionaires. The separate bids also could be compared as a basis for acceptability. Further, in case of default by one company, the other could easily take over the entire system.

The split into East and West zones could also have disadvantages. From the government viewpoint, it was more expensive to evaluate two bidding competitions and to administer two separate contracts. Moreover, political problems could arise if the two winning bids were far apart even though the two zones were technically very similar. Also, there was the chance that one company would win both bids, defeating the idea of two separate concessions.

Other complications of the two-zone system could develop. One was the need to interconnect the two adjoining systems under certain circumstances and how to handle intercompany charges where treatment costs differed. Another problem was how to split the rehired employees as well as splitting the customer database in accordance with the location of the customer.

3.2.5   Tariff Adjustments and the Regulatory Office

To ensure that the bidders could accurately estimate revenues from operations, the bid information had to contain a clear statement of tariff adjustment procedures over the life of the contract. With several variables affecting the terms and costs of service, determining a formula for tariff changes proved to be a daunting task.

The MWSS, with some reluctance, had to settle upon a "rate rebasing" scheme for rates that were required to be reviewed after five years. The formula was developed by the economic consultants from the National Economic Research Associates (NERA). The proposal was to establish a return on project investment. This predetermined rate of return was called the appropriate discount rate (ADR), and was defined as the prevailing rate for similar infrastructure projects in similar economies or countries.

MWSS was concerned about utilizing some type of fixed-return system. The company raised several issues:

1. Would it be possible to have a clear and objective rate escalator that would not be subject to reinterpretation?
2. Would it not virtually guarantee a profit that would remove the pressure to reduce costs?
3. Could some company submit a very low bid with the expectation of recouping losses when rebasing occurred?

With the help of the consultants, the MWSS incorporated a protective clause in the bidding rules. To avoid the possibility of underbidding, the rules specified that the first rebasing after five years could be canceled at the option of the government. A potential cancellation could make underbidding costly.

Nevertheless, a well-defined future rate escalation was essential to attract bidders. MWSS recognized that rate increases could offset several risks facing bidders, including the requirement for capital investment for upgrading, the potential risk of disapproval of future expenses by the Regulatory Office, and the political risk of future governments' adherence to the contract.

NERA came up with several approaches to the composition of the rate formula, some of which were very sophisticated and complicated. However, MWSS pressed for a simple, easy-to-understand approach. Eventually, the ADR approach was adopted. Given allowable costs, the concessionaire would be allowed a competitive return on his investment. Since the ADR would affect the value of the contract, bidders were asked to reveal the ADR used in formulating their bid. MWSS could check this figure based on costs and per capita water consumption figures provided in the bid.

The actual determination of the ADR would be made by the Regulatory Office. That office would be asked to consider the bidder's ADR estimate. In addition, the analysts would review the costs of debt abroad, the cost of equity for other utility businesses in the Philippines and abroad, and necessary adjustments for county, exchange rate, and other project risks.

The Regulatory Office was to be established and charged with not only establishing the ADR, but also monitoring the concession contract. The consultants recommended an independent agency following the British model. However, there were legal constraints on that configuration so the new agency had to be established as a "cooperative" partner of the MWSS.

3.3   End of Contract Capital Expenditures and Ownership of Assets

An important problem was the incentive for the concessionaire to make necessary capital improvements toward the end of the contract period. To handle that issue, MWSS would pay the firm the present value of any unamortized cost. Of course, that cost would be passed on as a required expense to the next winning bidder.

The general issue of the ownership and use of the current assets also had to be resolved. MWSS decided not to transfer the assets to the concessionaire and to allow him to use assets at no additional cost. Any assets not necessary to operate the business would be sold by MWSS, and the proceeds would be placed in a fund to be used in the case of default of the government. The value of the inventory of consumables at the beginning of the contract would be given to the concessionaire, but that value would be returned to MWSS at the end of the contract period. Vehicles and equipment existing at the end of the period would be returned to MWSS.

3.4   Termination

Another contract provision was developed relating to the possibility of default either by the government or by the concessionaire. If the government defaulted, it was obligated to make all payments to the creditors of the company as well as a lump sum payment at the time of default. If the concessionaire defaulted, MWSS would take over the operation and would limit compensation to the concessionaire to 70% of the depreciated value of the assets taken over, not to exceed the value of outstanding loans.

3.5   Relationship Between Concessionaires

Since there would be two companies operating using some common facilities, such as the basic raw water source, it would be necessary to have an agreement relating to these facilities. In particular, an interconnection agreement would be required.

3.6   Bidding and Approval

Borrowing from the Buenos Aires case, MWSS required two separate bid proposals: technical and financial. The technical proposal would be examined first and judged as to compliance with the terms of the contract. Those bidders in compliance would then be moved to the second stage, where their financial bids would be compared.

A predetermined set of approval steps was put in place. The approval steps would be as follows:

1. MWSS Management
2. MWSS Board
3. The Cabinet-level Special Advisory Committee to the President, which would involve going through their Working Group
4. President, who would likely seek the advice of the Committee on Privatization

4. THE BIDDING

Given the size of the proposed contract, it was clear that only large firms in the water industry were in a position to bid. MWSS desired as many bidders as possible, especially since there was a split contract and both water and sewerage service were involved. As expected, the major participants were large French and British companies: Compagnie Generale des Eaux (CGE), Suez Lyonnaise des Eaux, United Utilities, and Anglian Water Company.

The international firms, however, required local partners, who would retain a minimum of 60% ownership. Since local water firms were very small, the international firms had to recruit larger nonwater local firms. There could be a consortium of local firms, but the rules required that the dominant local firm hold at least 20% of the total equity.

Attracting local partners posed a problem. The local nonwater partners had to be convinced that the operation would be profitable, and were concerned that the capital requirements might be beyond their means. MWSS recruited large Philippine companies. It convinced them that this would be a very profitable, stable-cash-flow business for 25 years. MWSS also revealed that the actual capital required would be $220 million for two zones, and that the international partner would assume part of the burden and that the rest could be funded out of the cash flow.

4.1   Prequalification

Four groups were prequalified to bid. They were as follows:

International Partner	Dominant Philippine Partner
International Water (United Utilities and Bechtel Corp.)	Ayala
Suez Lyonnaise des Eaux	Benpres Holding
Compaignie Generale des Eaux	Aboitiz Equity Ventures
Anglian Water International	Metro Pacific Corp.

4.2   Bid Opening

Since all bids were in technical compliance, the contest was over the financial bids. The results were as follows, stated in terms of water rates proposed:

Zone	Winning Bidder	Bid
West	Ayala	P2.5140
	Benpres	P4.9688
East	Ayala	P2.3169

Since Ayala was the low bidder in both zones, it was awarded the contract in the East and Benpres, which came in second in the West, was awarded the contract there, since no company was allowed to have both zones.

4.3   Issues Raised

There were some concerns raised before and after the bidding. There was a legal challenge by a local contractor who had a contract from MWSS for a construction project. The contractor wanted to ensure that the current contract would be honored and wanted to prevent MWSS from privatizing in the hopes of securing future construction work. Even though the contractor secured a temporary injunction to the bidding, the restraining order was allowed to expire before the bid opening.

Examination of the bids caused some concern, since the bids represented rates from one-half to one-quarter of the MWSS water rates and a few critics thought this was too good to be true. The huge bid disparity between the zones concerned some observers, especially since a full explanation was not given. Also, the winning of the bids by the two largest Philippine companies upset others. However, the likely extension of water service and of the 24-hour availability of water was so attractive to most Manila residents that rate differences by zone did not became a major issue.

5. IMMEDIATE BENEFITS OF PRIVATIZATION

The anticipation of immediate benefits to ratepayers muted most of the criticism of the bidding results. The sharp cut in water rates was very popular. The promise of rapid service improvements in fixing leaks and installing new technology was enticing. There was an appreciation of the lowering of cost by shedding redundant personnel and the likely reduction of corruption, especially in the letting of contracts. The taxpayer subsidization of the water system that was required under government operation would now be unnecessary. A $7 billion infusion of new spending for water and sewer improvements was expected to boost the economy. It appeared likely that self-interest of the concessionaires quickly would lead to expanded service.

Of considerable importance were the service targets in the Concession Agreement. Table 10.1 gives the service targets.

Table 10.1   Service Targets (in Percent of Coverage)

	Year					
Service	*1996*	*2001*	*2006*	*2011*	*2016*	*2021*
Water service	67	87	98	98	98	98
Sanitation	0	39	40	36	33	29
Sewerage	8	7	15	26	38	54
Sanitation and sewerage	8	46	55	62	71	83

Determining the final set of service targets took considerable negotiation. The consultants had to link the service targets to capital and operations and maintenance budgets. In turn, the level of tariffs had to be linked to service levels.

IFC presented alternative scenarios to the MWSS Board that included conservative, moderate, and aggressive goals. The key element was the sewerage program. Given the concern about the level of sewerage service, the Board chose the aggressive scenario despite the fact that higher tariffs would be required.

With the choice of an aggressive set of service targets, it was expected that the winning bids would be relatively high. However, it turned out that the bids were much lower than expected. The analysis of these bids suggested that the bidders believed that the introduction of the most advanced technology would drive down costs from their present level. The strong competition among the large, international companies who bid for a potentially profitable contract also depressed the bid level.

6. CLEARING THE WAY TO PRIVATE TAKEOVER

It was expected that the actual takeover could take place by June 30, 1997, but some transition problems slowed the process. Most of the delay was related to the existence of two zones. Employees, properties, and the customer database had to be split between the two concessionaires. An interconnection agreement had to be negotiated.

In addition to the two-zone issues, the interim between bid and takeover also created some problems. MWSS was slow to repair leaks and to collect bills, and this situation caused concern that this would create problems for the concessionaires. Moreover, MWSS was uncertain about what power it had and what its responsibilities would be after takeover.

The biggest obstacle was the interworkings of the two concessionaires and the related interconnection agreement. After some difficult negotiations, these issued were resolved. Each company was allocated half the raw water from the Angat Reservoir, the main source of raw water supply. It was estimated that the initial water market demand would be split 60% to 40%, West to East.

The most vexing issue was the transfer price of water in the interconnection agreement. The parties agreed on two prices, depending upon the classification of the water transferred. The water transferred on a regular basis would be priced at direct cost, and the water transferred on a temporary or extraordinary basis would be priced at the retail price. The final price numbers had to be decided by arbitration. The arbitrator's decision set the direct or treatment costs at P0.40 per cubic meter and the opportunity or temporary costs at P10.0 per cubic meter. The final award was P1.80 per cubic meter.

7. CONCLUSION: PRIVATIZATION RESULTS

Before major benefits could begin to flow from the new operating systems, two serious events intervened. A very serious drought hit the Philippines, and there was a drastic devaluation of the Philippine peso. 98% of the water supply of the greater Manila area comes from the Angat Dam. At the time of the private takeover throughout 1998, the most severe drought since the dam was built reduced the water supply by 30%, even though no water was released for irrigation, which use typically represented half of the total releases from the dam.

At the same time that the drought took effect, the Asian financial crisis occurred, leading to a 60% devaluation of the peso. Since nearly all of the MWSS debt (assumed by the concessionaires) was in foreign exchange, there was a significant increase in the cost of debt service. By 1998, debt service was in the range of $100 million.

The cost of these twin disasters was significant. In the case of one concessionaire, the debt service exceeded total revenue. However, the financially secure concession companies were able to absorb these added costs. Consequently, water tariffs were not increased in 1998. They were forced up in 1999, but only by 15%.

Service problems, likewise, were handled by the concessionaires with minimal impact. For example, the drought situation created a number of problems in the face of the resultant water rationing. Operational problems had to be solved and water had to be delivered by tanker to some communities that completely lost their water supply.

Even though regulatory issues may arise as the concessionaires and the regulators interact over time, the privatization decision has, to date, proven to be beneficial to consumers and to the government. Rates are lower, service is better, and there is no longer a financial drain on the government's budget. The concessionaires have added more than one million new customers about half of whom are the poorest of the poor. These customers now have direct connections and their cost of water has been reduced by more than one-half at the same time their consumption has increased by four times, on average. So far, the concession method seems to be a winner for the Philippines.

REFERENCE

Dumol, Mark (2000). *The Manila Water Concession: A Key Government Official's Diary of the World's Largest Water Privatization.* Washington, DC: The World Bank.

Chapter 11

Government-Owned Public Limited Companies in the Dutch Water Sector

Klaas Schwartz and Maarten Blokland

The government-owned public limited company (PLC) enjoys considerable prominence in many parts of the world, where it can alternatively be known as a "public company," a "government-owned company," a "state-owned enterprise,"or a "state-owned company." The essence of the government-owned PLC is that the company is established and operates under company law while the shares of the company are in hands of national, regional, or local government authorities. Generally, there is no legal or organizational difference between a publicly owned and a privately owned PLC, apart from the government ownership of shares. In some countries, however, special laws may exist that regulate government shareholdings in PLCs.*

*In New Zealand, for example, it is a "constitutional convention that government requires explicit statutory authority before acquiring a shareholding in any company. . . . Accordingly, there are no cases of companies used as instruments of state action where the only statutory framework is the Companies Act" (McKinlay, 1998).

Klaas Schwartz is Lecturer, Department of Sanitary and Environmental Engineering Department, International Institute for Infrastructure, Hydraulic and Environmental Engineering. He is the coauthor of several studies in the water field. Maarten Blokland is the Vice-Rector Education, International Institute for Infrastructural, Hydraulic and Environmental Engineering, Netherlands. He is the coauthor of several studies on water issues.

Reinventing Water and Wastewater Systems: Global Lessons for Improving Water Management, edited by Paul Seidenstat, David Haarmeyer, and Simon Hakim.
0-471-06422-X   Copyright © 2002 by John Wiley & Sons, Inc.

In the past decades, the government-owned PLC has become increasingly popular in various branches of infrastructure services provision. The government-owned PLC has been promoted as a means of stimulating more efficient management of infrastructure by introducing corporate structures similar to commercial, market-oriented enterprises. One of the main reasons for introducing such a structure is that it allows burdensome civil service regulations, which may impede the performance of public utilities[†], to be bypassed. Also, the government-owned PLC has been used as a transitional stage in the move toward full divestment. Examples of Dutch PLCs that were at one time government-owned but have since been privatized include KLM Royal Dutch Airlines, DSM chemical company, the NMB/Postbank (which would later merge into the ING Group), the Postal and Telecommunications Service (PTT), and the Dutch Railways (NS) (Van de Ven, 1994).

For the drinking water sector, however, the Dutch Government recently decided against privatization of water supply companies. The decision to "secure government ownership of water supply companies" is based on the belief that the emergence of private monopolies in the water supply sector would be "undesirable" in the context of the Government's duty to ensure safe drinking water and to protect public health (Ministry of Housing, Spatial Planning and Environment, 2000).

In the water sector, government-owned PLCs or public water PLCs, as they are also known, are quite common in Western Europe, where they can be found in Germany, the Netherlands, and Belgium, as well as in the Scandinavian countries. They are also found in the United States under the name of municipal stock corporations, but are less prominent in low- and middle-income countries. In the landscape of modes of organization in the water supply and sanitation sector, government ownership and the legal framework under which it operates distinguish the Public Water PLC from other modes of organization. Figure 11.1 provides an overview of the main differences between the public water PLC and other generally accepted modes of organization in the water supply and sanitation sector.

This chapter examines the functioning of government-owned water PLCs in the Netherlands by looking at the PLC structure in general and by examining this structure for two water supply companies in greater de-

---

[†]In Indonesia, for example, one of the reasons that introduction of the PLC structure is being considered is the desire to mitigate the burdensome labor regulations that are stipulated in the civil service code.

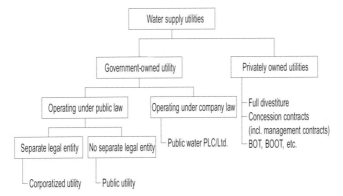

Figure 11.1 Distinguishing features of the public water PLC. BOT = build, operate, and transfer. BOOT = build, own, operate, and transfer.

tail. The chapter can be subdivided into five parts. The first part (Sections 1.1–1.3) briefly provides background information on the Dutch water supply sector in general and on the two companies subject of this case study in particular: the Water Supply Company Limburg[*] (WML) and Water Company Europoort[†] (WBE). The second part (Sections 1.4–1.5) examines the legal framework of the PLC. Relevant articles of company law are examined, and the role of the articles of association is discussed. The third part of this chapter (Section 2) focuses on issues relevant to the day-to-day functioning of the Dutch public water PLCs, the issues that distinguish them from other PLCs, both privately owned and publicly owned. The fourth part (Section 3) contains a performance comparison of various utilities in the Netherlands. The chapter ends with a conclusion (Section 4) concerning the functioning of the PLC as mode of organization in the water sector. In the conclusion, particular importance is given to the degree of autonomy that the PLC structure can provide to a utility.[‡] Also included in the conclusion is a discussion of the checks and balances between different stakeholders provided by the PLC structure.

[*]NV Waterleidingmaatschappij Limburg.

[†]NV Waterbededrijf Europoort.

[‡]The level of autonomy and the checks and balances greatly influence the performance of water utilities. A World Bank evaluation of water supply and sanitation projects between 1967 and 1989 even found that "institutional performance in the areas of operations and maintenance, unaccounted-for-water and human resource management has been good only where there was autonomy in utility management and an arms length regulatory regime" (World Bank, 1992).

1. WATER SUPPLY IN THE NETHERLANDS

1.1   Background

Since the early beginning of water supply in the Netherlands in the 1850s the sector has evolved continuously. This evolution is illustrated by the various modes of organisation that have been dominant since the first water supply company started operating in 1853. [See Figure 11.2.] Broadly three distinct periods can be identified.

—Blokland and Schwartz, 1999, pp. 35–36

1. *1853–1920.* In this era, the water supply sector was dominated by privately owned water supply companies (full divestiture). Toward the end of this era, the municipal public utilities were steadily increasing in number.
2. *1921–1974.* During this period, the municipal public utilities formed the predominant mode of organization in the Dutch water sector. The private utilities were declining further in number, and the public water PLC was emerging.
3. *1975–present.* During this phase, the public water PLC has become the most prominent mode of organization while the other ex-

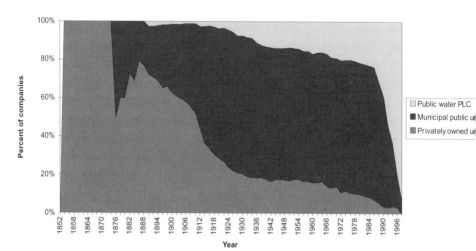

Figure 11.2   Mode of Organization of Water Supply Utilities in the Netherlands, 1850–1994. *Source:* Blokland and Schwartz, 1999.

isting forms were reduced further in number. At current, only two privately owned water supply utilities and one municipally owned public utility operate in the Netherlands.

The prominence of the public water PLC resulted largely from government regulation, most notably the Drinking Water Supply Act and amendments to this Act in 1975, which promoted the "[a]malagation of water utilities into larger vertically integrated units under the public water PLC format," while "horizontal integration with other utilities and other institutional forms were discouraged" (Blokland and Schwartz, 1999). The Drinking Water Supply Act was not without result, as the number of water supply companies dropped from 210 in 1950 to 27 in 1998. Of these 27, the overwhelming majority (21) are public water PLCs. Three water utilities are municipal public utilities, two companies are privately owned limited liability companies, and one utility is a privately owned PLC.

1.2   Characteristics and Responsibilities in the Dutch Water Sector

Two characteristics of the water supply sector deserve to be highlighted as distinguishing features of the Dutch water sector. Firstly, the production and supply of drinking water has been segregated from the treatment of wastewater. Water supply companies do not engage themselves in the collection and treatment of wastewater. The task of wastewater conveyance, treatment, and discharge is left to the water boards. Secondly, water supply companies are currently not subject to an external economic regulator. Despite plans from the Ministry of Economic Affairs to introduce economic regulation, the water supply companies, united in the Netherlands Waterworks Association (VEWIN), have been able to postpone such regulation. Instead, the companies rely on a system of economic self-regulation.* The quality of drinking water produced, however, is regulated. A Public Health Inspectorate ensures that drinking water meets hygiene standards that are set by national and European law. Table 11.1 outlines the division of responsibilities in the Dutch water sector.

---

*For a discussion of the self-regulating mechanisms in the Dutch water supply sector, see Braadbaart et al. (1999).

Table 11.1 Responsibilities in the Dutch Water Sector

Organization	*Responsibility*
Central government	Framework acts and general administrative measures Strategic national policy Operational policy and management for the North Sea and main rivers General supervision over provinces, water boards, and municipalities
Provinces	Strategic groundwater and surface ater policy Operational groundwater policy and management (and, in two provinces, surface water quality) General supervision over water boards and municipalities Reorganization plans for the water sector
VEWIN	10-year plans, drafted every five years
Water boards	Operational surface water management (quantity and/or quality), including sewerage treatment
Municipalities	Sewerage collection
Drinking water companies	Production and distribution of drinking water One-year plans

Source: Based on Perdok, table as cited in Mostert, 1997.

1.3   Water Supply Company Limburg and Water Company Europoort

The Water Supply Company Limburg (WML) supplies more than 78 million cubic meters of water to well over 1 million people in the southeastern Province of Limburg. A portion of the shares of WML are owned by the provincial government of Limburg, which owns close to 24% of shares. The remaining shares are in hands of 57 municipalities in the service area of WML.

Water Company Europoort (WBE) annually supplies 150 million cubic meters of drinking water to over 1.4 million people living in and around the city of Rotterdam in the Province of South Holland. The shares of WBE are owned by a group of 29 municipalities, which are all situated in the service area of WBE. In terms of water delivered, WBE is the largest water supply company in the Netherlands.

Statistics for the two companies are shown in Table 11.2, and their locations are shown in Figure 11.3.

Table 11.2 Selected Indicators for WML and WBE in 1998[a]

Indicator	Unit of Account	WML	WBE
Service coverage	Percent	99.9	99.9
Total water supplied	m^3 ($\times$ 1,000)	78,279	149,774
Connections	Number	414,417	461,957
Staff	f.t.e.	536	515
Net turnover from water sales	US $ ($\times$ 10^6)	96.4	174.4
Nonrevenue-water	Percentage of total drinking water supplied	5.5	6.3

Source: VEWIN, 2000.

[a]Currency exchange rate used: US $1 = 2.3 Dutch guilders.

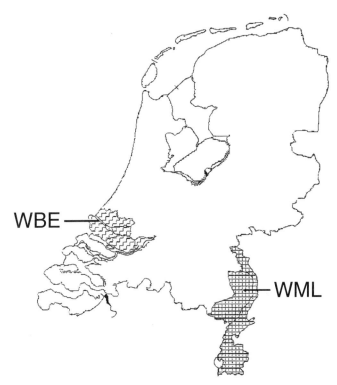

Figure 11.3 Location of WML and WBE in the Netherlands.

1.4   The Governance Structure of the Public Water PLC

The legal framework of the public water PLC consists of three complementary parts. The first part concerns legislation that regulates the drinking water sector in general and legislation that regulates government-ownership of shares in PLCs. In the Netherlands, the cabinet decision to prohibit privatization of the water sector is an example of such legislation. The second part of the legal framework concerns company law, which defines, in broad terms, the main characteristics of a PLC, as well as the main rights and obligations of the various actors in a PLC. In the Netherlands, company law concerning PLCs is mainly to be found in articles 64 to 174a of the second book of Civil Law or "Burgerlijk Wetboek" (BW:2). The third part of the legal framework concerns the articles of association that have to be adopted at the time the PLC is established. These articles of association specify in greater detail the obligations and rights of the various actors.

National and supranational legislation with respect to the drinking water sector will generally have precedence over company law, which in turn has precedence over the articles of association of a PLC. In this chapter, the focus is mainly on the company law and on the articles of association.

Dutch law identifies and attributes responsibilities to four main actors in PLCs. These four actors are the managing director or management board, the board of directors, the shareholders, and the workers' council. The first three actors are present in each PLC. The workers' council is usually only required in companies that have at least 100 staff members. These different actors are illustrated in Figure 11.4 and elaborated upon in greater detail* in Sections 1.4.1–1.4.4.

1.4.1   Managing Director

The managing director is responsible for the day-to-day management of the company. The managing director is also the company's legal representative unless stated otherwise by law. Decisions and actions taken by the management must be directed at realizing company goals. To ensure this, company law dictates that the managing director is personally responsible for any debt that cannot be covered by company assets, in case the PLC goes bankrupt and it is plausible that the bankruptcy is the result of mismanagement on the part of the managing director. Moreover, if the managing director has presented misleading figures about the (financial) state

---

*Based on articles 64–174a of the "Burgerlijk Wetboek" unless mentioned otherwise.

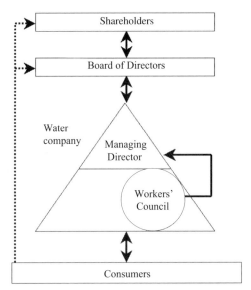

Figure 11.4   Structure of a Dutch PLC.
Source: Blokland and Schwartz, 1999.

of the company, he/she is personally liable for debts resulting from the misrepresentation of figures.

By law, the managing director is also responsible for presenting the annual accounts to the shareholders meeting and the workers' council within five months of the closing of the financial year. These accounts provide information about the company's financial performance over the previous year, solvency, assets, and liquidity. The annual account has to be approved and signed by the board of directors and by a reputable, independent accountant prior to being sent to the shareholders meeting.

Within the PLC structure, the managing director has a bridging function. The managing director cushions the impact of the board of directors on the work of the company's staff and interprets particular demands from the board of directors to the staff. At the same time, the managing director represents the interests of the staff before the board (Thynne, 1998). In addition, the managing director of a government-owned PLC must be able to display considerable political sensitivity. The government ownership of the company, as well as the fact that the company provides the essential service of drinking water supply, inevitably exposes the managing director directly or indirectly to the wider political environment. The managing director must be able to function effectively in that environment (Thynne, 1998).

1.4.2   Board of Directors

The board of directors is responsible for supervision of the management of the company and of the general functioning of the company. Article 139 of the "Burgerlijk Wetboek" specifically stipulates that, in performing their tasks, the board of directors is to be guided by company interests. Similar to the case of the managing director, company law dictates that the board members can also be held personally responsible for any mismanagement of the company.

The board has free and unlimited access to all company facilities and information and can advise the management of the company on any issue it considers relevant. Among the more important tasks of the board are the appointment, dismissal, and suspension of the managing director,* and approval of the annual plan. Dismissal of the managing director requires prior consultations with the shareholders meeting. As mentioned earlier, the board of directors also has to approve the annual accounts prepared by the managing director.

Members of the board of directors can be suspended by other board members. A request for dismissal of a board member on grounds of neglecting his/her responsibilities can be submitted to a Court of Law by other board members, a representative of the shareholders meeting, or the workers' council. The Court of Law in question ultimately decides if the request for dismissal is valid.

Similar to the managing director, the board of directors also fulfils a bridging function. The board of directors has to look outwards to the shareholders of the company and inwards to the management and staff of the company (Thynne, 1998).

1.4.3   Shareholders Meeting

The shareholders meeting, which is to be held at least once a year, is granted "all powers, within limits set by Law and articles of association, that are not bestowed upon the management or others" (art. 107 BW: 2). Although article 107 seems to attribute considerable powers to the shareholders, in most large PLCs, the shareholders have little direct control over the management of the company because considerable powers are bestowed upon other actors by the articles of association. Generally, the powers of the share-

---

*The appointment and dismissal can also be the responsibility of the shareholders meeting (art. 132 BW: 2).

holders are limited to approval or rejection of the annual accounts, proposals to amend the articles of association, and ultimately, proposals to dissolve the company. Although the shareholders theoretically have the opportunity to adjust the balance of powers and tilt it in its favor by amending the articles of association, the shareholders generally refrain from doing so.

The shareholders meeting, which is convened by the board of directors or the managing director, is entitled to receive all desired information from the management and board of directors unless disclosure of particular information poses a severe threat to company interests. All shareholders have the right to vote during the shareholders meeting. The number of votes that a shareholder has is dependent on the number of shares that the shareholder owns.

1.4.4   Workers' Council

The rights and obligations of the workers' council are stipulated in the Law on Workers' Councils.* Every company with over 100 employees must establish a workers' council. All Dutch water companies employ more than 100 employees, and as such, have a workers' council. The functioning of the council is mostly arranged in the statutes that every council must draft. These statutes relate to all matters that have been delegated to the council by the Law on Workers' Councils. The council and the managing director are to meet at least six times a year, during which meetings any issues considered relevant can be discussed. The council has the right to request all information that could be considered relevant in performing its tasks. The council has the right to inspect the accounts, annual budget, multiyear forecast, and other key strategic and legal information. In addition to the right to such information, the council has a range of advisory powers on issues such as changes in organizational structure, termination and expansion of business activities, company siting, major investment and lending activities, and tendering consultancy assignments. Moreover, the council has the right of approval on regulations pertaining to matters such as pension plans, work hours, holidays, recruitment policies, training, and handling of complaints.

The advisory powers of the council are only sparingly used. However, when offered, the advice tends to be accepted by the managing director. On average, the managing director adopts three out of four initiatives (Blokland and Schwartz, 1999).

---

*Wet op de Ondernemingsraden.

1.5   The Articles of Association

The articles of association are drawn up before a public notary and need to be approved by the government for compliance with private company law and public law. These articles may stipulate company objectives, maximum capital outlays or loans that the managing director can decide upon independently, the composition of the board, the ownership and transferability of the shares, the number of annual shareholder meetings, the financial result that is to be obtained, and how to utilize this. The articles also specify in greater detail the powers held by the board of directors, the managing director, and the shareholders. The restrictions and limits set in the articles of association are self-imposed by the shareholders meeting. The articles of association actually define in detailed terms the operating structure of the PLC. Below, the way in which the articles of association define the operational structure of the PLC is illustrated by looking at the articles of two water supply companies.

1.5.1   Articles of Association of Water Supply Company Limburg

The ownership of shares and the transfer of shares have been restricted to public bodies in the articles of association of WML. The articles of association limit shareholdership to the Province of Limburg and municipalities in that province, and dictate that shares cannot be transferred to other (government) organizations without prior approval of the board of directors. Municipal shares are to be owned in proportion to the number of inhabitants in the service area, and the board of directors updates the distribution of shares once every five years.

In the WML case, the 57 shareholders own a total of 500 (voting) shares. The largest shareholder is the Province of Limburg, which owns 118 shares (24%). The next largest shareholder is the Municipality of Heerlen, owning 35 shares (7%). As the decisions in the shareholders meeting are by common majority, the WML shareholders need to do quite some lobbying with coshareholders to have their opinions validated in the meeting. The board of directors consists of nine members who appoint new members by cooptation. Informal arrangements between the shareholders, which have evolved over time, have taken care of regional, provincial, and expert representation on the board, which now consists of six municipal representatives from three distinct regions of the service area, one provincial representative, and two experts. (See Table 11.3.) The board of directors of WML is empowered to set the limit for new short- and long-term loans to be taken

Table 11.3   Characteristics of WML Shares and Composition of the Board of Directors

Shares	
Exclusive ownership?	Yes, province and municipalities
Number of Shareholders	56 municipalities + 1 province
Share capital	US $2.6 million
Number of shares	500
Largest shareholding	118 (Province of Limburg)
Board of Directors	
Number of members	9
Membership prescribed?	No
Actual membership	6 Municipal representatives
	1 Provincial representative
	2 Professionals
Chairperson	Decided by board members

Source: WML Statutes, as cited in Blokland and Schwartz, 1999.

up by the company, and to approve key decisions, such as investments above a certain financial value, collective staff dismissals, modification of labor conditions, and the procurement by the company of financial interest in other PLCs.

With respect to the financial results that are obtained, WML limits the dividend to be paid to the shareholders to a maximum of 7% of share capital. The articles of association also specify the powers of the different bodies. Table 11.4 presents a general overview of the distribution of some of the decision-making powers over the three main actors, that is, the managing director, the board, and the shareholders in the cases of WML and WBE (the latter being discussed in more detail below).

1.5.2   Articles of Association of Water Company Europoort

Similar to the case of WML, ownership of shares is also restricted to public bodies by WBE's articles of association. Shareholdership is restricted to municipalities located in the service area, and shares can be owned only by the municipalities and by the company itself. Transfer of shares is only between these same parties. The distribution of shares is to be proportional to the water supplied to each municipality, with shares held by the City of Rotterdam fixed at 50% of the shares. This rule was set to limit the influence of Rotterdam, which consumes more than half the water sup-

Table 11.4   Powers of the Managing Director (MD), the Board of Directors (BoD), and the Shareholders Meeting (SM) for WML and WBE

Responsibility	*WML*	*WBE*
Bill and collect for services	MD	MD
Terminate service provision to defaulters	MD	MD
Determine tariff structure	SM	SM
Set water tariffs and connection fees	SM	BoD
Enter loan agreements	MD (limited), BoD	MD (limited), BoD
Procure goods and services	MD (limited), BoD	MD (limited), BoD
Procurement of assets	MD (limited), BoD	MD (limited), BoD
Hire and fire individual staff members	MD	MD
Promote and demote individual staff members	MD	MD
Determine salary and incentive structure	BoD	BoD
Determine structure of the Organization	MD	MD
Define internal work processes and standards	MD	MD
Hire and fire the managing director	BoD	BoD
Appoint and dismiss board members	BoD	Municipalities
Appoint external auditor	SM	SM
Approve rolling multiyear business plan	SM	SM
Approve annual plan	BoD	BoD
Approve annual report (budget/audit report)	SM	SM
Share issues	SM	SM
Participate in other enterprises	SM	MD (limited), BoD
Dissolve company	SM	SM
Amend bylaws	SM	SM

Source: WML and WBE Statutes, as cited in Blokland and Schwartz, 1999.

plied by WBE. Company self-ownership of shares is also limited, to a maximum of 10%.

In the case of WBE, the board also has nine members. Here, the distribution of seats on the board is prescribed in the articles of association. Although the appointment of directors to the board is regulated by company law, the articles of association of a PLC can demand that members of the

BoD be appointed by a government organization. In the case of WBE, the bylaws indicate that the BoD is to be named by a group of municipalities. The municipalities mentioned in the bylaws as the government organizations that appoint the BoD, are also the shareholders. So, de facto, the shareholders appoint the board members in the case of WBE. The City of Rotterdam appoints two board members, who decide between themselves which of the two is to be the chairperson of the board. The remaining 23 municipalities are divided into six groups, with each group appointing one board member. The members are appointed and dismissed by their constituent municipalities. The ninth board member, who is appointed by the other eight board members, is to be a professional with expertise in social, economic, and financial affairs. See Table 11.5.

In addition to the bodies demanded by company law, the WBE statutes require the institution of an "Advisory Council" that represents the Shareholders, and a three-member independent expert "Committee of Good Services." These committees are empowered to advise in particular cases of disagreement within the board of directors. Unlike the statutes of WML, no dividend limitations or guidelines are specified for WBE. The shareholders of WBE have decided to mandate the board of directors to water tariffs and connection fees, while the shareholders of WML have kept those powers to themselves.

Table 11.5   Characteristics of WBE Shares and Composition of the Board of Directors

Shares	
Exclusive ownership?	Yes, municipalities and company
Number of Shareholders	29 municipalities
Share capital	US $7.9 million
Number of shares	150,000
Largest shareholding	75,000 (City of Rotterdam)
Board of Directors	
Number of members	9
Membership prescribed?	Yes, 2 members from Rotterdam and 6 from other municipalities
Actual membership	8 Municipal representatives 1 Professional
Chairperson	1 of 2 Rotterdam representatives

Source: WBE Statutes, as cited in Blokland and Schwartz, 1999.

As may be seen in Table 11.4, most of the responsibilities at WBE have been assigned to the same actor as in the case of WML, although some significant differences exist. The point to be made is that, within the framework provided by company law, the articles of association of the company allow a degree of flexibility in the constitution of the various bodies and in the assignment of specific powers.

2. DISTINGUISHING CHARACTERISTICS OF DUTCH WATER PLCS

Publicly owned water PLCs in the Netherlands have certain characteristics that are not shared by other PLCs. These salient features result directly or indirectly from the fact that the water supply companies are owned by government agencies and from the nature of the drinking water sector in the Netherlands. These characteristics can be divided into managerial characteristics and financial characteristics. The management characteristics relate to mayors in the Netherlands being appointed instead of elected, the composition of the board of directors, the fact that the shareholders have limited their powers, and the fact that the issue of water supply lacks political appeal in the Netherlands. The financial characteristics are the low levels of equity, a history of full cost recovery, and limited profit sharing.

2.1   The Appointed Mayor

In the Netherlands, the mayor of a municipality is appointed by the national government. The relevance of the appointed mayor for provision of water supply lies in the fact that the members on the board of directors of public water PLCs tend to be mayors from shareholding municipalities. The only ways in which mayors can be removed from office are by a "vote of no confidence" from the city council (which rarely occurs) or if the mayor tenders his or her resignation. As such, mayors tend to remain in office for a long period of time. The long time horizon that the mayors can adopt and their (relative) independence from municipal politics insulate the board of directors of many Dutch water supply companies from political opportunism.*

---

*In discussing the constraints to improving water services in Latin America, Spiller and Savedoff (1999) have indicated that the main reason for the poor performance of water utilities there is the problem of governmental opportunism. Governments with short time horizons are likely to avoid unpopular decisions that may yield benefits only in the longer term.

2.2   Composition of the Board of Directors

A key question in the government-owned PLC is what role the board of directors should play. The role played by the board may be one of representing the government shareholders or it may be one of providing commercial expertise to the PLC. The membership of the board and how these members are appointed or elected will depend on how the shareholders define the role of the board. Historically, the Dutch water PLCs have placed emphasis on the role of representation of the government shareholders. Mayors from the shareholding municipalities and provinces generally populate the board of directors. As such, most board members are politicians with an administrative background.

The problem with a board of directors that emphasises representation of (government) shareholders is that the board members are often not well versed in the technical, financial, and commercial aspects of the drinking water industry. In fact, when one takes into account the likely expertise of the managing director, there appears to be a serious imbalance between the knowledge of the board members and that of the managing directors when it comes to running a water supply company. The result of such a knowledge imbalance is that the board members are unlikely to have the necessary skills to properly supervise the actions of the managing director. The dangers of inadequate supervision speak for themselves.

In recent years, this problem has been recognized and the role of the board as a source of commercial expertise is increasingly gaining importance. In other words, the board of directors is becoming increasingly professionalized. However, such a development is not without obstacles. Replacing a director with an expert-director requires a municipality to give up its seat on the board. Giving up a seat on the board means giving up influence. Needless to say, this is a very sensitive issue. Despite its sensitivities, the process of professionalization is slowly taking place. For example, the board of Limburg's WML decided to take additional expertise on board. The board invited a surface water expert into its midst. One geographical region has voluntarily given up one of its seats on the board to enable the water supply expert to join the board. In the long run, WML appears to strive for a 50–50 distribution: half private sector representatives with relevant expertise, half government officials.

A second characteristic that follows from the fact that the board members are overwhelmingly government representatives is that most board members have to return the fee they receive for sitting on the board. According to Dutch law, government employees who are also board members of a PLC have to return the fee they receive for being a board member. This means that the mayors who sit on the board of directors do not gain financially from their board membership.

A third issue that derives from the composition of the board of directors is the potential conflict of interests between the responsibilities as board member and those of mayor. Although formally, a board member must set aside other commitments (for example to the constituents of his/her municipality), it remains to be seen if such separation of responsibilities can always be achieved.

A fourth issue relevant to the composition of the board of directors in the Netherlands is the agreements (often specified in the articles of association) between the shareholders concerning the regional distribution of seats on the board of directors. Often, the board members will come from different regional areas of the service area, thus ensuring that all regions are represented on the board.

2.3   Limited Powers of Shareholders

As mentioned earlier, all powers not bestowed on the managing director and board of directors belong to the shareholders. Characteristic for Dutch water supply PLCs, however, is that the shareholders have substantially limited their powers in the articles of association. By way of these articles, the shareholders are largely limited to accepting or rejecting the annual accounts and to rejecting or accepting amendments to the company statutes.

2.4   Lack of Political Appeal

With almost universal service coverage and high customer approval rates,* the issue of water provision is largely lacking in political appeal in the Netherlands. Although the water supply sector did receive some media attention, due to a government decision to prohibit privatization of the water

---

*A benchmarking study undertaken in 1999 showed that the services provided by Dutch water supply companies are rated well. On average, a 7.7 (on a scale of 1 to 10) rating was given. This compared well with service provision by Dutch supermarkets (7.0) and Dutch transportation companies (5.9) (VEWIN, 1999).

supply sector and the Second World Water Forum in The Hague, the issue of water supply does not seem to be high on the agenda of the consumers. As long as drinking water quality is beyond reproach and prices are acceptable, the drinking water industry offers politicians little incentive to intervene for the sake of personal (political) gain.

2.5   Limited Profit Sharing

A fifth characteristic of the water supply PLCs is that a large number of these PLCs have limited the dividend payments to the shareholders. The limits are stipulated in the articles of association. Although some PLCs, like WBE, do not impose restrictions, most water PLCs either limit dividend payments, like WML, which restricts payments to 7%, or even entirely forbid dividend payments, like the Dune Water Company South Holland (DZH). All the profit earned above 7% in the case of WML, and any profit in the case of DZH, goes to the reserves of the water supply company to pay for future investments or production costs. Clearly, the limits to profit-sharing illustrate that the first priority of the shareholders is not the return on their investment but the continued supply of drinking water.

2.6   Low Equity

Another characteristic concerns the low amount of equity that the shareholders bring into the company. In the case of WML, for example, the 57 shareholders hold a total of 500 shares. Each share is worth approximately US $5,263.* The total amount of the shares thus amounts to just over US $2.6 million. Considering the capital-intensive nature of the drinking water industry, this is a very low amount. The shareholders are only liable for the capital they have put into company shares. Low equity levels thus also imply low (financial) liability levels for the shareholders. The low level of liability of the shareholders, in conjunction with the fact that the shareholders are municipalities and provinces, mean that the shareholders are not as adamant about performance efficiency as private investors would be. The shareholders have little incentive to press management of the PLC for greater efficiency.

*Figures date from 1997 and may have changed slightly due to exchange rate fluctuation.

2.7 Cost Recovery

As mentioned earlier, the origins of the water supply organization in the Netherlands lie in the private sector. This has meant that, since its early beginnings, the Dutch water supply sector has had relatively little to lose and, with some drinking water companies restricting their dividend payments, companies have operated on the principle of full cost recovery. Still today, no public funds are used for public water supply. Only for special innovative, pilot projects, such as deep-infiltration experiments, can companies request government subsidies. This has meant that the consumers are accustomed to paying the full cost of water.

3. PERFORMANCE COMPARISON WITH PRIVATE AND PUBLIC UTILITIES

In comparing the performance of public water PLCs with other public and private utilities, it should be emphasized that performance comparisons between different utilities are always a risky undertaking. The physical and social circumstances under which a utility operates are unique and hardly comparable. Those unique characteristics greatly cloud any performance comparison between utilities.

Below, performance indicators for three companies are presented. GWA is the municipal water supply company of Amsterdam. GWA is a public utility that serves approximately 415,000 connections in the capital city of Amsterdam. WML is a public water PLC. TWM is a privately owned public limited company of which TWM Holding owns the shares.* TWM is one of the smaller water supply companies in the Netherlands (68,000 connections) and annually distributes 13 million cubic meters of water in its service area in the southern part of the Netherlands.

The size indicators of these three companies (Table 11.6) reveal a glimpse of the particular unique circumstances under which these utilities operate. For example, the network density for GWA is relatively high, which is not surprising considering that the clientele of GWA resides in the capital city of Amsterdam. In the case of WML, which incorporates a large rural area in its service area, the density of the network is significantly lower.

The three utilities show comparable performance concerning the indicators stated in Table 11.7. Nonrevenue water is somewhat lower for TWN

---

*TWM Holding is a limited liability company.

Table 11.6   Selected Size Indicators for Three Dutch Water Supply Companies in 1998

	Unit of account	GWA	WML	TWM
Large consumers (>10,000 m³)	Percent	17	5	19
Small consumers (<300 m³)	Percent	65	63	66
Staff	Full-time employees	661	533	95
Length of mains	Kilometers	2,079	7,329	826
Total water sales	Cubic meters (× 1,000)	65,940	78,297	13,333
Connections	Number	415,955	414,417	68,374
Network density	Meters/connection	5.0	18	12
Water sales/connection	Cubic meters/connection	159	189	195

Source: VEWIN, 2000.

as compared to the other two utilities. However, TWM's smaller size and service area may explain part of this difference.

The financial ratios also are similar for the three utilities, as shown in Table 11.8. The low equity levels characteristic of the Dutch water supply companies result in very high debt-equity ratios, as is illustrated by all three companies, and by GWA in particular. All in all, a quick analysis of selected performance indicators and financial ratios of these three utilities does not show any strong performance contrasts between the different utilities. All three utilities show a comparable and good performance.

Table 11.7   Selected Performance Indicators for Three Water Supply Companies[a]

Indicator	Unit of Account	GWA	WML	TWM
Nonrevenue water	Percent of total water produced	3.5	5.5	0.0[b]
Labor productivity	Staff/1,000 connections	1.6	1.3	1.4
Net profit	Million US $	5.9	16.3	0.54
Price (excluding VAT)	US $/cubic meter	1.22	1.23	n.a.

Sources: VEWIN, 2000 (figures cover 1998); Annual reports 1999 of GWA, WMI, and TWM.

[a]Currency exchange rate use: US $1 = 2.3 Dutch Guilders.

[b]Of the 13 million cubic meters distributed annually, 4,000 cubic meters are not charged.

Table 11.8   Selected Financial Ratios for Three Water Supply Companies in 1999

Ratio	GWA	WML	TWM
Working ratio	0.63	0.54	0.68
Operating ratio	0.95	0.84	0.90
Debt-equity ratio	113.5	2.8	2.1
Return on fixed assets	0.02	0.04	0.02
Return on equity	2.6	0.16	0.03

Source: Annual reports 1999 of GWA, WML, and TWM.

4. CONCLUSION

The performance of Dutch public water PLCs is generally considered to be good. In trying to explain this performance, it is useful to examine the PLC structure in terms of the checks and balances it provides between the various stakeholders and the degree of autonomy it allows for. When examining the structure of the PLC, it is evident that, although the PLC structure itself can provide a general framework, much depends on the individual articles of association and on the characteristics of the water sector in general.

The following checks and balances can be identified as resulting directly from company law:

1. Company law provides a broad division of responsibilities and powers among the various actors. This division of powers limits the possibility that one of the actors is capable of getting the upper hand in the company.
2. All PLCs are required to file annual financial audit accounts at the chamber of commerce. These accounts, which are open to scrutiny by the public, contain information on the company's financial performance over the previous year, solvency, assets, and liquidity. The accounts are to be approved by the managing director, the board of directors, the shareholders, and by an impartial and reputable accounting firm.
3. The managing director and the members of the board of directors can be held personally responsible for any debts incurred due to severe mismanagement of the company.

The actual checks and balances present in an individual PLC and the degree of autonomy enjoyed by a PLC depend strongly on how the PLC framework is filled in. In other words, much depends on the content of the articles of association, which to a large degree determine the actual level of autonomy enjoyed by the utility. Features drawn up in the articles of association that increase the autonomy of Dutch water supply companies and add to the checks and balances in the governance structure include limited profit-sharing, low levels of equity, limited powers of the shareholders, the geographical distribution of board members, and limiting the number of shares that can be held by one municipality.

The limited profit-sharing and the low level of equity ensure that the financial interests of the shareholders in the utility are limited. On the one hand, limited profit-sharing means that the shareholders have little to gain financially from the utility, which means that profits from the company cannot be used to finance other public expenditures but have to be reinvested in the company. On the other hand, the low levels of equity, which imply a virtually risk-free investment, mean the shareholders have little to lose financially.

The self-imposed limits on powers of shareholders mean that most of the operational powers lie with the board of directors and the managing director. In many public water PLCs, the articles of association stipulate that board members represent different regions of the service area. By ensuring balanced regional representation, domination of the board of directors by one region of the service area is avoided. The limits on the number of shares that can be owned by a single shareholder provide a check by demanding that decisions in the shareholders meeting and board of directors can only be made by a coalition of shareholders.

In addition to the features in company law and the articles of association that provide the necessary checks and balances and autonomy, a number of characteristics of Dutch public law and the Dutch water sector also aid the performance of the companies. The first of these features is the appointment of mayors by the national government. The mayor of a municipality cannot be voted out of office and is only indirectly dependent on the support of the electorate of the municipality by way of the municipal council that is elected by the constituents of a municipality. Considering that a mayor is not dependent on elections in order to remain in office, the incentive for the mayor (as a board member) to influence the company for political interests is small.

The second of these features relates to the lack of political appeal of the water supply sector. Very little political mileage can be gained from issues relating to drinking water, as consumers appear to be satisfied with the quality of water and find the price acceptable. As a result, the degree of autonomy that the water supply companies enjoy is relatively large.

REFERENCES

Blokland, Maarten W., and Klaas Schwartz (1999). "The Dutch Public Water PLC," in Maarten W. Blokland, Okke D. Braadbaart, and Klaas Schwartz (Eds.), *Private Business, Public Owners: Government Shareholdings in Water Enterprises.* The Hague: Ministry of Housing, Spatial Planning and the Environment.

Braadbaart, Okke D., Maarten W. Blokland, and Bjorn Hoogwout (1999). "Evolving Market Surrogates in the Dutch Water Supply Industry: Investments, Finance, and Industry Performance Comparisons," in Maarten W. Blokland, Okke D. Braadbaart, and Klaas Schwartz (Eds.), *Private Business, Public Owners: Government Shareholding in Water Enterprises.* The Hague: Ministry of Housing, Spatial Planning and the Environment.

Gemeentewaterleiding Amsterdam (GWA) (2000). *Year Report 1999.* Amsterdam.

McKinlay, Peter (1998). "State-Owned Enterprises and Crown Companies in New Zealand," *Public Administration and Development,* Vol. 18, pp. 229–242.

Ministry of Housing, Spatial Planning and Environment (2000), "Eigendom Waterleidingbedrijven," *Staatcourant,* August 29.

Mostert, Erik (1997). *Water Policy Formulation in The Netherlands.* Water 21, Phase 1 report, Research Report number 6, RBA Centre, Delft, The Netherlands.

Spiller, Pablo T., and William D. Savedoff (1999). "Government Opportunism and the Provision of Water," in Pablo T. Spiller and William D. Savedoff (Eds.), *Spilled Water: Institutional Commitment in the Provision of Water Services.* Washington, DC: Inter-American Development Bank.

Thynne, Ian (1998). "Government Companies as Instruments of State Action," *Public Administration and Development,* Vol. 18, pp. 217–228.

Tilburgsche Waterleiding-Maatschappij (TWM) (2000). *Financial Year Report 1999,* Tilburg, The Netherlands.

Van de Ven, Adrian, T.L.M. (1994). "The Changing Role of Government in Public Enterprises: Dutch Experiences," *International Review of Administrative Sciences,* Vol. 60, No. 3, pp. 371–383.

VEWIN (1999). "Transparantie en Doelmatigheid Voorop bij Waterleidingbedrijven," *Water Spiegel,* No. 1, March, pp. 2–8.

VEWIN (2000). *Waterleidingstatistiek 1998*. Rijswijk, The Netherlands: Vereniging van Exploitanten van Waterleidingbedrijven in Nederland.

Waterleiding Maatschappij Limburg (WML) (2000). *Financial Report Magazine,* NV Waterleiding Maatschappij Limburg, Maastricht, The Netherlands.

World Bank (1992). *Water Supply and Sanitation Projects: The Bank's Experience—1967–1989*. Washington, DC: Operations Evaluation Department, World Bank.

ADDITIONAL READING

Schwartz, Klaas. "The Policy Environment," in Maarten W. Blokland, Okke D. Braadbaart, and Klaas Schwartz (Eds.), *Private Business, Public Owners: Government Shareholdings in Water Enterprises*. The Hague: Ministry of Housing, Spatial Planning and the Environment, 1999.

Part V

Restructuring Operations in Selected US Cities

Chapter 12

Innovative Contracting for New Facilities

Seattle's Use of Design-Build-Operate for Implementation of Water Treatment Plants

Elizabeth S. Kelly, Scott Haskins, and Rodney Eng

The City of Seattle, in the State of Washington, has used an innovative public–private partnership approach for development of its Tolt River drinking water treatment facility. In contrast to a conventional design-bid-build public works procurement method, Seattle has been successful in implementing a high quality and reliable facility on schedule and at about 40% below estimates, by using an alternative contracting approach called design-build-operate (DBO). Under this method, a single integrated team of designers, constructors, and operators work as a single entity for development of the project, while the City maintains public ownership and financing. This facility was put into operation in December 2000, almost

Elizabeth S. Kelly is Program Manager, Seattle Public Utilities, Seattle, Washington, and a professional engineer. She manages large capital improvement projects for the agency. Scott Haskins is Resource Management Branch Director, Seattle Public Utilities, Seattle Washington. He oversees watershed management, water quality and supply, resource planning and development, and solid waste management. Previously, he served as Finance Director and Deputy Director of Seattle Water. Rodney Eng is Assistant City Attorney, Seattle, Washington.

Reinventing Water and Wastewater Systems: Global Lessons for Improving Water Management, edited by Paul Seidenstat, David Haarmeyer, and Simon Hakim.
0-471-06422-X   Copyright © 2002 by John Wiley & Sons, Inc.

four years after the project initiation. The Tolt facility is a 120 million gallon per day (MGD) filtration and ozonation plant for treatment of Seattle's Tolt River source of supply, providing about a third of the water for Seattle and its wholesale customers. The success on the Tolt project prompted the City to use the same approach for its Cedar Treatment Facility, which will provide treatment for the remaining two-thirds of the water to Seattle and its regional customers, and which is expected to be operational by the end of 2004. In addition, other cities—including Phoenix, Houston, Detroit, Tampa Bay, Vancouver, Canada, among others—have followed Seattle's lead by using similar approaches for their major water treatment projects.

This chapter presents the key features of Seattle's innovative DBO approach, describes how it was implemented, and summarizes lessons learned, which can be applied to other similar projects.

1. BACKGROUND

Seattle Public Utilities (SPU), a department of City government, is responsible for providing water supply, drainage and wastewater, and solid waste services to the City and, for some services, to the surrounding region. The water system provides water for a population of about 1.3 million people in the City of Seattle and in surrounding cities and water districts in the region. SPU takes 99% of its water from two surface water sources: the Cedar River in southeast King County and the South Fork of the Tolt River in northeast King County (see Figure 12.1). The Cedar River currently provides about two-thirds of the water to the system, with the South Fork of the Tolt River providing the other third. The Tolt and Cedar sources originate in pristine, unoccupied watersheds in the Cascade Mountains, east of Seattle. Seattle has been very fortunate that, since the development of the Cedar source in the late 1890s and Tolt source in the late 1950s, the water has provided public health protection, and met regulatory requirements without major treatment facilities.

In the early 1990s, Seattle began the process of developing a filtration project for the Tolt source. The project needed: (1) to increase Tolt system reliability, particularly during stormy periods when turbid water could require a shutdown; (2) to meet increasingly stringent federal drinking water regulations (with regard to the Tolt source, this primarily has to do with regulations for *Cryptosporidium* and disinfection by-products); and (3) to increase system yield by allowing greater drawdown of the South Fork Tolt Reservoir.

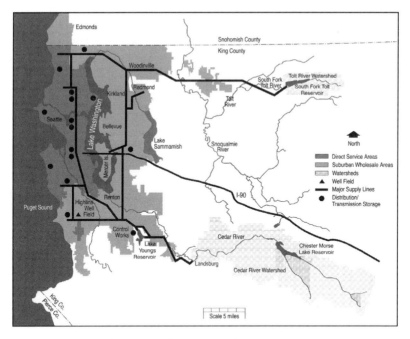

Figure 12.1   Seattle's regional water supply system.

Upon completion of the siting assessment, pilot studies, and environmental reviews, Seattle began to seek the most cost-effective and reliable approach to completing development of the project. The traditional model for developing and operating major treatment plants in the United States has been the design-bid-build-operate model. This model involves three basic actors—the designer, the constructor, and the operator/agency responsible for supplying water from the plant to its customers—and a serial approach to the design, construction, and operation steps. In the first step in this process, a private design firm is hired under contract. Their responsibilities include determining plant requirements for the municipality, including (implicitly) many of the risk elements of the problem. They then design the plant and develop bid specifications. In the second step, bids are tendered to meet the specifications and the low bidder is tasked with the job of constructing the plant in accordance with the specifications. In the third step, the municipal operator is responsible for operating the plant in accordance with performance standards.

While the conventional design-bid-build-operate approach to public works contracting is well understood, and most appropriate for many applications, its shortcomings are numerous and well documented:

1. Segmenting the design and construction process often leads to excessive design costs, due to undesirable risk allocation and mixed incentives, and a failure of the designer and builder to collaborate in ways that would lead to reductions in construction costs.
2. Implementation budgets expand unpredictably, and the low-bid method of selecting constructors has high potential for costly change orders.
3. Schedules become extended due to difficulty in managing and staffing large, technically complex project-delivery systems.
4. Responsibility for cost overruns associated with design or construction are typically borne by the owner, and risks associated with the failure of the plant to operate in accordance with the municipality's intent are often ultimately borne by the municipality/operator because of difficulty in establishing liability between the designer, contractor, and municipality.
5. The low-bid, fixed-price method of selecting constructors heightens the risk of performance failures, particularly in a circumstance of ambiguous responsibilities for a failure.
6. Municipalities are not always the most qualified party to implement new technology to mobilize, train and start up new plants or to operate them most cost effectively.

In today's dynamic environment, most of the traditional roles are beginning to be reconsidered. A changing point of view about municipal versus private sector capabilities, the increasing prominence of integrating design and construction processes, and the emergence of capable private operators have all contributed to a broader set of opportunities for formulating a new and better approach to the conventional model.

2. DEVELOPING A NEW PROJECT APPROACH

Seattle's initial willingness to consider an innovative approach to developing the Tolt Filtration Plant, and the ultimate selection of the DBO model derived from the overall effort to find more efficient ways of doing business. A variety of specific factors led to the viability of using an alternative approach over the traditional approach to design, construction, and operations of the Tolt facility: Seattle's lack of experience in filtration plant design, construction and operation; a growing body of demonstrated private experience in plant operation; the emergence of capable design, build, and operations consortia; potential cost savings associated with an alternative approach; a statute allowing design-build procurement within Washington State.

In implementing the DBO model, Seattle sought to ensure that the Tolt facility would perform to a standard at or above conventional implementation and operation, but at minimum cost to the City within a preferred risk allocation framework. Seattle's process of needs and objectives setting resulted in a "project philosophy" built around the following concepts:

1. That the City describe its needs in terms of facility performance requirements and other key outcomes, and allow vendors to propose solutions to achieve these requirements;

2. That technological innovation be encouraged within the range of proven technology, and competition be used to achieve *both* technical innovation and lower cost;

3. That an integrated contractual responsibility be provided for design, construction, and long-term operations with a single point of accountability, and that all phases of the contract be secured by a single financially strong guarantor;

4. That risk be allocated between the City and the vendor in a manner that minimizes overall project costs (i.e., assign the risk to the party best able to manage it);

5. That a long-term service agreement be established for operations, thereby assuring performance of the facility beyond construction completion; and

6. That, as the entity ultimately responsible for providing a safe and reliable water supply, the City should maintain ownership of the facilities (an economic analysis and market feedback also concluded this to be the least-cost approach over the long term).

2.1   A Decision Process for Approval of DBO

In addition, having made a preliminary decision to pursue a DBO approach for development of the Tolt Filtration Plant, it was clear to the project team and elected officials that there was not a sufficient base of experience from which to firmly project cost savings or market interest in the project. The City was interested in proceeding with the DBO alternative only if sufficient savings could be obtained from capable teams, while meeting Seattle's stringent performance requirements. At the same time, the vendor community was concerned about a fair and efficient decision process, particularly considering the large up-front expense that would be required to develop proposals.

These factors led the City to develop a process by which final adoption of the DBO approach would be subject to a *threshold determination of its viability*. This process was intended to protect both the interests of the City and those of prospective proposers. Under this threshold determination process, the City agreed to proceed with the DBO approach only if one or more proposers could guarantee:

1. Cost savings of at least 15% compared to the life cycle cost of the benchmark plant that Seattle would have had to build under conventionally sequenced design, construction, and City operation process;
2. Water quality, quantity, and cost performance through the application of proven technology; and
3. Agreement with the City's published risk-allocation requirements.

2.2 Communications Protocol

Finally, by early 1996, it was apparent to the City that proposal teams were already being created for the Tolt project that consisted of consortia of designers, construction contractors, contract operators, and support firms. The City was committed to a fair, open process for interested parties to receive information about its DBO projects and the competitive solicitation processes. For the Tolt DBO project, the City issued a series of quarterly project updates that were mailed to all parties who had expressed interest in the project or who were on existing City consultant mailing lists. The first of the project updates outlined the solicitation process, schedule, and communications protocol. Because of the unique nature of the project, its high industry visibility, and the fact that new firms or teams were interested in making proposals to the City for the first time, the City committed to:

1. Achieve the fair distribution of relevant information to anyone interested.
2. Avoid the appearance of unfair, or "insider" information going to any firm.
3. Provide regular informational updates about the project.
4. Provide prompt, thorough responses to parties with questions or concerns on any information.

There was a single contact person identified within the City for communication with interested parties, response to questions, and receipt of requests for information and clarifications. Reference materials, including policy and technical background documents were made available to interested parties. In addition, the City initiated the use of an Internet web site for making project information generally available.

3. IMPLEMENTING THE NEW DBO APPROACH

Seattle implemented the Tolt project in phases: defining the project, selecting a DBO partner, negotiating a long-term service agreement, designing and permitting the project, and completing the project. Following is a brief description of each of these important phases.

3. Step One: Defining the Project

In late 1996, the City selected a Support Services Consultant to assist in the selection of a Tolt DBO partner and to provide additional technical expertise necessary to define the project and related performance requirements. A team of R. W. Beck and Malcolm Pirnie was selected for this role bringing technical expertise related to the design of major treatment facilities, and local aspects of project permitting, as well as expertise in the area of alternate contracting processes. This augmented a very high-level City project team, comprised of water quality, engineering, and policy and administrative leadership in SPU. Additional specialty skills were brought on board with representatives of the City Council Staff, Mayor's office, and other independent technical, legal, and financial advisors. From the beginning, this high-level involvement and process buy-in throughout the City's bureaucracy was a major factor in the success of the project.

Project definition for the Tolt project included establishment the specific project objectives, development of project performance requirements, and definition of a desired risk allocation.

General Facility Objectives

1. The facility must provide the necessary unit processes, process control, monitoring and control, and system redundancy in order to continuously meet the treated water quality requirements, and structures, piping, and equipment must be designed such that they can be maintained while the facility continues to meet the supply and quality performance criteria.
2. The facility, project features, and operations must be environmentally compatible with the site(s), and the site layout must provide space and flexibility for future process and capacity additions, taking into consideration impacts of future expansion on wetlands and other sensitive areas.
3. Project features must be designed and functions must be grouped together to provide for easy and efficient operation, monitoring, and collection of operating data.
4. The facility layout must provide for efficient and safe access into and throughout the site.
5. The facility must be designed and constructed using materials and equipment consistent with a 50-year life for buildings and structures.
6. Materials and equipment must be of sufficient quality to assure long life expectancy, a low incidence of failure, manufacturer support and service, and compatibility with Seattle's system.
7. Equipment, buildings, landscape, and other facilities must be designed to require infrequent maintenance while presenting a well-kept and pleasing appearance.

The Facility Design and Construction Objectives

1. Optimization of present and future water treatment processes.
2. Minimization of design and construction costs.
3. Use of existing facilities and infrastructure where beneficial and feasible.
4. Sound design and quality construction for long-term operational reliability.
5. Design for optimal power efficiency.

6. Sustainable design to ensure prudent, long-term use of resources.
7. Development of a project schedule that achieves project efficiencies and quality.

Plant Operation and Maintenance Objectives

1. Integrated operation with the City's water system.
2. Reliable, efficient water treatment services to the public.
3. Ongoing compliance with all applicable regulations.
4. Effective response to both standard and unforeseeable or emergency operating situations.
5. Achievement of the lowest reasonable operational costs.
6. Ongoing minimization and mitigation of environmental impacts.
7. Facility operations that obtain optimal power efficiency.
8. Prudent management and protection of public resources.
9. Maintenance of highly qualified and trained Treatment Facility operating personnel.
10. Provision of necessary facility maintenance and rehabilitation to protect the City's investment and assure its functionality over the useful life of the project.

A wide range of issues were addressed in the course of developing the performance specifications, contract elements, and risk allocation matrix. The major allocation of risk is shown in Table 12.1. Seattle carefully evaluated risk allocation and circulated a summary to proposal teams for review and comment.

Table 12.1 Risk Allocation Matrix

Risk / Responsibility	City	Company
Financing	X	
Site acquisition	X	
Permitting		X
Design		X
Construction		X
Operation		X
Raw water quality	X	
Treated water quality		X
Change in law	X	

During this phase of the project, regulatory agencies were consulted to discuss the DBO approach and encourage a cooperative arrangement for successful project implementation. It was determined that flexibility would be necessary relative to typical permitting processes, and through a consultive process, it was determined that sufficient flexibility existed. This was a significant turning point for the project.

3.2   Step Two: Selecting a DBO Partner

Step two included conducting the request for qualifications (RFQ) and request for proposal (RFP) processes, including development of the shortlist and final selection.

3.2.1   Conducting an RFQ/RFP Process

The City requested that those submitting statements of qualifications (SOQs) comment on certain concepts presented in the RFQ, including treatment processes, financing options, risk allocation, guarantor requirements, and confidentiality restrictions. The City's preliminary risk-allocation matrix—a much more detailed version than that shown in Table 12.1—was included as an appendix to the RFQ.

In addition, the RFQ discussed the concepts of the threshold decision to proceed and benchmark comparison, listed the evaluation criteria, and specified that the City would pay an honorarium, or stipend, in the amount of $100,000 to firms submitting proposals responsive to the RFP who were not awarded the DBO contract. The RFQ also stated the City's intent to require a single, financially strong guarantor for all phases of the project. The primary reasoning behind this was that the City did not want to find itself in the center of disputes among the members of the DBO consortium. Continuous operations are imperative, and resolution of problems must occur in a manner that would insure this.

The RFP presented proposal requirements and provided proposers with information to facilitate preparation and submission of proposals, and included 25 schedules that set forth baseline conditions and requirements for the treatment process, facilities, and operations. The City issued several addenda to the RFP and provided written responses to questions submitted by proposers. Responses to questions were issued in writing to all proposal teams.

The Tolt RFP required proposers to submit two separate proposals. One was intended to result in a facility that met both currently enacted regulations and those that could be reasonably anticipated as of the projected facility on-line date, and the other was to result in a facility that met current regulations, took a more expansive outlook on potential enhanced surface water treatment rule requirements, and addressed enhancements that could be desired in a long-term operational period for the facility. It was expected that the second proposal would require additional treatment processes, such as ozonation or membrane filtration.

3.2.2   Making Final Selection

Seattle's DBO proposal evaluation was based partly on financial criteria (40%) and partly on the technical strength of the proposal and team qualifications (60%). The financial criteria included cost effectiveness and financial qualifications. Team and technical criteria included project implementability, technical reliability, technical viability, environmental considerations, proposer and staff past performance, and women and minority business enterprise (WMBE) utilization.

The evaluation committee included individuals with expertise or responsibility for utility and project management, water quality facilities development, DBO contracting, financial analysis, permitting, engineering, and construction. Technical teams consisting of evaluation committee members and advisors were created for detailed investigation into the various proposal components.

In addition to reviewing technical information contained in the proposals, the evaluation committee conducted a process of thorough reference checking, site visits, interviews, financial strength assessment, and request for final offers from each proposal team.

3.3   Step Three: Negotiating a Long-Term Service Agreement

The evaluation committee reached a unanimous recommendation to select the contract team for initiation of negotiations. The Committee also determined that their recommendation would not be materially affected by the choice of whether to develop the base or alternate facility.

In March 1997, the Seattle City Council formally authorized SPU to proceed with the DBO process based on their findings that all Best and Final Proposals met the City's technical requirements and substantially exceeded the required 15% savings over the benchmark. Based on the Tolt predesign work completed in 1995, the City anticipated total costs of $156 million dollars (net present value, 1998 basis)—construction costs of approximately $100 million dollars and operations and renewal costs estimated at $56 million dollars over a 25-year period. This estimate was for a filtration-only facility (no ozonation or sedimentation), and was based on awarding a competitively bid construction contract with an independent design engineer's specifications, and City operation.

The cost range for the four Best and Final A Proposals was $83–$108 million and for the B Proposals was $103–$124 million (net present value, 1998 basis). Although the Proposal B facility was not used to create the benchmark cost estimate of $156 million, Seattle estimates a Proposal B facility would cost $171 million using a conventional design-bid-build process. (This estimate was produced by addition of the Proposal A benchmark of $156 million with the average cost differential between Proposal A and Proposal B facilities received from the proposers.) Thus, with the final negotiated contract valued at $101 million, the City estimates savings of 40% ($70 million) over a conventional approach to design, construction, and operation of a similar facility.

The objective of the City at the initiation of negotiations was for each party to develop a strong understanding of the others' interests and to develop a basis of trust that would sustain the long-term partnership. The negotiations were conducted in three phases. The first phase consisted of the City's identification of major issues and concerns, and the resolution of these issues through the initial collaborative work session. The second phase involved several intensive joint work sessions for facility refinement, and exploration of various facility enhancements. This phase proved to be very beneficial to the project, resulting in an improvement over the proposed facility *and* a reduction in cost. Facility improvements included enhanced facility redundancy, increased water storage capabilities, greater flow commitments, more incentives for efficient chemical use, and a more comprehensive definition of raw water quality parameters. The third phase included finalization of all service agreement terms and conditions, and other related documents. As contract negotiations were completed, CDM Philip and SPU had developed a solid basis of trust, cooperation, and partnership from which to launch project implementation. In May 1997, the City Council Utilities and Environmental Management Committee advanced an ordinance authorizing SPU to enter into a DBO service agreement with CDM Philip for the Tolt Treatment Facilities.

3.4   Step Four: Completing the Project

3.4.1   Design and Permitting

The service agreement designates the period of time between contract execution and initiation of construction as the "Development Period." During this time, Azurix CDM (CDM Philip was renamed Azurix CDM in January 1999) prepared site plans and permitting documents, submitted permit packages, and completed the facility design (to the degree necessary to receive site permits). These activities were completed and site clearing began in May of 1998. The accomplishments of the development period were an extraordinary attribute to the cohesive structure of the Azurix CDM team. The contractor and the designer worked with the permitting team to develop a streamlined approach and address concerns and potential concerns of the regulatory agencies—both for the site and building permits and the department of health approval.

The City submitted several hundred comments to the contractor related to each package. The comments were not representative of concern with the quality of design or design progress, but rather, were offered in the spirit of partnership and information sharing. However, the City did not "approve" the facility design, as is typically done on a conventional design-bid-build project, because of the risk and responsibility for facility operations as a contract obligation. In fact, problems arising out of faulty design or difficult construction are the responsibility of the contractor, and thus the contractor maintains the obligation to correct problems. For example, during the construction of the filter underdrain system (designed by a supplier), the fiberglass forms used to support the concrete during curing failed, causing the floor of the filter beds to become slightly deformed. If left uncorrected, the potential resulting problems would have ranged from barely noticeable operational inefficiencies to major operations and maintenance problems, and possibly even system failures. Any of these operational complications would have been the responsibility of the contractor. Clearly, a benefit of the DBO contracting approach is that, when such situations arise, the contractor must consider long-term implications when deciding on an action to take. Azurix CDM chose to remove and replace the substandard filter bed floor. If this situation had occurred in a conventional design-bid-build contracting project, in all likelihood, it would have resulted in an unhappy client, cost and schedule overruns, and probably litigation.

3.4.2   Construction

Site clearing was initiated at the beginning of what is typically the less rainy season in the Puget Sound region—late spring. Unfortunately, the summer and fall of 1998 combined to become one of the wettest seasons in recorded history for the area. Because no cost or schedule relief was afforded the company due to rain, the contractor had incentive to accommodate the situation with innovative approaches to reducing sediment runoff from the site and to keeping dry spaces for concrete curing.

To further complicate the first season of construction, culturual artifacts were located on one of three distinct construction areas. Such a finding is explicitly listed as an "uncontrollable circumstance" in the service agreement. For this reason, cost impacts of the artifact recovery were borne by the City, and fortunately, the location of the finds was in an area for which construction was not on the critical path.

The City's approach to design and construction oversight for the project was to ensure compliance with the service agreement with primary focus on areas that:

1. Have plant longevity considerations
2. Could cause major maintenance problems
3. Would be extremely difficult to repair, replace, or rebuild
4. Have major facility reliability considerations

Design review included a pilot testing confirmation process, and review of four major design packages. The design review was conducted to confirm conformance with service agreement, and not to convey design preferences.

3.4.3   Start-Up and Operations

During December of 2000 and January of 2001, the contractor and operations staffs worked together with SPU for transition of the facility to operations. In early February 2001, the facility successfully completed the very demanding 14-day acceptance test. The test is a contractual obligation for the DBO vendor to show—on a 24-hour per day basis—that the facility can meet various challenges while complying with the rigorous performance requirements.

4. LESSONS LEARNED

Now that Seattle has successfully implemented the Tolt project, and developed new practices surrounding this effort, a number of lessons emerge. The following are some of the key focus areas where attention should be focused.

4.1 Carefully Consider Your Desired Degree of Flexibility

Seattle described its water quality and supply needs in the RFP and developed performance requirements detailing expected outcomes and then asked proposers to provide solutions for the achievement of these standards. Thus, in a DBO context, the service agreement is primarily a *performance-based contract*. Our agreement stipulates output requirements for water quality and quantity, but does not provide detailed specifications for facility design and construction. Water quality standards are higher than those required by current and anticipated state and federal regulations, and supply obligations are conservative enough to allow the City ample flexibility for conjunctive use with its other supply sources. For each project, the owner must decide the degree to which it is comfortable with a purely performance-based approach versus prescribing specific features or methods to be employed. Following are service agreement features we considered critical to us, along with an explanation of how they were addressed.

4.1.1 Liquidated Damages and Incentive Payments

One of Seattle's fundamental water supply obligations is to provide safe drinking water to its direct service customers and wholesale purveyors. In this context, a major challenge in crafting the RFP and draft service agreement was to definitively convey the essence of this public trust to the contractor. The City used the service agreement as the primary method of meeting this objective. The service agreement includes modest financial incentives for exceptional facility performance and substantial liquidated damages for nonperformance (addressing regulated and enhanced water quality and water supply). Eighteen treated water quality parameters are to be monitored for which noncompliance liquidated damages vary, depending on the water quality or public health consequences and severity of the infraction. All of the more serious infractions include grounds for termination after a specified period of noncompliance. In addition, the contract includes significant liquidated damages for supply of unacceptable treated water, which may or would result in drinking water regulatory actions up to and including a "boil water" notice.

Incentive payments are to be made to the contractors by the City (in the amount of $5,000 per month) for any calendar month in which the total number of noncompliance events is zero. We consider financial incentives to be a significant and effective tool to bring about the type of behaviors we desire.

4.1.2   Project Liability

The Tolt Service Agreement includes no provisions limiting the liability of the company or the guarantor. In Seattle's view, this helps assure that the facility will be designed and constructed to the highest standards, and that the company will reliably operate to meet the City's high standards of performance. It has been an important principle to the City that our DBO partner, and ultimately the financial guarantor, accept full responsibility for any damages resulting from its negligent actions or inactions, including injuries or death to consumers and damage to the facility and the property of others that may result from the acts, errors, or omissions by the company occurring in connection with, or arising out of, the design, construction, start-up, testing, operation, or maintenance of the facility.

As a matter of public policy, it has been important to the City that it not share liability for negligence by the DBO Company. Some operating companies are reluctant or unwilling to enter into agreements that expose their entire balance sheet to this type of stipulation.

4.1.3   Structure of Capital Payments

The capital payments to be made during the development and construction phases of the project were structured in two different methods. Payments to be made during the development phase—the time during which contractor completes design and permitting for the facility—totaled no more than 2% of the capital component of the contract. The remainders of the capital payments were to be paid out based on a predetermined milestone payment schedule, with 5% retainage held out until final acceptance testing is complete.

4.1.4   Structure of Operation and Maintenance Service Fee

The structure of the service fee is a very important component in meeting the City's objective of allocating incentives appropriately. The service fee was designed to mirror the City's foremost water supply objective of reliable delivery of high quality water to customers. Other objectives, including minimization of chemical use and environmental impacts, and aggressive maintenance practices for the long-term health of the facility, were conveyed in the service fee structure as well. The operation and maintenance (O&M) service fee is divided into two primary components—a fixed base fee and a variable component. The fixed base fee includes the base operations and maintenance service charge (for labor, facility maintenance, including costs for equipment repairs under $10,000, and all other nonchemical and non-pass-through costs of operating the facility) and the base chemical charge (the negotiated fixed cost of chemicals used in the treatment process). The variable component consists of pass-through costs primarily for insurance and electricity, which are "passed through" for City reimbursement, without markup. Careful attention to service fee elements and incentives can be useful in assigning risks to parties best suited to managing them and in creating cost efficiencies.

4.1.5   Renewal and Replacement

An important City objective for the facility included the requirement that it be designed and constructed using materials and equipment consistent with a 50-year life for buildings and structures, and that materials and equipment be of sufficient quality to assure long life expectancy, and a low incidence of failure. Another City objective was to assure that the facility be maintained over time to a consistent high standard of appearance, longevity, and functionality. Accordingly, the City wanted to ensure, through a carefully crafted service agreement, that: (1) the initial design, construction, and equipment specification be conducted with long life expectancy; (2) preventative maintenance be performed to specified standards; (3) appropriate investments would be made in minor and major equipment replacement, both on an ongoing basis and especially toward the end of the contract term.

The City's approach in meeting the first of these objectives was to evaluate proposals based on design and equipment quality. In addition, proposal teams were evaluated based on past experience in development and construction of high quality facilities.

The approach in meeting the second objective was to require the contractor to include in their O&M plan schedules listing all major items with replacement frequency and type of each, and to require that the facility be maintained to that schedule unless the City authorizes otherwise. In addition, a requirement of the service agreement was that third-party facility audits be performed semiannually throughout the term of the agreement in order to assure proper maintenance, repair, and replacement.

The approach for meeting the third interest was to require proposers to submit, as part of their proposal, a plan for major equipment renovation throughout the term of the agreement, including preventative maintenance plans, related operator training, and a Renewal and Replacement Plan for replacement of major items (over $10,000) over the term of the service agreement. The City will make annual payments as designated in the service agreement to a Renewal and Replacement fund to cover major equipment cost replacements that have been identified in the plan. Higher or unforeseen maintenance or replacement costs are the responsibility of the contractor, and if the contractor maintains equipment to lives longer than those specified in the plan, cost savings will be shared between the City and the contractor. Repair and replacement of items of $10,000 or less are considered ordinary maintenance activities and are covered within the fixed base fee.

The City's primary internal staffing objective during the construction phase of the project was to maintain an owner's presence on the job, ensure implementation by the contractor of their assured quality/quality control (AQ/QC) program, and to be available to assist in interface issues if necessary. The City expected that its role on site would evolve as experience dictated, and was mostly interested in observing the execution of quality construction practices.

In fact, problems arising out of faulty design or difficult construction are the responsibility of the contractor, who is obligated to correct problems. The problem that arose during the construction of the filter underdrain system (designed by a supplier), when the fiberglass forms used to support the concrete during curing failed and caused the floor of the filter beds to become slightly deformed, has already been discussed. Clearly, the DBO contracting approach benefited the City in this instance. After considering the long-term implications of the possible solutions, Azurix CDM chose to remove and replace the substandard filter bed floor. The project schedule was maintained, and substantial completion was accomplished on December 1, 2000.

4.1.6   Address Public Versus Private Financing Issue

From the outset, the City desired to own the Tolt facility and believed that utility tax-exempt financing would achieve a lower cost of capital for the project. The City did invite proposers, as part of the RFP process, to submit alternatives for private financing or suggest other alternatives. No workable responses were received, and the market supported our basic assumptions. Private financing may be appropriate in other situations where the owner does not regularly finance large capital construction or where the owner does not have sufficient debt capacity to cover capital costs.

4.1.7   Resolve Potential Labor Issues

Seattle is fortunate not to have had any major labor conflicts during the procurement process or execution of the contract. This is primarily because new technology is being introduced in this facility, because existing employees are not being displaced, and because from the beginning, the city worked closely with unions to explain the project and labor implications. The contract includes a provision for joint hiring of operators prior to initiation of operations. In the interim, between contract execution and start-up of facility operations, this provision allowed Seattle to fill open operator positions with personnel selected jointly with the contractor and with the intent that these operators would transition to the new facility once it became operational. The city also worked closely with union officials and the DBO contractor to assure the salary/benefit package afforded treatment operators met or exceeded that which would have existed for regular City employees. Our experience suggests these types of efforts and provisions are necessary to assure a smooth transition to a DBO approach.

4.1.8   Assure Owner Oversight During Operations

It is essential for the owner to maintain an appropriate level of oversight during the operations phase, including monitoring of key performance criteria, review of operations, attention to regular reporting, involvement in facility planning and maintenance, and review of personnel actions and transitions.

For the Tolt project, water quality standards are strictly defined in the service agreement and are higher than those required by state and federal law. The City maintains a "window" into operations of the facility through monitoring of instrumentation and control at Seattle's Operations Control Center, and through sampling and review of operational reports. City staff can participate in facility operations training at the discretion of the City, and monitor water quality parameters within the facility at its discretion.

4.2   Create and Maintain an Effective Teaming Relationship

While the owner and vendor team have distinct and separate roles and responsibilities, all phases of the project can be more effectively conducted if a solid partnering relationship is formed during project negotiations. If done well, this allows the parties to build on a quality service agreement, identify common interests, and understand divergent goals and objectives. Ongoing communications and issue resolution must occur, and tools must be created to manage the complex DBO relationship.

Seattle and Azurix CDM conducted formal partnering initially and then maintained a structure for regular communication and reporting. The entire team identified critical success factors and continually monitors them with open meeting dialogue to address issues and concerns, and celebrate successes.

5. CONCLUSION

A frequently expressed concern when considering alternative procurement strategies is that a public–private partnership like the DBO process may lead to a loss of control by the owner and a sacrifice of quality. Seattle's DBO contracts are structured to allocate each responsibility and risk to the party best able to manage it. We endeavored to build provisions into the service agreement that fairly allocate the risks, clearly provide guidance regarding the issues to be managed, and actively monitor performance against the expectations. *Appropriate allocation and fairness* are the keys. A service agreement must be capable of applying to a wide range of real situations, many of which cannot be anticipated prior to project initiation.

The backlog of water quality and other infrastructure requirements makes it essential for public utilities to deliver high-quality projects at the lowest life cycle costs. New contracting practices and performance methods are being implemented for procurement, selection, risk allocation, permitting, design, construction, transition to operations, and ongoing maintenance of new facilities. Thus far, our experience demonstrates that DBO projects are capable of achieving high-quality results at a lower life cycle cost where a well-planned process is used and allocation of risks, roles, and responsibilities is thoughtfully established.

Early development of an overall project approach and process is crucial to the success of any project, as is thoughtful development of project procurement documents, legal agreements, and technical requirements. Our objectives determined the specific approach, project staffing, decision-making and communication process, evaluation criteria, and implementation strategy. However, in the use of an alternative approach to project development, like the DBO model for the Tolt project, it is also critical that there be a very well-designed and well-executed process of implementation. Since the alternative approach is new, utility staff and management, elected officials, and the public at large are not entirely sure of all of the steps. The process provides a roadmap of goals, activities, expectations, decisions, and points of involvement.

In reflecting on the Tolt DBO procurement process, Seattle can turn to the underlying objectives and core features of the project in evaluating its success. The City maintains ownership of the facility, and the contract includes strong financial guarantees for the three phases of the project. The service agreement appropriately allocates risk, is fair to both parties, and sets forth a clear roadmap for project implementation. The length of the agreement is important in ensuring the viability and long-term reliability of the process design and facility construction. The contract is performance based, but is clear about those water quality, supply, maintenance, and engineering standards that must be incorporated into the facility. The performance-based nature of the contract has allowed proposers the rare (for public works projects) opportunity to do a project in the most efficient and practical manner—fostering cooperation among site permitting specialists, designers, constructors, and operators, and allowing them to push the design envelope and be innovative in their proposals. A single point of accountability has been established within the contracting team, rather than segmenting responsibility and liability. This same consolidated accountability is driving the team to coordinate design, construction, and operations phases to achieve better quality, a faster schedule, and lower cost.

We observe a market that is ready to develop projects using a DBO process and similar approaches. Since many consider Seattle's Tolt project to be a test case (or a controlled experiment) for implementation of projects using alternative contracting approaches, it is important to recognize that the keys to Seattle's success with the development of the Tolt DBO project have been multifaceted.

The project's diverse and multidisciplinary team was critical in bringing in diverse perspectives and better solutions to complex issues. The project philosophy, developed early in the process, was key to balancing performance versus prescriptive interests, choices between innovation and traditional approaches, and trade-offs between control and costs. The project partnership between the utility and Seattle elected officials ensured that stakeholders were on board throughout the process, and clear points of involvement for elected officials were identified early in the process.

A relentless and continual emphasis on maintaining process integrity provided the assurances needed by the private sector in participating in the competitive process. Thoughtfully developed performance specifications, contract incentives, technical requirements, and other contract provisions have provided the basis for an approach of *trust, cooperation, and partnership* with the contractor for project implementation.

READINGS

Westerhoff, G. P., D. Gale, P. D. Reiter, S. A. Haskins, and J. B. Gilbert, "The Changing Water Utility—Creative Approaches to Effectiveness and Efficiency," American Water Works Association, Washington, DC, 1998.

Kelly, E. S., S. Haskins, and P. D. Reiter, "Implementing a DBO Project—The Process of Implementing Seattle's Tolt Design-Build-Operate Project Provides a Road Map for Other Utilities Interested in Alternative Contracting Approaches," *Journal AWWA,* June 1998, pp. 34–46.

Chapter 13

The Challenges and Benefits of Privatizing Wilmington's Wastewater Treatment Plant

A Case Study

James H. Sills, Jr.

This chapter will serve as a case study analysis of the dynamics associated with the three-year (1995–1998—see Table 13.1) privatization of the management of the Wilmington (Delaware) Wastewater Treatment Plant. The principal intent is to provide information and some direction to those persons who may be interested in learning how the process of the privatization of wastewater treatment plants might impact governmental planning and decision-making in their local communities.

It should be stressed at this point that privatization can have several definitions. For the purpose of this chapter, the word privatization may refer to the handing over of either ownership, management, operations, or capital investment to the private sector. When appropriate, the word privatization will be clarified for the benefit of the reader. If the word is not otherwise defined, it is to be assumed that privatization refers to the sale of the wastewater treatment plant.

The factors that led the City of Wilmington to consider plant privatization and to the political, legal, and regulatory challenges encountered dur-

James H. Sills, Jr., was formerly mayor of Wilmington, Delaware. He is Associate Professor Emeritus at the University of Delaware. He was formerly cochairman of the Urban Water Council of the U.S. Conference of Mayors.

Reinventing Water and Wastewater Systems: Global Lessons for Improving Water Management, edited by Paul Seidenstat, David Haarmeyer, and Simon Hakim.
0-471-06422-X   Copyright © 2002 by John Wiley & Sons, Inc.

Table 13.1   Wilmington: A Privatization Timeline

Summer 1994	The City begins discussing possibility of privatizing the ownership or operations of the Wastewater Treatment Plant.
August 1994	City Council passes resolution urging administration to study possible privatization of Wastewater Treatment Plant.
September 1994	Mayor Sills announces that the City will explore options for privatizing the Wastewater Treatment Plant.
January 1995	A consultant's report suggests that privatization based on the Franklin, Ohio model (sale and repurchase after 20 years) could benefit the City.
May 1995	A two-part Request for Proposal is published.
June 1995	Proposals from four companies are received.
July 1995	New Castle County proposes to buy the Wastewater Treatment Plant.
August 1995	City reviews the four proposals and ranks Wheelabrator EOS first.
September 1995	City begins negotiations with Wheelabrator.
October 1995	A taxpayer suit challenges the City's selection and decision process.
November 1995	Delaware Chancery Court rules that the City's procurement and selection process adheres to State law and makes "perfect sense."
December 1995	City performs further analysis of options while negotiations with Wheelabrator continue.
January 1996	Analysis suggests that a long-term management agreement provides the best economic benefit to the City.
February 1996	New Castle County solicits proposals to privately operate the Wastewater Treatment Plant.
March 1996	Mayor Sills announces in his budget address, "Based on analysis of the City's options, I am convinced that the best option for the City and our customers is to privatize operations while retaining ownership of the plant."
	County Executive Dennis Greenhouse states his commitment to working out a new City/County service agreement in the next several months. New Castle County staff is included in review of proposed agreement.

Table 13.1 *Continued*

June 1996	State Senator Tom Sharp introduces Senate Concurrent Resolution 68, urging the City and New Castle County to wait until January 1997 before concluding an agreement on the Wastewater Treatment Plant. SCR 68 passes both houses of the General Assembly.
August 1996	City officials meet with EPA to discuss review of the proposed agreement.
October 1996	The City joins in lobbying the federal Treasury Department for changes in tax treatment of municipal bonds to make contracts with private operators easier to implement.
December 1996	Mayor Sills is appointed Cochair of the Urban Water Council of the US Conference of Mayors.
January 1997	New Castle County Executive Tom Gordon promises cooperation with the City on sewage issues.
February 1997	US Filter announces that it will acquire Wheelabrator EOS.
October 1997	Negotiations are concluded on a proposed service agreement with US Filter EOS. The proposed service agreement is presented to City Council for consideration.
	The City and New Castle County agree on the substance of a new City/New Castle County service agreement.
November 1997	Mayor Sills meets with EPA officials regarding federal review of privatization proposals.
	The final version of the City/New Castle County service agreement is presented to both governments for review and acceptance.
December 1997	Mayor Sills signs privatization contract with US Filter EOS.

ing the privatization process are discussed in detail in the pages to follow. Examples of these challenges encountered by the City will be developed. These challenges involved: successfully defending a legal challenge to our procurement process in the Delaware Chancery Court; working with the United States Conference of Mayors to lobby the Federal Treasury Department for changes in tax treatment of municipal bonds in order to make long-term contracts with private operators easier to implement; and, working with the US Environmental Protection Agency (EPA) to create a fair review process for wastewater treatment plant privatization and sale projects.

Before delving into any particulars, it is important to have a basic understanding about the City of Wilmington. Wilmington, a city of nearly 73,000 residents located in the northern county of Delaware, is central to several large northeastern metropolitan areas; yet, possesses many of the qualities of small towns and medium sized cities. The City, which encompasses 15.86 square miles, is the largest city in the State of Delaware and is the focal point for most of the State's commerce, economic, and governmental activities. Included among this mix are several notable business firms that call Wilmington their international headquarters; among these are the worldwide chemical firm, the Dupont Company, and the credit card company, MBNA America.

Wilmington is home to less than 10% of the state's population, but generated 23% of the wages in Delaware from July 1997 to June 1998. An estimated 75,000 people work in Wilmington on a daily basis.

Wilmington's sewer and water systems measure over 500 miles combined. The wastewater treatment plant, constructed in 1954, has a capacity of 105 million gallons per day and serves approximately 400,000 City and County residents combined, or nearly two-thirds of the entire Delaware state population.

The treatment plant, originally constructed in 1954, was constructed and expanded at a cost of $93,780,237. Of this amount, $37,680,237 was provided by the City, $11,200,000 was provided by the State, and $44,900,000 was provided by the Federal Government. Prior to privatization of the wastewater treatment plant, it had an annual operating budget of $18.5 million and had a work force of 55 employees. The plant had generally enjoyed a favorable reputation for the scope and quality of its services over the many years of its operations. There were few publicly expressed citizen concerns and complaints.

1. THE ROAD TO PRIVATIZATION

1.1 Circumstances Leading to Privatization

The feasibility of privatizing Wilmington's wastewater treatment plant, and the final decision to privatize, were largely brought about by circumstances connected with a city facing an $11 million annual operating deficit and a more than 40-year-old wastewater treatment plant with a failing infrastructure. While certain periodic capital upgrades had been made to the wastewater treatment plant during its 44-year history, other improvements, along with daily maintenance and management, continued to be necessary and costly realities for the City of Wilmington.

Relative to finances, it is to be stressed that, with the beginning of a new City administration in January 1993, the City of Wilmington had posted annual operating budget deficits in three of the previous four fiscal years, and had come to rely on short-term borrowing to meet its operating expenses. Due to these budget problems, the City had forgone its regular biannual capital bond issue in fiscal year 1994, and the City's infrastructure needs were not being met. The inability to resort to the bond market to borrow money for capital improvements was forcing the City to make painful choices as to which infrastructure needs would be met and which ones the City would forgo.

In addition to these interior constraints, the City of Wilmington was also facing an exterior barrier to alleviating its deficit problem: The Clean Water Act of 1972. This federal act, which required cities to improve their water treatment facilities, placed added federal pressure on local governments to enhance their wastewater treatment facilities, and to improve the quality of water being released from such plants back into local streams and rivers. The lack of federal funds, however, to assist cities in meeting the required federal standards caused an additional budgetary constraint on the City of Wilmington.

Monetary and external federal factors were not the only pressures affecting Wilmington. The City, like many other urban areas with a disproportionate number of minority, elderly, and low-income residents, had lost population in the 1960s and the 1970s. The City began to grow again in the 1980s and the 1990s; but the need to provide expanded services for its residents was also growing.

The City's necessity to reduce the operational costs of its services to its budgeted revenues and the need to stop annual budget deficit spending became primary concerns of a new City administration in 1993. In this context, privatization via the sale of the wastewater treatment plant or through operational privatization became two viable options in restructuring city government.

1.2   The Privatization Process

In keeping with the aforementioned needs and problems, Wilmington had three fundamental objectives in exploring privatizing the City's wastewater treatment plant:

1. Utilize private funds to make much needed capital improvements to an aging plant, which had a net book value of approximately $52 million dollars in 1994.
2. Improve management operations of the plant and reduce the annual cost of service, which entailed an annual operating budget of $18.5 million, and a staff of 55 workers.
3. Replenish the City's cash reserves by securing a major up-front payment for the sale of the plant.

With these objectives in mind, the City retained a nationally recognized consulting firm, Raftelis Environmental Consulting Group, to examine the feasibility of privatizing the plant. The consultant's report, completed in January 1995, concluded that the City could benefit from privatization based on the model being tested at the time in Franklin, Ohio. The City of Franklin was considering a plan to sell its wastewater treatment plant to Wheelabrator Environmental Operating Service (EOS) for 20 years, with the option of buying back the firm at the end of the contract.

In May 1995, a request for proposal (RFP), based on the Franklin, Ohio, model, was advertised in newspapers on the Northeastern Seaboard by the City of Wilmington. The RFP asked for two proposals: a technical proposal outlining how the privatizer would run the plant, and a cost proposal defining the cost of selling the plant and managing operations of the plant. Evaluation and consideration of the technical proposals were to be based on the following criteria:

- Corporate profile
- Corporate experience and expertise
- Regulatory experience
- Key management and operational personnel
- Financial strength
- Utilization of disadvantaged business enterprises and Equal Employment Opportunity Act (EEO) compliance
- Employee considerations
- References and reputation
- Completeness and responsiveness of proposal

The cost proposals would be required to meet the following criteria:

- First, the amount and structure of the proposed Service Fee

- Second, was the completeness and responsiveness of the cost proposal to the requirements outlined in the RFP

Additionally, the Wilmington RFP requested proposals concerning the sale of the plant with a repurchase after 20 years, once again based on the Franklin, Ohio, model. Alternate proposals for privatizing the plant without selling it were also sought. This provided the City important flexibility in analyzing and negotiating varying contractual options with interested private business firms.

By reviewing the different options, including privatizing the plant without actually selling it, provided Wilmington the opportunity to view the internal management capabilities of the prospective privatizers. These firms were also invited to present the expenses associated with their management of the wastewater treatment plant. This enabled Wilmington to compare its current management costs with those of the prospective firms in order to determine the amount of operational savings associated with each privatization option.

1.3   Critical Appraisal

Following the receipt of several proposals, each was then evaluated by a staff committee with the help of the City's consultants. While the consultants did not have a voice in the selection of the privatizer, they were able to contribute valuable expertise gleaned from their experience in the private marketplace and from contractual relationships with other governmental outlets. To ensure that the privatizer selection was not based on price, the technical and cost proposals were required to be submitted as two distinct documents.

As an end result, the City of Wilmington ranked Wheelabrator Environmental Operating Systems (EOS), the private partner in the Franklin, Ohio, project, first among the submissions. It was only after the rankings were announced that Wheelabrator's cost proposal was then opened. The City's possession of the cost proposals from four companies provided a crucial advantage in negotiating an economically sound agreement with Wheelabrator.

1.4   Negotiations

Wilmington, from its evaluation of the technical and costs proposals, first concluded that it should sell its wastewater treatment plant rather than lease the plant. In late 1995, the City was negotiating a tentative agreement to sell the wastewater treatment plant to Wheelabrator for about $52 million. The advantages of selling the wastewater treatment plant were twofold:

- One, the sale of the wastewater treatment plant would provide a large infusion of money to the City of Wilmington at a time of budgetary need for the City.
- Two, privatization without selling the plant had not been previously tested. At that time, no city on record had entered into a 20-year agreement with a private operator for management of an existing facility.

While we were negotiating the terms to sell the plant, we were also asking Wheelabrator for details about the alternate option of privatizing management operations without selling the plant. Upon further analysis of this alternate option, we decided that the best choice for the City would be to retain ownership of the plant and privatize the operations with Wheelabrator for the following reasons (see also Table 13.2):

- First, the greatest long-term operational savings for the City and the County would be created in this way. While the City would not receive as great an initial payment as compared to the sale of

Table 13.2   Evaluation of Privatization Options

	Option 1: Sale to Private Firm, Repurchase after 20 Years	Option 2: City Ownership, 20-Year Management Contract
Benefits to City		
Up-front payments to City	$23,749,243	$0
Cost savings, 20 years	2,808,252	35,136,135
Property tax to city	5,963,024	0
Total benefits	$32,520,519	$35,136,135
Benefits to New Castle County	36,700,079	65,406,702
Total benefits	$69,220,598	$100,542,837

the plant, annual operational costs would be further reduced. We expected that privatization would reduce annual plant operational costs from approximately $18.5 million to approximately $15 million.

Under the terms of the contract with US Filter (which acquired Wheelabrator in 1997), the City is responsible for paying an Annual Professional Service Fee to US Filter for their provision of services. This fee of $8.18 million will result in reduced expenditures and stable costs for 20 years. The fee provides for all costs of management, operation, and maintenance of the Wastewater Treatment Plant. Included in said fee are limited adjustments for the Producer Price Index. The fee further includes adjustments based on the price of electricity at the Treatment Plant and can be adjusted for large increases or decreases in flows to the Plant.

The Service Fee is equal to the sum of the Base Component, the Electricity Component, the Adjustment for Biosolids, the Adjustment for Flow, and the Extraordinary Items Component:

$$ASF = BC + EC + ADJB + ADJF + EIC$$

where: ASF = Annual Service Fee
   BC = Base Component
   EC = Electricity Component
  ADJB = Adjustment for Biosolids
  ADJF = Adjustment for Flow
   EIC = Extraordinary Items Component

The Base Component is adjusted for each Fiscal Year. The first calculation was based upon the Fiscal Year beginning July 1, 1997. Mathematically, the base is determined by multiplying the Initial Base Component by the Index Factor:

$$BC_n = BC_i \times I_n$$

where: BC_n = Base Component in Fiscal Year n
   BC_i = Initial Base Component
    I_n = Index Factor Applicable to Fiscal Year n

- Secondly, we believed that federal review of our proposed privatization of our plant would simplify the review required under Executive Order 12803.* While the EPA had informed the City that review was still mandated under 12803, we did not expect that review to be a significant hurdle in this long process.

- Thirdly, we believed the public would be more accepting of the option of retaining ownership of the plant and privatizing the management of operations.

- Finally, retaining ownership of the plant—which serves a majority of Delaware citizens—enhances Wilmington's political leverage in dealing with the county and state governments on other governmental matters affecting the welfare of the City.

We therefore decided, as a City, to pioneer a new model for privatization in a wastewater treatment plant—one that called for the privatization of the management of operations while maintaining City ownership of the plant. We decided the best policy decision was to structure an agreement that would do the most to generate long-term savings for the City.

2. POLITICS AND PRIVATIZATION

While negotiating with Wheelabrator EOS, we were also talking with New Castle County government officials regarding the future of the plant. New Castle County residents in the county surrounding Wilmington provide about 70% of the flow to the City's treatment plant. County officials expressed concerns that the City would receive an initial payment for the sale of the plant that was not to be shared. These concerns raised certain issues relating to the sale of the treatment facility: mainly, that the County, in its view, had a right to purchase the plant.

New Castle County had paid for the treatment of its sewage since the construction of the plant in 1954. The County claimed that these monies were analogous to shares in a company; these shares, therefore indicated partial ownership in the plant. The City of Wilmington argued that the arrangement between the County and the City was parallel to a landlord

---

*The federal government's Executive Order 12803 of April 30, 1992 called for privatization initiatives to ensure the most beneficial economic use of [American] resources. The order also maintained that [a] adequate and well-maintained infrastructure is critical to economic growth. Under Section 2b of the Order, the Federal Government reserved the right to recoup in full the amount of Federal grant awards associated with the infrastructure assets. The Order also outlined a multistep privatization review process for local governments.

and a tenant: a tenant pays rent each month in exchange for use of a percentage of the space in the building and a portion of the services provided by the landlord. Upon termination of the tenant's lease, said tenant does not have an explicit, nor an implied, right to purchase the building based on status as a tenant.

Concerns were also raised by officials of both the County and the State as to rate control if the wastewater treatment plant was sold to a private firm. The fear was that sewerage fees and rates to consumers would automatically increase with private ownership of the plant. With the City deciding, however, to maintain plant ownership, these factors were no longer major concerns.

State officials also raised questions as to whether the State would be reimbursed $9,500,000, the amount it had contributed in the form of grants for expansion of Wilmington's plant, if the plant were sold to a private firm. One other concern raised by State officials was whether or not local government officials should have the authority to finalize long-term agreements affecting the wastewater treatment plant right before the November 1996 elections. The view was expressed that new people elected to City and County positions in the 1996 elections might object to privatization plans of the plant. With this concern in mind, in June 1996, the Delaware General Assembly passed Concurrent Resolution 68, which called for New Castle County and the City of Wilmington to wait until after January 1997 (following the November 1996 elections) before concluding any agreement on the wastewater treatment plant. Both City and County elected officials abided by the spirit and intent of this particular resolution.

The possibility of the loss of the valuable partnership that existed, and that continues to exist, between the City, the County, and the State was also at issue. This partnership, coupled with the County and State's use of the City's wastewater treatment plant, provided an important political ally for the City of Wilmington, Delaware. Wilmington, like many other cities, depends on funding from other levels of government to meet many pressing social and economic needs such as transportation, housing, and education. County and State officials aided the City in realizing its need to continue to consider various options beyond that of sale of its plant. It is important to a city to maintain ownership of this vital asset in order to remain a full partner with the County and the State in providing for residents.

3. LEGAL, REGULATORY, AND TECHNICAL CHALLENGES

When it was finally decided that long-term leasing to a private firm was the best option for the City, we faced a series of legal, regulatory, and technical challenges in realizing this privatization model.

3.1   Regulatory and Legal Challenges

3.1.1   Creation of a Satisfactory Review Process

First, there was a challenge working with the EPA to create a review process for a privatization structure that the federal government had not seen before. There were a number of instances in which cities had entered into short-term (up to five year) agreements to operate water and wastewater facilities. No city on record at that time, however, had entered into a 20-year agreement with a private operator for management of an existing facility. While Executive Order 12803 explicitly provided for the sale of existing wastewater treatment plants, it was ambiguous on the issue of long-term leasing agreements.

With no explicit federal regulations to guide us, we began discussing our proposed private operator agreement with the EPA representatives in Washington, DC. Our discussions with the EPA followed two lines: The first set of discussions involved gaining regulatory approval of Wilmington's proposed deal. In the second set of discussions, the City worked with the US Conference of Mayors in presenting ideas to the EPA for interpretation of Executive Order 12803 that would expedite the review and approval of novel privatization structures for other cities.

The US Conference of Mayors had created a task force called the Urban Water Council to promote the interests of cities in addressing issues of water and wastewater management. As Wilmington became more involved in these issues, Mayor James H. Sills, Jr., was invited to serve as cochair with Pat McManus of Lynn, Massachusetts, of the Urban Water Council. Serving as cochair of the Urban Water Council provided two key advantages: we were able to learn and therefore benefit from the combined expertise of the Urban Water Council while sharing Wilmington's situation and ideas for privatization.

3.1.2   Possible Tax Consequences for Municipal Bonds

The second regulatory challenge involved the federal treatment of municipal bonds used to fund the construction of wastewater treatment plants and other public facilities that are later privatized. As Wilmington began negotiating a 20-year operating deal, we were forced to look at the City's outstanding bonds. A comprehensive analysis of the City's bonds used to fund the construction and development of the treatment plant, based on complex federal tax regulations, raised questions as to whether all City bonds would continue to enjoy tax-exempt status if we entered into a 20-year management agreement with a private firm. Working with the Urban Water Council, Wilmington joined in lobbying the federal Treasury Department for changes in tax treatment of municipal bonds that would make long-term contracts with private operators easier to implement. Fortunately, in January 1997, the Treasury Department agreed with and acted upon our argument that service agreements of up to 20 years could be necessary to allow private firms to make long-term investments.

3.1.3   City Charter Limitations

Another challenge was the limitation of Wilmington's City Charter. According to the Charter, the city government could not be held liable for economic damages resulting from a contract that exceeded four years. It was therefore necessary to change the Charter to allow a 20-year agreement with US Filter. A solution was reached to solve this dilemma: US Filter agreed to enter into a contract with the City of Wilmington for a four years, while Wilmington sought approval for a 20-year agreement from the Delaware General Assembly. US Filter's willingness to sign such an agreement was based on its interest in getting started on such a nationally high profile project. The agreement was signed in December 1997, and the change in Wilmington's City Charter was granted by the Delaware General Assembly in June 1998.

3.1.4   Sale of the Bidding Company

In February 1997, as the City of Wilmington was nearing the end of its negotiations, another challenge arose. Wheelabrator EOS was sold by its parent company, WMX Technologies, to US Filter. WMX Technologies was under pressure from its shareholders to improve its financial performance by focusing on its core business (primarily, trash removal and landfill operations).

Wheelabrator was then sold intact to US Filter, which allayed some of the City of Wilmington's concerns. Once US Filter acquired ownership of Wheelabrator EOS, the economic strength of US Filter became a primary issue for the City of Wilmington.

Evaluation of the management and financial strength was an essential part of our ranking process of the original proposals. As a result, the City had to conduct a review of the financial strength of US Filter in order to be sure of its ability to live up to the terms of the agreement, because US Filter must continue to meet the financial strengths required for the duration of the 20-year agreement.

3.2   Preserving Employment and Protecting Employees

Managing dictates making good choices. Using privatization, therefore, as an indirect way to reduce staff is, in our view, not fair to employees and not good management practice. In this case, we consulted with our employee unions before issuing the RFP, which included provisions that, "The Privatizer shall offer employment to present City employees. The Privatizer shall provide total compensation (i.e. wages plus benefits) without lapse in benefits, comparable to that currently available to City employees . . . for a period of two (2) years" (RFP Appendix B-19). In short, the City of Wilmington employees would lose nothing in moving from the public sector to the private sector.

To further encourage and ease the privatization transition for the employees, US Filter offered $2,000.00 to each employee who accepted a position with its firm. Building confidence in the workforce was further assisted by the fact that US Filter had already been operating Wilmington's sludge dewatering facility for approximately five years. This history of working with US Filter, including union representation, played an important role in creating confidence in the proposed privatization.

3.3   Citizen Opposition and Public Apathy

A group of five citizens filed a lawsuit against the City of Wilmington in 1995, claiming that the selection process of prospective privatizers was improper. In November 1995, Delaware Chancery Court ruled that the City's procurement and selection process did adhere to State law and "made perfect sense." Such citizen opposition was the exception.

Two large-scale public meetings concerning the privatization process were held at different stages. At neither meeting, nor at any time in the interim, did a public outcry arise. Mayor Sills, reflecting upon the public apathy to the privatization process, observed, "It is clear that costs will go down, and that is a story that most people can understand and accept. . . ."

Continuing, the Mayor further observed, "The public does not have the same emotional attraction to waste-water plants that it does to garbage pick-up, public works, and safety."* Retaining ownership of the plant while providing continued service and reinvestment into the wastewater treatment plant served to reassure the public.

4. THE FINAL AGREEMENT: DECEMBER 1997

4.1 City Council Approval

Prior to signing the final agreement, Wilmington's City Council needed to approve the Service Agreement because the City operates under a Mayor-Council form of government. Under this system, the Mayor and the 12 Council Members are each elected to serve four-year terms. The Mayor is therefore not a member of City Council. In accordance with our City's Home Rule Charter, the final agreement was submitted to City Council. Following City Council's approval in November of 1997, the agreement was also reviewed and authorized by the Delaware Department of Natural Resources and Environmental Control (DNREC) and the EPA. The success of the agreement put forth to City Council was due in part to the job and union security offered to the City's 55 employees working at the wastewater treatment plant.

4.2 The Nuts and Bolts

Under the terms of the approved agreement, the City will continue to own the plant, including all future capital improvements. The City will continue to set water and sewer rates for all customers, and will retain oversight of operations and capital improvements at the Wastewater Treatment Plant. The City will retain all rights to negotiate electric rates for the Plant while continuing the right to make decisions regarding future expansion of the Treatment Plant. US Filter EOS pays an annual Site Access fee of $465,000 and reimbursed the City $1 million for its transaction and administrative costs associated with this project.

The economic value of our agreement with US Filter lies not only in reducing our annual wastewater treatment costs, but also in the use of private capital to make needed investments in the plant. The agreement with

*Originally quoted in the *New York Times,* May 5, 1996.

US Filter includes $15 million in capital investment for the first two years to make the plant more efficient; which reinforces the value of using private investment to replace aging public infrastructure. By not having to borrow money to make plant infrastructure improvements through our capital bond program, the City can thus give greater priority to other pressing capital development needs.

It is also important to mention that during the 20 years, as outlined in the agreement, the service fee as paid by the City of Wilmington to the privatizer will not increase, except under one or more of the following conditions:

- *A Change in Law.* Should a change occur within the law that affects the cost of operations of the wastewater treatment plant, the Service Fee may be adjusted.
- *An Increase in the Producer Price Index (PPI).* Historically, the PPI has not risen as much as the Consumer Price Index (CPI).
- *Natural Disaster.* Should damage occur to the plant through uncontrollable circumstances, the Service Fee will be adjusted accordingly.
- *A Change in Flow.* If the facility flow, in any fiscal year, is greater than 110% of the base flow, or less than 90% of the base flow, then a readjustment will be made to the Service Fee.

5. CONCLUSIONS AND RECOMMENDATIONS

This was a long, difficult process. The City has had to forge new ground along the way. There is, as yet, no prepackaged model for privatization that will work best in every situation. But we hope other cities can learn from our experience and make more efficient use of the private sector to manage and invest in public infrastructure facilities. We encourage city leaders to explore the possible use of private enterprise to manage facilities like wastewater treatment plants. The rewards are attainable for those who will take the risks. We have already observed one such reward: the City's operating expenditures have been reduced by $1.4 million in the first year of operation. We anticipate plant annual operational savings of $20 million over the life of the 20-year contract.

While we believe our experience justifies highlighting major issues that other cities need to consider and be aware of before privatizing their wastewater treatment plants, we would like to caution other city governments not to use this case study analysis as a blueprint for future directions in their city government. Our experience is so new that some time needs to elapse in order for us to make final judgments about some of the decisions that were made and their public policy implications for the future of Wilmington and other cities. With these factors in mind, we do believe our experience to date warrants our urging cities to at least consider the following:

- *Timing Is Ripe for Public / Private Partnerships.* The economic climate in cities is undergoing some positive changes. The fact that many people are moving back into downtown urban areas, coupled with a growing antidevelopment sentiment in the suburbs, is attracting more private firms to consider cities as a place to do business. An already in-place infrastructure also makes cities more attractive to the private sector because this convenience for firms could represent a cost savings. New opportunities for public/private business partnerships are growing. We would therefore urge cities to pursue privatization in the context of this changing pro-city climate.
- *A Multipurpose RFP Is Essential.* The request for proposal should be written to obtain a variety of information about the privatizer and options for privatization. The more information a city receives from the privatizer, the more likely a city will be able to adequately evaluate the marketplace. Such an evaluation is important when determining which privatizer is best and which form of privatization will best meet the political and economic needs of local communities.
- *Do Not Be Unduly Enticed by Up-Front Cash Payments.* Our limited experience would indicate that cities should give second thoughts to accepting up-front cash payments for the sale of their plants. By selling its wastewater treatment plant, a city is likely to give up a considerable amount of money in the long-term. A city government may receive a greater economic benefit as a result of reduced annual operational costs over a 20-year period in the long run, through privatizing the management and operations of its wastewater treatment plant. In short, cities should weigh carefully the benefits of long-term investment and not be unduly influenced by large up-front payments associated with plant sale.
- *Capitalize on the Market for RFPs.* After authoring an RFP, as described above, the city concerned would be advised to derive a mon-

etary benefit from having done so. Cities should be more entrepreneurial in their pursuit and receipt of RFPs. Atlanta, Georgia, for example, requested that interested firms pay $50,000 to submit their proposals to the City.* Drafting RFPs can prove to be time consuming and expensive for cities, and charging a fee to the privatizers for submitting their proposals is one way of recouping a portion of this expense. We agree with Atlanta's approach and would urge other cities to consider doing the same.

- *Protect the Interests of Public Employees.* One key concern in privatization is the employees. Protecting their positions, their benefits, and their rights is paramount in negotiations with the privatizer and in gaining public acceptance and support from political decision-makers. This is a goal best achieved when employees are involved from the onset in the planning and decision-making process.

- *Bear in Mind that Wastewater Treatment Plants Represent Political Leverage.* In considering our options as a city government, we discovered that the wastewater treatment plant's value as political capital was as important as the plant's economic value. The fact that Wilmington's wastewater treatment plant serves a majority of Delaware homes and businesses creates important political and economic leverage when it comes to securing funding support for other local, social, and economic needs. For this reason, we saw continuing City ownership of our wastewater treatment plant to be in the long-term strategic interest of the City of Wilmington. We therefore suggest that cities should carefully evaluate proposals to transfer ownership of their wastewater treatment plants to private firms, as they may lose valuable assets that help promote their political influence on other matters.

We are pleased with the outcome of this wastewater treatment privatization process. It is true that there were risks, and even doubts, during this process. We hope our experience is instructive to other cities. We believe the economic rewards are there for those cities that are willing to take the risks.

---

*Information shared at a Conference of the Urban Water Council in Chicago in fall, 1998.

Chapter 14

Houston's Strategy to Secure New Water Supply, Infrastructure, and Maintenance

Lee P. Brown

1. OVERVIEW

With the dawn of the twenty-first century, the energy capital of the world finds itself at the forefront of another challenge: securing and managing an adequate supply of drinking water for future generations. Houston is facing the same challenges that confront many of the communities in this country: rapid growth and development, escalating demands for water, increasing pressures to maintain water rates, regulatory requirements that promulgate higher standards for drinking water, an aging infrastructure, environmental concerns, subsidence caused by the depletion of groundwater, rising construction costs for improvements to its water treatment and distribution system, and competition for existing water supplies. These factors have mandated a new approach to an old problem: doing more with less.

Lee P. Brown is Mayor of Houston, Texas. He has a Ph.D. in Criminology from the University of California, Berkeley, with extensive service in the law enforcement field. He headed police agencies in Atlanta, Houston, and New York, and held a cabinet position as Director of the National Drug Control Policy. Prior to being elected mayor, he was Professor of Public Affairs and a Scholar at the James A. Baker III Institute for Public Policy at Rice University.

Reinventing Water and Wastewater Systems: Global Lessons for Improving Water Management, edited by Paul Seidenstat, David Haarmeyer, and Simon Hakim.
0-471-06422-X   Copyright © 2002 by John Wiley & Sons, Inc.

2. INCREASED DEMANDS FOR WATER

The City of Houston supplies reliable drinking water to 1.95 million Houstonians and 19 other communities in the region. In fact, the Houston Primary Metropolitan Statistical Area has 4.1 million residents. Houston has been supplying treated surface water for more than 40 years. Its position as the regional supplier of surface water is evidenced by its ownership, either in whole or in part, in three major reservoirs that have a combined capacity of 1.1 billion gallons per day of supply. While this amount of firm capacity is sufficient to meet the city of Houston's needs for the next 75 years, it only represents a 35-year supply when the entire region that could potentially rely on Houston's water supply is taken into consideration.

Houston's population has increased dramatically in the past decade, from 1.5 million people to 1.9 million. It is projected that Houston's population will grow by 1.6 million by the year 2020. In order to keep pace with the current demands, and plan for the future, Houston has embarked on an aggressive program to secure additional surface water supplies and to involve the private sector in delivering new facilities, as well as reaching out to its neighbors to develop a partnership aimed at a regional solution to the water challenge before us. To echo a common refrain in the space exploration program: failure is not an option.

3. SECURING ADDITIONAL WATER SUPPLIES

As previously stated, Houston currently has interest in three reservoirs: Lake Houston (100%), Lake Conroe (67%), and Lake Livingston (70%). We are reaping the benefits of tough decisions made by our predecessors many years ago. In order for us to pass on a legacy of an adequate water supply to future generations, it is now our turn to make some tough decisions.

Houston is in the process of finalizing its efforts to secure a permit for a $200 million reservoir on the Brazos River. The Allen's Creek reservoir will provide almost 100,000 acre-feet of water per year, or 90 million gallons per day. Houston and the Brazos River Authority have a 70/30 partnership in this reservoir that, at the time of this writing, is one of only four permitted sites in the State of Texas. The difficulty of finding suitable sites for future reservoirs, coupled with increased environmental concerns, have made the construction of new reservoirs an arduous task. Of course, at some point in time, someone has to step forward with the monies to fund the design, construction, and operation of the reservoir and raw water transmission system. This last aspect alone is a significant challenge facing many of the smaller communities in the state. Fortunately, the Texas Water Development Board is exploring alternative options to make financial assistance available to those communities that have a need for additional surface water supply, but lack the financial capacity to bring the project to fruition.

Once in place, the Allen's Creek Reservoir will give Houston access to 60 million gallons per day (MGD) in raw water supply for Houstonians in the western portion of our community. We will also be in a position to provide treated surface water to the surrounding communities. Because a water treatment facility will be needed, we are examining our options regarding a new water purification facility near the Allen's Creek Reservoir. We will be working with the Gulf Coast Water Authority to identify a site for the plant, and to plan for its design and construction. The cost of the Allen's Creek plant is estimated to be approximately $180 million. All water providers understand the need to plan for new facilities, and to maintain existing plants. Houston is no different.

4. IMPROVEMENTS AT EXISTING FACILITIES

Houston's water supply infrastructure is extensive. We provide both groundwater and surface water to our citizens and the surrounding community. Our groundwater system is comprised of 198 wells that have a combined capacity of approximately 260 MGD. To keep this vital resource in top operating condition, we have a defined program to rehabilitate these wells. Over the past 5 years, we have invested approximately $20 million in well refurbishment. We systematically address the needs of approximately 30–35 wells every year via major capital improvements. Our surface water system is no less important.

We currently own and operate four major surface water facilities. Our East Water Purification Plant Complex is actually comprised of three separate, but wholly integrated, water treatment plants. While the East Water Plant currently has a firm capacity of approximately 310 MGD, we are in the process of upgrading that facility to a firm capacity of 350 MGD. We also are a coowner, and managing partner, in the Southeast Water Purification Plant (SEWPP). This 80-MGD facility is also being upgraded at this time, to a firm capacity of 120 MGD. We are also in negotiations to expand this plant to a possible 240 MGD. This water plant provides reliable drinking water not only to the City of Houston, but also to 10 other communities, or coparticipants, who own 75% of the plant. The SEWPP is a perfect example of the alliances that will need to be formed in order for communities to work together to solve a common challenge.

5. REGIONAL COOPERATION: "WATER AIN'T NECESSARILY FOR FIGHTIN'"

While we are all familiar with Mark Twain's famous remark that "whiskey's for drinking and water's for fightin'," the Southeast facility presented an remarkable opportunity for Houston to work with the surrounding communities to join in a mutually beneficial partnership for the design, construction, and operation of this water plant.

When the price of oil plunged in the early 1980s, the greater Houston area was confronted with both a challenge and an opportunity. As the economy mandated that we be fiscally prudent, new approaches had to be explored to address the very real need for additional treated water supplies. Rather than build numerous new facilities, Houston joined with the surrounding communities in a cooperative effort to build a regional facility: the Southeast Water Purification Plant (SEWPP). Though Houston is the managing coparticipant, at this time we only have a 25% share of the treated water. The balance is used by the other 10 coparticipants. Not only are the capital costs shared, but the annual expenses as well. Overall, our relationship with the coparticipants has been a very positive experience. Since it came on line, the SEWPP has been operated by a private entity.

6. HARNESSING THE PRIVATE SECTOR

In a procedure that can be referred to as "outsourcing" or as "privatization," the SEWPP coparticipants decided to solicit bids for the operation and maintenance of the facility. While Houston submitted a competing offer to manage the facility, the private sector was the successful bidder.

The city's Water Production Branch is gaining more and more experience and knowledge in the area of "outsourcing" the operation and maintenance services of its drinking water facilities. The city's experience in the outsourcing of water production facilities began in 1991, when the operation and maintenance of the SEWPP was first outsourced for a five-year contract. The contract was rebid in 1996, and most recently in 2001. As stated earlier, the city's Water Production Branch submitted a bid in 1996 but was not the successful bidder.

The city has also outsourced the operation and maintenance of the Kingwood and Forest Cove groundwater plants, as well as the distribution systems. This experience began in 1996 when these areas were annexed by Houston. Essentially, the contracts that existed with these systems were extended through 2001 and are currently in the process of being rebid. While the two subdivisions were previously operated by two different contract service providers, the new contract will cover both of the subdivisions as well as two additional systems.

The following observations follow from Houston's experience in the area of contract service operations of water production facilities.

6.1   Benefits of Outsourcing

There are many aspects of outsourcing that are beneficial to any utility that is considering this option. First of all, it allows a utility to compare its cost of service with the private operating firms. This gives the utility the opportunity to increase its efficiency. This process is known in the industry as "optimization." It is important for utilities to see outsourcing as an opportunity to recognize and obtain efficiencies, which the private firms may have to offer in their treatment techniques, or the way in which they do business.

Cities are generally criticized because of their bureaucratic processes of hiring and purchasing. Private firms use these terms when they are marketing their services. "We can do it better, faster, and cheaper because we do not have to go through all of the red tape that city government deals with." By recognizing this reality, Houston has enacted a "P-Card" (purchasing card/credit card) program where workers can immediately purchase nonwarehouse materials for repair and operations of utility facilities. The P-card limit of $2,000.00 allows operations and maintenance staff to "cut through the red tape" of purchasing for the price range of the most frequently needed items. This is an example of different divisions working together to improve efficiencies.

Because of the fact that the utilities of the City of Houston are spread out over a 617 square mile area, there are efficiencies in having private operators in the outlying areas, as it lessens the difficulty of the logistics of travel and supply from the city's already spread-out operations area. This has really proven to be a benefit to the city's operations by allowing us to focus our resources in areas that are centralized in quadrant staging areas; thus allowing for more operation and wrench time rather than windshield time.

6.2   Potential Problems with Outsourcing

In the highly competitive market of operation and maintenance service providers, the bidders have become expert at bidding exactly what they feel the contract requires. Many of the people who work at public utilities have been public servants the majority of their career and work with the understanding that whatever you need to do to serve the customers, both internal and external, is performed without question. While contract service providers are willing to provide such services, their patented statement is that these services are "outside the scope of the contract" or are "additional services" and will require additional payment.

As new contracts have been developed for future contract periods and facilities, the contract language has increased in volume to attempt to cover areas that were found through previous contract experience, to be less than clear and concise. The problem we have experienced is that the contract administration requirements are increasingly more intensive. Legal assistance and the involvement of higher-level management are becoming more routine.

While it may appear that hiring a contract service provider will lessen the amount of time that utility staff spends in a service area, there may in fact be a shift from operational-intensive involvement to an administration-intensive involvement. Generally speaking, the larger contracts have bigger problems with contract administration than the smaller contracts.

6.3   Relationship Between the Participants

While it is certainly not a requirement of contract administration, there is a tendency for parties to posture into roles of buyer versus seller. While no one really plans for this type of relationship; it may be naïve to think it can easily be otherwise. The nature of this relationship can make for a very long contract term.

It is very important to develop the best relationship possible and to attempt to establish what the utility industry terms a "Public–Private Partnership." Such a relationship is definitely a win-win situation, where the utility gains from the service provider and vice versa.

7. THE LEGISLATIVE HORIZON

The 77th Texas Legislature ended its session in May 2001 with little significant water-related legislation affecting Houston being enacted. The session did, however, include proposed legislation that will, in all likelihood, be introduced in some form in the 78th Legislature in 2003. Some of the items that would affect Houston are:

1. Changes to existing law related to the transfer of surface water from one river basin to another that strengthen the need to show that the benefit to the receiving basin outweighs any detriment in the basin of origin. Currently, those transfers make the water transferred a junior water right to all other water rights in that basin. This provision will need to be rescinded to make water marketing a reality.

2. Two additional Groundwater Conservation Districts were created by the 77th Legislature in the Houston area in Brazoria and Montgomery counties. These districts will be developing water management plans that will look at aquifer depletion rates, ground subsidence, and transfer of groundwater outside the district. Other Groundwater Conservation Districts may be created in future legislatures.

3. A Texas State Water Plan will be completed in 2002 that combines the approved plans for the 16 regions of the state. It is expected that the plan will identify water projects with a cost of about $18 billion that will need to be funded over the next 50 years. The recent legislature considered fundraising plans that ranged from a "head tax" to a "sales tax on water" and included increased permit fees. These types of fees and taxes add disproportionate cost to large municipalities for the benefits received.

4. Additional protection for water rights for municipalities have been proposed in previous legislation and need to be enacted in future legislatures. These include no cancellation or reassignment of an unused surface water right where the water right results only because of the construction of a reservoir or other facility by the water right holder. Other protections could include assigning definitive priorities to types of water use, marketing tools to allow a workable system for sale and use of water supplies, and assistance in planning and developing new sources and technology for water supplies.

5. Legislation at the federal level could include funding for water supply projects that could be in the form of assistance to state and local governments. In addition, the U.S. Army Corps of Engineers
could be funded and directed to engage in development of new
water supply sources and transfer of existing sources to areas of
high demand.

Caution will need to be exercised in proposed legislation that may have
the impact of reducing the flexibility for transferring, treating, developing,
and using the current and proposed water supplies.

8. CONVERSION TO SURFACE WATER

The effects of the overpumpage of groundwater are ever present. Subsidence in some areas surrounding Houston has been extreme. To combat
this issue, the Texas Legislature mandated that this area convert from relying upon groundwater to using surface water. The Harris Galveston
Coastal Subsidence District (HGCSD) has been tasked with enforcing the
deadlines. Should a groundwater permittee fail to convert as required, the
HGCSD will assess a "disincentive fee" of $ 3.00 per 1,000 gallons. This fee
will be calculated against 80% of the permittee's annual groundwater
pumpage.

The schedule is aggressive. For example, each groundwater permittee
must have an approved Groundwater Reduction Plan (or GRP) no later
than January 1, 2003. Once approved, any needed construction must begin no later than January 1, 2005. The potential disincentive fee for fail-
ing to convert that area to surface water is sufficiently high to encourage
this activity. Indeed, Houston has already embarked on an aggressive program to ensure compliance with the regulations.

9. CHARTING A NEW COURSE

The "frontier spirit" is alive and well in Houston. In March 2000, we so-
licited input from a broad range of private sector companies as to how best
to deliver a state-of-the-art water treatment plant to assist us in addressing the need to convert to surface water. The recommendations were almost unanimous: implement an alternative project delivery method. For
us, that meant design-build-operate (DBO). The aim was to contract with
a single entity to design, construct, and operate the 40-MGD Northeast

Water Purification Plant (NEWPP). We were confident that we could realize significant saving using DBO instead of the traditional method of design-bid-build.

Unfortunately, in Texas, the public sector is precluded from utilizing DBO for a municipal project such as a water treatment plant. Rather than admit defeat, we chose to find a way to accomplish our mission. Houston created a local government corporation (or LGC): the Houston Area Water Corporation (HAWC). The HAWC's Board of Directors is comprised of seven Houstonians who are nominated by the Mayor and confirmed by the City Council. They act as an independent entity to deliver the NEWPP and assist the city in developing a GRP for the entire region. It has truly been a remarkable success.

Following an extensive submittal and review process, the HAWC Board of Directors selected Montgomery Watson Constructors, Inc. to deliver the NEWPP. The proposed Service Agreement calls for a 40-MGD facility and 61,000 linear feet of 84-inch and 42-inch transmission line. We also have a guaranteed price for an additional 40 MGD of treatment capacity, should the HAWC exercise the option within the first three years of the Agreement. Lastly, the Agreement provides for an initial 10-year operating period, with an option for two 5-year extensions. The capital savings realized by utilizing DBO exceeded our expectations: a phenomenal $80 million. We are on schedule to have the 40 MGD facility completed by December 2003.

Houston is proving that the water challenge being faced by so many cities can be overcome through vision, planning, and courage. The right thing to do is not always the easiest. However, when the issue is as fundamental as a safe, reliable, and adequate water supply, there are no alternatives. Failure is not an option.

Chapter 15

A New Standard for a Long-Term Lease and Service Agreement

Michael A. Traficante and Peter Alviti, Jr.

It is a scenario that is familiar to many mayors. Twenty years ago, the City of Cranston, Rhode Island, built a shiny new wastewater treatment facility that was generously subsidized with state and federal grants. Over the past 20 years, improvements were made to the facility only on an as-needed basis, but even these improvements added up to $26.2 million of Sewer Enterprise Fund Debt. Sporadic sewer user fee increases generated revenues that barely met the operational expenses as the city government attempted to keep fees affordable for ratepayers. From time to time, the City's General Fund loaned money to the Sewer Enterprise Fund to meet overruns in operational expenses, resulting in the Sewer Enterprise fund owing $8.6 million to the City's General Fund. Stricter air and water quality mandates imposed by the federal government require $30 million of dollars worth of upgrades to the wastewater facility, while the state and federal grant subsidies have disappeared.

Michael A. Traficante is Government Affairs Liaison, Laborers International Union of North America. He was formerly the mayor of Cranston, Rhode Island. Peter Alviti, Jr., is an engineer with Laborers International Union of North America. He was formerly the Public Works Director for the city of Cranston, Rhode Island.

Reinventing Water and Wastewater Systems: Global Lessons for Improving Water Management, edited by Paul Seidenstat, David Haarmeyer, and Simon Hakim.
0-471-06422-X   Copyright © 2002 by John Wiley & Sons, Inc.

The citizens of the City are becoming increasingly resistant to approving capital improvement bond referenda that may increase user fees or taxes, leaving public works officials uncertain as to how they will comply with the state and federal mandates. Sewer rates would have to increase nearly 100% during the next five years in order to meet these financial and capital improvement obligations. On a separate front, the City's General Fund suffered $6.9 million cumulative deficits during the past four years, mainly due to two years of record-breaking snowfalls and the loss of a labor-related lawsuit.

On September 11, 1997, with a stroke of his pen, Mayor Michael A. Traficante solved all of these problems and more by signing an innovative $400 million 25-year Wastewater System Lease and Service Agreement with Triton Ocean State L.L.C. The agreement allowed Cranston to meet all of its financial and capital improvement obligations and insures compliance with all air and water quality mandates over the 25-year period. It provides the most cost-effective means of providing wastewater services and long-term stabilization of sewer user fees for the citizens of Cranston.

Certainly, a deal such as this did not happen without a considerable effort conducted over a two-year period. It involved several carefully planned parallel efforts that addressed the technical, legal, financial, political, regulatory, labor, public sector, and private sector issues related to the transaction.

1. HISTORICAL BACKGROUND

Cranston, located immediately south of Providence, is the state of Rhode Island's third largest city, with a population of 78,000. The city operates under a home rule charter adopted in 1962. It has a mayoral form of government with a nine member City Council headed by a Council President. Cranston's annual General Fund operating budget is $134 million with a separate Sewer Enterprise Fund budget of $12.7 million. Prior to the privatization, the City's bond indebtedness was $75 million, including $26 million related to the wastewater system.

The City's wastewater treatment plant was originally built and placed in service in 1942. At that time, it had an average daily flow of 5.5 million gallons per day (MGD). Construction of a new facility in the 1970s brought the operating capacity of the facility to its present 23 MGD and included 190 miles of sewers and 21 pumping stations. Current operating flows average approximately 12.4 MGD. This flow comes from 28,700 residential units, 2,300 commercial units, and 268 industrial units, representing 92% of the city's existing population.

When the treatment plant facility construction was completed in the late 1970s, the City initially operated it. During the early 1980s, the City employees had considerable difficulty in meeting the discharge limitations in the plant's permit. In the late 1980s, the City retained an outside firm to manage the City employees operating the facility and optimize performance. This approach lasted several years, but the treatment performance violations were not corrected.

In 1989, the City retained Professional Services Group (PSG), a subsidiary of Air & Water Technologies Corporation, under a short-term five-year operation and maintenance agreement. PSG operation and the City's funding of major repairs and replacements have improved the plant performance over the last several years, but the increased cost of repairs and replacements of plant equipment required additional bond debt to be incurred. Further, the user rates charged by the City were not adequate to meet annual operating expenses. The City's General Fund regularly loaned money to the Sewer Enterprise Fund in order to meet operating budget shortfalls.

In 1994, Mayor Traficante commissioned a study to determine the long-range plans for the city's wastewater facilities. The plan included paying back the General Fund the amounts due to it from the Sewer Enterprise Fund over a 10-year period. It also included paying the existing bond debt and installing advanced wastewater treatment facilities, along with other mandated improvements that would cost another $30 million. The traditional solution would include increasing the user fees nearly 100% over a five-year period to cover the cost of the plan. That would result in the average residential sewer fee increasing from $231 to over $400 annually. Instead, the Mayor and his administration decided to pursue other alternatives that could accomplish the same results at a lower cost.

2. THE REGIONALIZATION OPTION

The first alternative studied was regionalization. In 1995, Cranston entered into negotiations with the Narraganset Bay Commission, a regional operator of wastewater facilities for three other municipalities in Rhode Island. Regionalization offered economies of scale that could potentially lower the cost of wastewater services. After a year of negotiations, which resulted in a Memorandum of Understanding, the City determined that the transfer in ownership to the Bay Commission would not meet all of its objectives.

The regional commission would not cover the cost of the $8.6 million due to the City's General Fund, nor would they pay for the defeasance of the entire $26 million Sewer Fund debt. This would leave Cranston responsible for addressing the shortfall, most likely by raising taxes. Further, the Bay Commission could not assure that Cranston sewer customers would be afforded stable user fees in the future, especially in light of the fact that the Bay Commission's own system needed $500 million worth of combined sewer overflow upgrades. In early 1996, the City became aware of Executive Order (EO) 12803, which enabled municipalities to sell or lease their wastewater facilities to private companies.

3. THE PRIVATIZATION OPTION

During 1996 and 1997, Cranston conducted a procurement process that involved three private companies competitively negotiating the contract provisions and costs associated with leasing the wastewater facilities for 25 years. The private company selected would be responsible for all wastewater system activities, including operation and maintenance, capital improvements, major repairs, financing, odor control, performance guarantees, air and water quality compliance, Industrial Pretreatment Program administration, and costs associated with the City's wastewater treatment plant, 21 pumping stations, and the 190-mile sewer collection system. To assure proper decision-making, the Mayor assembled a team, headed by his Director of Public Works, Peter Alviti, Jr., P.E., and including: the New York law firm of Hawkins Delafield and Wood serving as legal counsel; HDR Engineering, Inc., of White Plains, New York, serving as technical and economic advisors; and the New York office of Bear Stearns, serving as financial advisors.

While considering long term privatization, the team needed to include the views and concerns of several agencies and organizations, each having unique and separate priorities. The US Environmental Protection Agency (EPA) and the White House Office of Management and Budget had oversight of the EO 12803 approval process, with particular concern for the disposition of grants that funded the original facility, federal permits, the City's use of the contract payment, and ratepayer impact. The State of Rhode Island Department of Environmental Management reviewed and approved the water quality, operating, and performance criteria, various consent decree issues, and the capital improvement program. The State Public Finance Management Board, along with the Rhode Island Clean Water Finance Agency, provided the Private Activity Bond Cap Allocation and the financial structure necessary to provide the private contractor with tax-exempt financing for capital improvements. The Laborers International Union of North America was consulted with regard to the disposition of City employees who would transition to employment by the private contractor. The City Council was continuously involved in the political process and the review and approval of the privatization contract. Also, the City's residents who were concerned with the disposition of a major City asset and the future rate impact were given ample opportunities to voice their concern and be briefed during numerous public hearings. The concerns of each entity were given careful consideration and were addressed in the formulation of the City's objectives during the procurement process.

4. THE PROCUREMENT PROCESS

In March of 1996, Cranston developed its request for proposals (RFP). Using the results of the study that had determined the long-range plans for its wastewater facilities, the City created an initial set of objectives. The RFP was intentionally kept general in nature and contained only Cranston's statement of objectives and historical data on the financial, technical, regulatory, and operating status of the wastewater facilities. This approach provided the companies basic information and encouraged them to be creative and to use their private sector entrepreneurial skills to determine how they would accomplish the City's objectives. Three companies responded to the RFP, each one presenting an array of innovative solutions.

The Mayor's team then reviewed the responses, interviewed each company, and held discussions with the City Council, regulatory agencies, and other organizations. The team decided that a 25-year lease and service agreement would accomplish all of its objectives and that the most beneficial way to proceed with the procurement should involve transitioning into a competitive negotiation with all three companies. On that basis, the City's team developed a request for resubmittal that defined a set of basic contract principles, the capital improvement program, financing guidelines, and a response format that included fixed price bids on the capital improvements, finance costs, and service fees. After submitting their responses, negotiations with the companies were conducted, with care taken to negotiate the best deal between each company and the City independently. The team benchmarked each company's offer against projections of the City's original plan for operation of the wastewater system.

At the end of the year-long process, Mayor Traficante and his team selected Triton Ocean State, L.L.C., as the company offering the best overall contract. Triton Ocean State, a subsidiary of Poseidon Resources Corp. of Stamford, Connecticut, will be responsible for financing and management of the lease and service agreement. Triton's team includes Professional Services Group of Houston, Texas, which will provide services for the operation, maintenance, major repairs, and replacement of the wastewater facilities, and Metcalf & Eddy, Inc., of Wakefield, Massachusetts, which will design, engineer, procure, construct, and test the capital improvements.

5. THE PRIVATIZATION CONTRACT PRINCIPLES

After signing the contract in March 1997, the initial phase of the contract included the Approval Period. During this time, Cranston was obligated to secure all regulatory agencies' approvals and help the company secure its project financing. Triton was obligated to secure financing of company debt for the $48 million contract payment and for the tax-exempt $30 million capital improvement debt. Triton provided a $15 million letter of credit guarantee to insure that they would secure this financing once the City obtained all regulatory agencies' approvals. Both the City and Triton completed these obligations, which resulted in the contract commencement on September 11, 1997.

On the commencement date, Triton provided the City with a contract payment of $48 million and an additional $30 million of private financing for the state and federally mandated capital improvements. The $48 million contract payment enabled Mayor Traficante to defease the $26 million Sewer Fund debt, repay the General Fund $8.6 million owed by the Sewer Fund, eliminate the $6.9 million General Fund deficit, and establish a $6 million General Fund Surplus. Triton's $30 million private tax-exempt financing of the capital improvements guarantees that Cranston will meet all state and federal mandates without incurring additional city debt.

The financing for the contract payment and the financing of the capital improvements were provided by Triton as private company debt with no recourse to Cranston. If the company defaults on the terms of the contract, it will be terminated. Upon default, Cranston is not obligated to repay any of Triton's debt. These terms and conditions help guarantee that the City's wastewater system will receive uninterrupted operation and maintenance during the 25 years, in accordance with the contract principles. A parent company guarantee and a performance bond also back Triton's performance.

Beginning on the commencement date, under the terms of the agreement, Triton Ocean State is responsible for the operation, maintenance, major repairs and replacements, all mandated capital improvements, air and water quality regulation compliance, and Industrial Pretreatment Program administration for all of the city's wastewater facilities, including the treatment plant, collection system, and 21 pumping stations. The city will pay Triton a monthly fixed price Service Fee that escalates over the 25 years in accordance with the Consumer Price Index and Chemical indices. The contract service fee paid to Triton provides the City with long-term user rate stability and will save City ratepayers $35 million when compared to the City's continued operation projected.

Triton will be responsible to design, construct, and acceptance test 10 planned system improvements and the advanced wastewater treatment system for a fixed price and in accordance with an agreed schedule. The company will pay delay damages if it does not complete the projects within the contracted schedule. Triton will also be liable for the payment of any regulatory agency fines resulting from the delays. If the company delays result in contract default, the contract will be terminated. These contract provisions provide the necessary guarantees that the city will remain in compliance with state and federal mandates. Triton will also be responsible for operation and maintenance performance guarantees to meet effluent standards and sludge and grease processing. If Triton fails to meet the effluent standards, it will constitute an event of default for which the company must pay fines and liquidated damages and will face possible termination.

During the 25-year contract period, Triton will perform all major repairs and replacements at the wastewater facilities, except for those resulting from uncontrollable circumstances or changes in law. For those circumstances, the City has assumed financial responsibility in the future. An independent engineer will perform an initial audit of the wastewater facilities' condition. This will serve as a baseline that the company is responsible for maintaining during the 25-year contract. Triton is obligated to return the facilities to the Cranston in the same condition at the end of the 25-year contract. During the 25-year period, the company is also obligated to abide by an odor guarantee that assures the City residents of an acceptable odor level in the community around the facilities. The contract also contains other general provisions governing the company's performance, reporting requirements, city termination rights, parent company guarantee, and performance bond requirements.

6. LESSONS LEARNED

Several important lessons were learned by Cranston in pursuing this comprehensive privatization effort that may help other municipalities venturing into public–private partnerships.

1. *It Pays to Be Innovative.* Even if you believe your present method of operation can work effectively, a strong commitment to abandon traditional methodology is necessary in order to reap the benefits of innovation.

2. *Make Sure Your Team Is Committed to Change.* This is important from the top down. Privatization efforts will draw scrutiny and criticism from citizens, politicians, unions, and even the private companies. Unless there is a commitment to withstand the pressures that radical change will precipitate, the process will not succeed.

3. *Don't Fear the Loss of Your Job.* This may be a primary reason that many municipal officials resist the idea of privatization. The fact is that private companies can do some things better than governments. It is in the public's best interest that these alternatives be pursued, and, if a public official is doing a good job in the public sector, there will be just as many career opportunities in the private sector.

4. *Members of Your Procurement Team Should Be Detached from Past Involvement.* Traditional relationships may hinder innovation and credibility. Involving existing municipal consultants or employees who would be affected by the privatization, may, in a worst case, cloud proper judgment and, at best, lessen the credibility of the procurement process in the eyes of the public.

5. *Get Experts on Your Negotiating Team.* You can be assured the private companies will have them on their side of the negotiation table.

6. *Form Partnerships with Regulators and Agencies.* Many of the privatization contract principles will deal with passing regulatory compliance responsibilities from the municipality to the private company. There is a common benefit when the municipal and regulatory agencies work jointly to negotiate contract provisions that will insure the company's performance and regulatory compliance. For once, you are both on the same side.

7. *Give the Private Companies an Opportunity to Use Their Entrepreneurial Skills.* Detailed specifications in your request for proposals may result in predefined responses from the companies submitting proposals. Create the RFP in a manner that will encourage private companies to propose innovative solutions. That is the whole point of privatizing.

8. *Allow the Company to Make a Profit.* How much money the company makes for a profit is not your business. Only the savings provided from privatization is. In fact, if the company is not allowed to profit from the venture, then the contract is doomed to failure in the long term.

9. *Make the Contract Self-Sufficient.* The contract should contain incentives for good performance by the company, as well as disincentives for bad performance. Properly written contract provisions can help minimize the expense of municipal oversight.

10. *Include the Public in the Process.* It is their money you are saving.

7. THE BENEFITS OF PRIVATIZATION

Cranston has met all of the objectives that it set out to achieve. The 25-year lease and service agreement with Triton Ocean State will:

1. Create an economical and cost-effective means to provide wastewater treatment services and meet financial objectives through increased private sector participation. Savings to city over 25 years is more than $35 million.
2. Provide economies of scale and increase operating efficiency of the wastewater facilities, which will result in an improved environment.
3. Require the contractor to assume responsibility for operating and maintaining all wastewater facilities and capital improvement construction.
4. Assure compliance to facility plan and all other state and federal regulations and environmental mandates.
5. Pay in full $8.6 million in Enterprise Fund debts to the City's General Fund.
6. Pay in full $26 million in existing sewer-related bonded indebtedness.
7. Pay in full $6.9 million in the City's General Fund deficit.
8. Create a $6 million General Fund surplus.
9. Eliminate the City's interest expense for short-term borrowing.

10. Increase the City's long-term investment income.
11. Eliminate the need for future City bonded indebtedness for capital improvements and all future improvements or mandates.
12. Insure company compliance with the contract provisions.
13. Provide long-term stability of sewer user fees.

Chapter 16

North Brunswick's Model Water and Wastewater Public–Private Contract

Paul J. Matacera and Frank J. Mangravite

Like many municipalities, the Township of North Brunswick, New Jersey, was faced with the problem of paying for the increasing cost of providing services for its residents without increasing taxes. Decreasing state and federal aid contributed to the problem. For over 10 years, the Township had contracted with a private firm to operate its water treatment plant under a succession of one-year professional service contracts. The positive experience of this public–private partnership included decreased costs and improved water quality. With confidence in the partnership concept, the Township extended the private firm's scope of services to the management of its water distribution crew.

Throughout this succession of one-year contracts, it became increasingly apparent that additional savings and other financial and technical benefits could be achieved by a long-term contract. After evaluation of several possible alternate revenue sources, North Brunswick decided to focus its search on the realization of benefits from a long-term public–private partnership for the operation of its water and wastewater systems.

Paul J. Matacera is Principal with GluckShaw Group of Trenton, New Jersey, and former mayor of North Brunswick, New Jersey. Frank J. Mangravite is President, Public Works Management, a consulting firm in Morris Plains, New Jersey.

Reinventing Water and Wastewater Systems: Global Lessons for Improving Water Management, edited by Paul Seidenstat, David Haarmeyer, and Simon Hakim.
0-471-06422-X   Copyright © 2002 by John Wiley & Sons, Inc.

In April of 1995, North Brunswick issued a public advertisement for proposals. The scope of requested services included the management, operation, and maintenance of North Brunswick's water and wastewater treatment systems and the billing and collection of user fees for 20 years. The Township would retain ownership of the systems, and no asset transfer or lease would be involved. An up-front concession fee and annual "concession" payments were to be paid to the Township. The private firm was to obtain its fees from water and sewer user fee revenues. Capital improvements in excess of $50,000 remained the responsibility of the Township, although the private firm was free to invest capital at its own expense.

Three proposals were received. Contract negotiation with the selected private firm, US Water L.L.C., was completed in February 1996. Project start was July 1, 1996. The contract had a value of about $214 million ($110 million present value) over its 20-year term.

The North Brunswick water and wastewater systems serve about 33,000 residents through approximately 13,000 connections. The wastewater system is a collection system that flows into a regional treatment plant. The water system consists of a 10-million-gallon per day surface water treatment plant and a distribution system. Current flow is about 6.5–7 million gallons per day. The treatment plant employs coagulation, powdered activated carbon addition, clarification, pressure filtration, chlorination, sludge thickening, sludge dewatering, and pH control. The water source is the Delaware and Raritan Canal.

1. LONG-TERM CONTRACT BENEFITS

Short-term operation contracts, typically one to five years in term, can provide certain benefits such as those North Brunswick experienced over the last 10 years. They cannot, however, capture the benefits of a private firm's financial or technical resources that take longer periods of time to implement and to recover their costs. The four primary advantages of a long-term contract are:

- *Access to Private Capital.* Under a long-term contract, the private firm can invest a significant amount of capital because the debt can be amortized over a long contract term, which minimizes the impact on utility user fees. Therefore, the municipality can consider the use of private capital as an alternative to long-term public debt to finance utility projects, such as the construction of a new facility. In addition, the private firm can use the extended contract term to invest its own capital in improvements or technologies that require longer periods of time to recover their costs.

- *Employee Considerations.* A long-term contract increases the opportunity to implement comprehensive employee training and certification programs. Staff reductions can be achieved gradually through attrition, employee incentive programs, or specific gradual schedules. Surplus employees can more easily be transferred to other positions within the municipality, the project, or other projects of the private firm. This is much harder to do within a one- to five-year contract that must simultaneously offer significant savings.

- *Access to Technology.* Technologies that provide immediate benefits or that require little capital or human resources can be applied to both short- and long-term contracts. Other technologies take years to implement, require careful planning, design, and installation, require capital investment on behalf of the private firm, or provide long term rather than immediate benefits. These technologies become available to the project only if there is a long-term contract.

- *Compounding of a Lower Rate of Annual Cost Increases.* Typically, the private firm pegs its annual fee adjustment to changes in the Consumer Price Index. Historically, municipal utility budgets often increased at a rate of one to several percentage points above the Consumer Price Index. For example, if the municipality's projected annual budget increases are 5% and the private firm's annual increases are 3%, after 20 years, the municipally operated budget would be about 45% greater than the private firm's fee. This calculation does not take into consideration any immediate savings. It assumes both budgets are the same in year one. Obviously, if the private firm's first-year fee was less than the projected municipally operated budget, the differences after 20 years would be much greater.

More specifically, North Brunswick's contract's benefits and savings include:

1. Private capital investment in the form of an up-front "concession" fee and guaranteed annual "concession" payments.

2. Private capital investment for retirement of the outstanding tax-exempt debt, which increased the Township's bonding capacity.
3. Private firm's assumption of most of the inflation risk.
4. Savings from the compounding of the private firm's projected lower rate of annual cost increase as compared to continued public operation.
5. Savings of about $45 million (present value $21.9 million) over the projected cost of continued municipal operation.
6. Private firm's greater ability to market and sell through sharing of new wholesale water sales.
7. Savings from the private firm's investment in long-term cost reduction programs including reduced energy usage, improved technology, automation, and decreased labor costs.
8. Elimination of periodic procurement costs for the operation of the facilities. For the Township, they are procurement and contract negotiation costs. For the private firm, they are marketing, proposal preparation, and contract negotiation costs.
9. Private capital investment for the purchase and installation of new water meters.
10. Cost decreases from the private company's assumption of nonlabor operating costs that were not included in the previous one-year contracts.
11. Cost savings from broadening the scope of services to the operation of the Township's sewerage collection system.
12. Cost savings from broadening the scope of services to the full operation of the water distribution system, including all personnel and operating costs.
13. Cost savings from broadening the scope of services to the billing and collection of user fees.
14. Savings from the private firm's assumption of a specified portion of repair and replacement costs.
15. Reduced liability or insurance premiums as a result of the private firm's comprehensive general liability and other insurance, provision of a performance bond, and indemnification of the Township for certain circumstances, including negligence, willful acts, omissions, breach of contract, and regulatory noncompliance.

Items 1 through 8 are the result of the longer contract term. Items 9 through 14 are due to the broadened scope of the project as compared with the township's previous operations contract. However, some of these, such as items 9 and 14, would be difficult to incorporate into a short-term contract. Item 15 is a benefit that is representative to almost all private contracting for these services.

2. ENABLING LEGISLATION

State legislation enabling long-term public–private partnerships for water supply and wastewater treatment either does not exist or varies widely from state to state. In 1993–1994, two laws, each designed to end after six months, enabled long-term water system public–private partnerships in Hoboken and Sayerville, New Jersey. These were negotiated, sole-source contracts. Their success, along with the support of Governor Christine Todd Whitman, created awareness that new broadly enabling legislation was needed. These new laws, the Water Supply Public–Private Contracting Act (P.L.1995, Chapter 101) and the Wastewater Treatment Public–Private Contracting Act (P.L.1995, Chapter 216), were passed by the legislature and signed by the Governor in 1995. North Brunswick's contract used both laws for the first time.

The laws allow customization for each municipality's individual circumstances and needs. This is balanced by a comprehensive process, which includes public procurement, a public hearing, and review or approval by state agencies. Any combination of the following services is allowed, provided that private capital investment is required: financing, designing, construction, improvement, operation, maintenance, and administration. There is no specification as to the amount of private capital investment, but the amortization of that debt is the justification for the long contract term, which may extend 40 years. Asset sales are prohibited by both laws. Briefly, the two laws require:

1. Public advertisement for proposals.
2. Selection of a private firm.
3. Negotiation of a contract containing specific language and provisions covering:
   - Private firm's charges and fees
   - Allocation of risks
   - Default and termination
   - Employment of the current employees
   - Any performance bond requirement

- Private firm's authority relative to bulk water sales
- Directions for payment of concession fees
- A specific dispute-resolution procedure
4. Bond counsel opinion.
5. Notice of public hearing.
6. Public hearing in which the municipality explains all of the terms and conditions and answers questions.
7. Contract adoption by the municipality.
8. Submission of the contract, public hearing records, and responses to state agencies. For both water and wastewater contracts, the Department of Environmental Protection reviews the contract and the Department of Community Affairs approves or disapproves the contract. For water contracts, the Board of Public Utilities also approves or disapproves the contract, regardless of whether the system is regulated or unregulated.
9. Review, approval or conditional approval by the agency/agencies.
10. Resubmission if conditionally approved.
11. Approval or disapproval.

Additionally, if tax-exempt debt is involved, review by the Internal Revenue Service is required to ensure compliance with applicable IRS regulations, such as the need to retire recent bonds via a tender offer.

The most innovative provision of the laws is the allowance of "concession" fee payments to the municipality. Concession fees are "payments that are exclusive of, or exceed, any contractually specified reimbursement of direct costs incurred by the public entity." They can be looked at as a loan on future savings or the recovery of excess revenues. The allowed uses of the concession fees differ between the two laws. In the water act, which was written earlier, only tax relief is allowed. In the wastewater act, they can be used for reducing or offsetting taxes or service rates, capital asset expenditures or one-time nonrecurring expenses. Concession fees can occur at any time during the contract. The current administration can recover some of the value of the contract without having to sell it using up-front concession fees.

The balance between capturing the net present value of future savings or revenues as an up-front fee versus providing savings or limiting rate increases throughout the contract life is one of the most important local decisions. Although North Brunswick will receive up-front and annual concession fees and other capital investments for debt repayment and new water meters, significant savings are still projected over the contract's life as compared to continued municipal operation. The projected savings of $45 million represents about 21% of the overall contract value.

3. PROFESSIONAL CONSULTANTS

Procuring a long-term public–private contract for water and wastewater systems is a complex process requiring specialized financial, legal, and technical knowledge. Usually, the municipality employs one to four professional consultants, depending on the size and complexity of the system and on the amount and complexity of required services and financing. Other factors are the capabilities and availability of the municipality's professional staff. Of 10 municipalities in Northeastern United States that have either completed or are in the process of such procurement, all but one used specialized consultants.

In North Brunswick's case, the financial and technical complexity and the inherent liabilities of the utility operation required three types of consultants. DeCotiis, Fitzpatric & Gluck were legal advisors, Ernst and Young were financial advisors, and CME Associates, the township's existing engineering consultants, were technical consultants. Their services included: review and recommendation of the optimum contract structure and enabling legislation; drafting the request for proposals; supporting the procurement process; reviewing and analyzing the proposals; recommending the selection of a private firm, and contract negotiation. These consultants served a valuable role in the undertaking of this procurement.

Two consultants provided continuous oversight of the contract. CME Associates provide technical oversight and developed the capital improvement program, which the Township chose to exclude from the contract. Lapercq Financial Advisors furnished financial oversight. Several key municipal managers provide ongoing direct oversight and daily administration of the contract.

4. NORTH BRUNSWICK'S CONTRACT PROVISIONS

The major provisions of this contract appear in the introduction. The most important provisions are the term, the breadth of services provided, the exclusion of capital improvements, the payment of an up-front concession fee, and the private firm's derivation of its revenues from user fee receipts. The payment of annual "concession" payments is not unexpected, since the utility already generates an annual surplus. Following is a summary of the contract's provisions:

1. Twenty-year term with four additional five-year renewal terms.
2. Private firm to operate, maintain, and manage the water treatment, water distribution, and wastewater collection systems, and to institute an infiltration and inflow reduction program.
3. Private firm to bill and collect user fees on behalf of the township.
4. Private firm's revenue is the user fee receipt after distribution by the trustee. The Township bills and collects connection fees and pays the private firm a tapping fee.
5. Private firm to install new automatic read water meters over five years ($2–$4 million).
6. Affected Township employees, primarily some of the water distribution and sewer collection crews and the user fee billing and collection staff, are guaranteed employment for two years. During this time, both the Township and the private firm will seek to find new positions for those employees who cannot be utilized.
7. Township is responsible for capital improvements and repairs over $50,000 and receives user fee revenues allocated specifically for such expenditures.
8. Township retains ownership.
9. Township can terminate the contract without cause by paying a prorated termination fee (facilities are not leased).
10. Private firm assumes inflation risk up to the occurrence of 6% or greater inflation in two consecutive years.
11. Revenue from additional water sales is shared. The private firm receives 10% of the gross revenues of water sales between 7.3 and 8.3 MGD (millions of gallons per day) and 15% of gross revenues of water sales above 8.3 MGD.
12. Private firm may seek and propose additional wholesale water sales opportunities.
13. Private firm responsible for all normal operating costs including raw water purchase costs from the New Jersey Water Commission and sewerage conveyance fees to the receiving treatment facility, Middlesex Utility Authority.
14. Up-front concession fee of $6 million.

15. Annual "concession" payments of between $0.6 and $2.0 million, totaling $23.9 million in place of less certain utility surpluses.
16. User fee rate increases for the services covered by the contract are set for 20 years.
17. Annual or more frequent system audits to be performed by the Township's consulting engineer in addition to other oversight mechanisms.
18. Retirement of the Township's existing tax-exempt debt, $22.8 million, by defeasance and tender offer.
19. Forfeiture by the private firm of the nonrecovered portion of the up-front concession fee in the case of its termination for cause.
20. Private firm furnishes a $3.0 million performance bond.
21. Private firm provides the following insurance coverages in millions:
    - Workers Compensation and Employer's Liability in accordance with state law;
    - Public Liability—$5.0 each occurrence/annual aggregate with XCU coverage of $1.0 CSL;*
    - Automobile Liability and Property Damage—$5.0 each occurrence/annual aggregate;
    - Excess Umbrella Liability—$10.0.

5. RATE INCREASES AND ANNUAL PAYMENTS

The water and wastewater rates in the year prior to the first contract year were $19.47 and $23.98 per 1,000 cubic feet, respectively. The rate increases for the services included in the contract for both rates are:

- *Year 1.* 5.75%
- *Year 2.* 5.50%
- *Year 3.* 5.25%
- *Year 4.* 5.00%
- *Year 5.* 4.75%
- *Year 6.* 4.50%
- *Year 7.* 4.25%
- *Years 8–20* 3.00%

*XCU is insurance coverage that involves Standard Blasting or Explosion Coverage, Standard Collapse Coverage, and Standard Underground Coverage. CSL is liability insurance written with combined single limits. It is the maximum that an insurance company will pay.

The water and sewer rates in the twentieth year of the contract, 2016, are $40.23 and $49.55 per 1,000 cubic feet, respectively. These rate increases are separate from increases for capital improvements. The annual "concession" payments paid to the township by the private firm in each year of the contract are as follows (all payment values in millions):

- *Year 1.* $1.0
- *Years 2–10.* $0.6
- *Years 11–15.* $1.5
- *Years 16–20.* $2.0

6. PROCUREMENT AND SELECTION

The procurement process took North Brunswick about 10 months from advertisement for proposals to contract signing, and it took four more months until start-up. These time frames are about twice as long as needed because the laws were not passed until about four months after advertisement for proposals. As a result, the original request for proposals had to be modified. It was the first use of both laws. The procurement process required the submission of a second set of proposed fees based on a specific up-front concession fee and a more detailed description of required services. Selection of the private firm involved negotiation first with one firm and then with the final firm. Six firms requested the proposal package, and three submitted proposals.

One lesson learned from the process is that the city must be specific about the amount of up-front concession fee desired. The original request for proposals did not do this. Comparing 20 years of rates, along with varying up-front and annual concession payments, added unnecessary complexity to the financial evaluation portion of the selection process. Alternately, the rate increases or annual payments could have been fixed and the up-front fee left variable. Another lesson is to be specific about the scope of services. This applies particularly to the collection and distribution systems, where there is a wide range in the level of service practiced. Equalization of the services to be provided was one of the reasons negotiations with the first firm ended. After such equalization, their proposal's relatively small advantage in net present value of savings became a slight disadvantage. A contract was successfully negotiated with the second firm, US Water L.L.C., which also happened to be the firm that had operated the water treatment plant for ten years.

An important consideration is whether to base selection on a single financial criterion. The original proposal documents identified the sole financial evaluation criterion as the net present value of savings using a specific set of discount rates, Treasury Note, Bill, and Bond closing interest rates on the day of proposal submission. As mentioned, the difference in net present values between the two lowest cost proposals was relatively small. Since the selection of discount rates for the net present value calculation is somewhat arbitrary, it may be better to use more than one set of discount rates to evaluate each proposal. If a proposal has the best net present value under a range of discount rates, it clearly should be ranked first financially. If two proposals have the best net present value with some but not all the discount rates, other financial criteria should also be used to determine their financial ranking, or they should be ranked equivalently.

7. CONCLUSIONS

Long-term contracts for the operation of municipal water and wastewater systems are relatively new. Previously, a municipality that sought to recover financial value from its utilities had only one option, the sale of the systems. Not only does the municipality lose control forever, but also it loses the ability to recover the value of future advances in automation, design, operation, and management. Sale to a private firm also results in water-rate increases from the application of franchise, gross receipts, and state and federal income taxes.

As the first application of two new state laws, North Brunswick Township's long-term contract for the private operation of its water and sewer utilities is a model for other municipalities that desire to:

- Capitalize on the capabilities of the private sector;
- Recover as "concession" fees part of the value of future savings;
- Not lose control over the system and user fee rates; and
- Not involve a lease, mortgage, or lien to secure the private firm's project financing.

Chapter 17

Privatization of Operation and Maintenance of Wastewater Control Facilities

City of West Haven, Connecticut

H. Richard Borer, Jr.

At the end of our fiscal year 1991–1992, the City of West Haven faced a looming financial crisis with a deficit balance of over $17 million. This deficit had accumulated over the course of two years prior to the Borer administration coming into office. The City of West Haven was literally out of cash, and due to mismanagement in previous administrations, the city was unable to access the capital markets. As this administration struggled to manage the financial crisis, we also had other battles to fight in other sectors of city government operation, most notably, upgrades to our wastewater treatment plant facilities.

West Haven, population 52,000, is a densely populated residential community, comprising only 10 square miles, on Long Island Sound. The city had experienced increasing problems with its wastewater services over

H. Richard Borer, Jr., is Mayor of West Haven, Connecticut. He is recognized as a public policy expert in the water field.

Reinventing Water and Wastewater Systems: Global Lessons for Improving Water Management, edited by Paul Seidenstat, David Haarmeyer, and Simon Hakim.
0-471-06422-X   Copyright © 2002 by John Wiley & Sons, Inc.

the years, primarily due to long-deferred maintenance of its aging sewer lines (some installed in the early 1900s), 13 pump stations, and secondary treatment facility.

1. WASTEWATER TREATMENT PLANT UPGRADE

West Haven considered privatization of the water pollution control plant because of a history of mismanaged operations. Decades of underfunding for maintenance and capital replacement had left the City with an obsolete and worn-out wastewater treatment plant that was routinely cited by the State Department of Environmental Protection for being out of compliance with permitted effluent requirements.

The treatment plant was viewed by city workers as the most undesirable place to work in city government. As a consequence, good employees routinely left the plant when openings occurred elsewhere in city government. City managers at the plant were never able to effectively cultivate a professional, motivated workforce that took pride and satisfaction in their work. Rather, the plant was viewed as a "dumping ground" for employees who did not function well in other departments. A few workers had even used intimidation and sabotage to destroy earlier attempts to improve the operations at the plant. The "culture" of the organization was largely self-destructive.

When the Borer administration came into power in December 1991, the City was being required by the Connecticut Department of Environmental Protection (DEP) to invest $25–$30 million to adequately protect against water pollution. Due to past violations, the City was fined $500,000 by the DEP for noncompliance. This was the largest fine ever imposed on a municipality by the DEP. As a result of continuing noncompliance, the courts had imposed a stipulated judgment on the City, which required us to make substantial improvements to the plant. The state order had also found that the City was required to hire a Class IV certified wastewater operator to manage the system.

It is estimated that the City will need to borrow $23 million to complete the project according to DEP specifications. The City has already borrowed almost $12 million. These major problems cropping up at the facility were due in large part to the City's past poor management of the wastewater treatment facilities. The facility had not been maintained well on a daily basis, and the City had not invested in the long-term capital requirements to insure that the machinery was replaced before it wore out or had already broken down.

Breakdowns in the City's ability to properly manage our facilities resulted in the state fining the City because of continuous sewer overflows. We also lacked the ability or failed to provide the necessary budgetary support to comply with state and federal permits. Our facility was in need of a comprehensive maintenance management program. In order to insure that the City improved the operation and maintenance of the wastewater treatment system, the state order also mandated the adoption and implementation of a direct user charge system, removing day-to-day funding for the system from the city's general government budget.

The Stipulated Judgment required the City to operate within the limits of its NPDES (National Pollution Discharge Elimination System) Permit in terms of effluent quality and treatment practices. The Stipulated Judgment also required the City to meet tight timetables with respect to the elimination of sewer overflows, which had been a chronic problem in West Haven.

Other provisions of the Stipulated Judgment covered: standards with respect to staffing levels and shift coverage; educational levels and certification requirements of the staff; operator training; preventive maintenance activities; reporting requirements; and method of funding the operation of the plant through a sewer user charge.

The City proceeded to get its wastewater treatment system in order by taking the following steps. In 1993, it signed an agreement with Wheelabrator Environmental Operating Service (EOS), Inc. to supply interim management services for the facility. The purpose of retaining the interim management services was to properly staff and maintain the facility to assure compliance with the state permit and to comply with the requirements of state regulations. The interim contract with Wheelabrator ran from July 1993 until October 28, 1994.

Wheelabrator's scope of services during this interim period provided for:

1. Daily management of facility operation, process control, maintenance needs, and safety practices.
2. Weekly written reports regarding system operations, process performance, and maintenance needs/accomplishments.
3. Establishment and implementation of basic hands-on training for process control and maintenance management for plant operation

and maintenance (O&M) staff. This included safety training, maintenance management system training, maintenance training, and process control training.

4. Establishment of a Hazardous Materials (Haz-Mat) Team to respond to plant emergencies, particularly chlorine handling.
5. Setup of a maintenance management system and spare-parts inventory, and evaluation of present staffing needs.

Wheelabrator would also provide that, in accordance with DEP requirements, a Class 4 operator would manage the West Haven facility and Class 3 operators would supervise each shift. Not meeting this requirement could have resulted in a DEP fine of $25,000 a day.

During this interim period, we found that there were certain bureaucratic constraints involved in having a privately managed but publicly staffed operation. These constraints further limited the City's ability to reach the full potential for improvements to the operation and maintenance of the wastewater system. In addition, since we were in a financial crisis, it also lessened our ability to provide cost savings to the citizens.

The City began pursuing a contract for full operations, maintenance and management (OM&M) for the system once the interim contract period was up in October 1994. The City's rationale for this move was in part that full contract operations would provide us with the technical expertise needed to maintain regulatory compliance for our entire wastewater system, as well as reduce our operating costs.

The city first issued a request for proposals (RFP) for full contract operations to several national OM&M firms in the spring of 1994. Developing a clear and comprehensive RFP is crucial to the success of a privatization effort. Among the items we considered in formulating our RFP were the following:

- Lump sum versus cost plus agreement.
- Single, all inclusive proposal document or separate technical and cost proposals.
- Handling of repair and maintenance items. Would these be included in a lump sum bid or paid on a pass-through or cost plus basis?
- Handling of capital improvements. Which party would be responsible for improvements to the plant?

- Impact of capital improvements on future operating costs and how to reflect this impact in a multiyear contract.
- Engineering and other technical services required under the contract.
- Operating standards. Who would be responsible for compliance with all regulatory requirements, odor control standards, and so on.
- Provision for contingency: is that included in contract price?
- Reporting requirements, both operational and financial. What reporting arrangements would be required?
- Liability issues. Which party would assume liability for what exposures? Areas of concern included pollution liability, liability for DEP fines, sewer back-up claims, chlorine leaks, and so on.
- What selection criteria would be used to select a vendor?

Three proposals were received, and they were evaluated by a city committee composed of the city's Public Works Director, Director of Finance, City Engineer, and representatives from the city's Sewer Commission and Corporation Counsel's office. The committee met several times to review the responses and to formulate questions of the respective firms. Committee conducted two-hour interviews with each of the three firms (the Evaluation Questionnaire is shown in Section 5).

The key information that the Committee was attempting to determine included: technical capabilities of the firm to manage the plant in a thorough and professional manner; financial stability of the firm; satisfaction of current and former clients; and completeness of the firm's proposal.

The Committee also sent a team of City officials to treatment plants that were being operated by each of the three firms. The officials toured the plants unannounced to get a feel for the general operating conditions of each plant and to speak with plant personnel, including the line operators. They also spoke with government personnel in the cities where the plants were located to assess the satisfaction of each firm's respective clients.

2. SELECTION CRITERIA

The selection criteria used by West Haven were established before the City issued its RFP, and price was not the only criterion. The criteria, with their assigned weight, were as follows:

Technical merit	30%
Experience and responsibility	25%
Capability to complete the contract	25%
Price	20%

Since all the firms were major players in the industry, the Committee rated all three firms very highly on all criteria. Although the Committee found differences among the firms, there is little doubt that any of the firms we interviewed could have done a credible job of running our plant. Most of the distinctions between the firms tended to be rather subtle.

Partly because all of the firms were technically so strong, price played a large role in the selection process, and the Committee ended up recommending the firm Professional Services Group, Inc., (PSG) Houston, ranking it highest with all criteria, price among them. Following subsequent approval from myself and the state's Financial Review Board, PSG assumed operation of West Haven's system in October 1994.

3. OTHER CONSIDERATIONS

There were other considerations that went into the selection process. As part of its contract, West Haven required the contract operator to hire all existing City employees at the plant. This reduced disruption from union bumping in the rest of the organization, reduced unemployment costs, and eased the transition to contract operations. Existing union contract provisions had to be considered. Until a new labor contract was in place, our collective bargaining agreement did not allow the City to subcontract bargaining group work. This is a major impediment to privatization in strong labor union states.

Any privatization contract that is ultimately signed must contain very clear language on the subject of whether and under what circumstances the issue of contract price may be reopened by the contract operator. In our case, the only reason the contract can be reopened is due to a change in the scope of the project, which is defined as a 10% or more increase in total flow or suspended solids, or a 10% or more adjustment in the Consumer Price Index (CPI), or in energy costs.

As in any contractual arrangement, there are always going to be disputes over the intent of the language. The City and contractor cannot always agree on every issue, in particular the change in scope. Under our contract, instead of going to court and enduring long, and usually, costly legal battles, the parties agreed to arbitrate differences through procedures of the American Arbitration Association.

4. CONCLUSION

Our experience in this process has been good. West Haven has been able to reduce both operating and overtime costs, and to prevent overflows; employees have gained pride in their work performance. We had already utilized contract operations for our tree department and yard waste compost site, but wastewater services are by far our largest public–private partnership. I am a proponent of putting services out to bid to see if the private sector can compete on both cost and delivery.

After privatization was initiated, we saw a significant improvement within only a matter of months at our treatment plant due to adopting full contract OM&M. The move to contract operations has definitely been in the best interest of the taxpayers, the employees, and the users of these services—our citizens and businesses.

5. THE EVALUATION QUESTIONNAIRE: OPERATIONS AND MAINTENANCE PROPOSAL EVALUATION ITEMS

1. Baseline flow deviation allowance? Also include Biochemical Oxygen Demand (BOD) and Suspended Solids (SS).
2. How are extra services handled? How are they billed?
3. Right of Town to audit? Performance audit?
4. What is your approach to identifying and solving process problems?
5. What experience do you have with budget preparation?
6. Requirements to maintain personnel as stated and/or agreed to. No changes in personnel without written authorization of owner.
7. How will NPDES permit changes be handled?
8. Evaluation weighted factors:
   - Technical merit—30%
   - Experience and responsibility—25%
   - Capability to complete contract—25%
   - Price of proposal—25%
9. How are reimbursable expenses handled, if any?
10. Project organization? Name key personnel or person to be on-site.
11. Ownership of proposer: Who? How much? Traded stock?

12. Central or main office location? Nearest office location?
13. Bank reference?
14. References for current operations in Connecticut and/or adjacent states? Names and phone numbers
15. Names and references for five completed operations contracts in Connecticut and/or adjacent states?
16. Audited financial statements for past three years.
17. Detailed technical approach to West Haven's O&M.

Chapter 18

Phoenix Water Services Reengineers to Improve Practices and Save Money

Skip Rimsza

Citizens of Phoenix, Arizona, are giving the City's elected officials and the Water Services Department (WSD) high marks for service and water quality. One reason is a new program designed to help the department improve customer service and stay competitive. Begun in 1996, the WSD Reengineering Program is based upon a cooperative program between labor and management to achieve higher levels in excellence in all of its services, including being more responsive to the customers' needs in both water quality and support functions, and cutting costs through a change in work practices and workforce culture. Since the program's inception, the City Council has strongly supported it, and the program is achieving for the City a combined cost savings and cost avoidance in excess of $10 million per year.

The Phoenix WSD operates one of the nation's largest water and wastewater utilities, annually producing more that 100 billion gallons of water for about 1.3 million residents living in a 530 square mile service area. The department's five water treatment plants have a capacity of 630 million gallons daily (MGD). The two wastewater treatment plants can process

Skip Rimsza has been Mayor of Phoenix, Arizona, since 1994. He previously served as vice mayor and city councilman and is an innovator in the field of urban management.

Reinventing Water and Wastewater Systems: Global Lessons for Improving Water Management, edited by Paul Seidenstat, David Haarmeyer, and Simon Hakim.
0-471-06422-X   Copyright © 2002 by John Wiley & Sons, Inc.

about 225 MGD. One of the plants is operated by the department for Phoenix and the cities of Glendale, Mesa, Scottsdale, and Tempe.

Because of the size and successes of the municipally run WSD, it naturally became a target for privately held utility companies—both domestic and internationally owned—attempting to privatize public utilities. Over the past 10 years, the utility marketplace has grown increasingly competitive, due to privatization efforts led by foreign-owned companies. This effort is forcing public utilities to take a hard look at their operations and determine where they can enhance operational efficiencies to stay competitive with the privatizers when they come to lobby for time with the Mayor and City Council.

Cities such as Wilmington, Delaware, Arvin, California, Atlanta, Georgia, and Honolulu, Hawaii, are just a few of the locations where privatizers have taken over operation of some public services. With full knowledge of what the privatizers were capable of, and without full knowledge of where it stood competitively, the WSD began its Reengineering Program by conducting a competitiveness assessment to see how it compared with the most efficient utilities, public or private, using best practices.

Based on the results of that assessment, the WSD closely examined its organizational structure, work practices, and information and control technology infrastructure, as well as the opportunities presented by upgrades and expansions planned for treatment facilities. Then, with the assistance of a consultant, the department developed a unique partnering relationship with its unions, beginning the most ambitious change process in its 92-year history.

The partnership with labor, called PALM for the Participative Association of Labor and Management, is achieving a new level of cooperation and teamwork that benefits labor, management, the City, and, most of all, consumers. With representatives from the American Federation of State, County, and Municipal Employees Union (AFSCME locals 2384 and 2960) the Administrative, Supervisory, Professional and Technical Employees Association (ASPTEA), and department management, PALM provides oversight for the reengineering effort. The high degree of collaborative participation by both labor and management has created a powerful and uncommon partnership. Labor and management work together to ensure appropriate staffing levels are maintained for safety, customer service, and environmental protection, while increasing sensitivity and responsiveness to the community's wants and needs.

Together, management, the labor unions, and a consultant looked at strategies for enhancing the department's productivity levels and overall operation. Through this process, the WSD management developed realistic goals for the program, some of which went far beyond just saving money. Included among those key goals are: achieve excellence in regulatory compliance and environmental protection, achieve excellence in service levels and customer service, minimize staff increases due to expansions and plant upgrades, keep safety an utmost consideration, and limit future rate increases to low single digits. All of these goals, and much more, add up to the chief objective of reengineering in the WSD: to become a "Best of Class" utility.

Early in the program, management promised the enhancement to productivity would not occur through a traditional downsizing in the department. The WSD opted instead to undergo a hiring freeze, to retrain its employees to work smarter—not harder—and to change the work culture of a single-function, ivory-tower workforce to a multifunctioned, team-based culture. Throughout the entire process, the department sought and received approval of the Mayor and City Council. So far, the hiring freeze has resulted in an average of about 145 vacancies over the past four years, all due to normal attrition. These positions created an inventory of jobs that could be used for employees if their existing job classifications were not needed. Consequently, jobs were eliminated without eliminating employees.

Before any major changes occurred, though, the WSD management and labor partnership agreed to break the program into three distinct phases. Phase I focused on quick savings. Phase II called for the workforce culture change and for testing teamwork principles at the facilities with "pilot" programs. And last, Phase III targeted major changes to be implemented throughout the divisions undergoing reengineering.

In Phase I, department employees used their experience and knowledge to make recommendations that could be implemented quickly to result in some quick savings. This "bottom-up" approach to making changes is a cornerstone of the department's reengineering program. Labor and management agreed early on that employees doing the job know best how to change the operation. The employees' commitment has made the program particularly effective, because they are implementing their own ideas and concepts.

The success of the "bottom-up" approach was exciting. In the fall of 1997, the reengineering team recommended to the Mayor and the City Council measures to optimize electrical and chemical usage and to eliminate some already vacant job positions. The City Council approved those measures, which saved the WSD some $3.1 million per year and produced momentum to energize Phase II of the program. In addition to those Phase I savings, the hiring freeze was creating salary savings of more than $2 million per year.

Not just for Phase I, but throughout the ongoing reengineering program, the WSD has continued to follow the same sequence of bringing recommendations to Mayor and the City Council for approval before implementing changes, no matter how small the change.

In Phase II, or the pilot phase, a team of employees began to design a new mode of operating a water treatment plant that would test the concept of combining operations and maintenance functions while establishing a team-based work culture.

The team started with the city's 140-MGD Squaw Peak Water Treatment Plant (WTP). First, the team analyzed the work being performed and then they designed new operational practices that would increase efficiency, maximize cost savings, maintain safety, and still meet all regulatory standards.

Historically, an operations staff member would hand off work to a maintenance staff member and then go on to perform another operations-related task. Under the new combined function of an Operations and Maintenance (O&M) Technician, an employee takes ownership of a problem or situation and sees it through to completion, sometimes with the assistance of a fellow O&M team member. And if a problem goes beyond the scope of what the team can handle, the team goes directly to the plant administrator to resolve the problem. The middle level of supervision is eliminated in favor of the team concept.

The first pilot began in July 1998 at the Squaw Peak WTP. After some basic training in team building, the team, comprised entirely of volunteers, set out on the journey that would show that change was possible in the operation of the department. Changing a group culture is difficult, but the team at the Squaw Peak plant persevered despite facing training deficiencies and pressure from others who hoped to maintain the status quo.

In December 1998, the team completed the pilot and then began to identify benchmarks that could be used in subsequent pilots. The analysis revealed that a savings of about $800,000 per year could be realized by operating the plant under the new mode. Most of the savings were due to the decrease of staff (from 30 to 14 employees), and by the team working smarter, not harder. The decrease in staff affected all levels.

The next pilot, held at the 23rd Avenue Wastewater Treatment Plant (WWTP), was approached in similar fashion, but the staffing level of 38 (reduced from 80) proved to be a challenge when it came to steering the pilot in the right direction.

The pilot design differed slightly from the one used at Squaw Peak because of the type of operation. Where the team at Squaw Peak had received only limited team building (sometimes called soft-skills) training before it started the pilot, the staff at the 23rd Avenue plant received much more team-building training. An analysis showed an imbalance: too much team building and not enough on-the-job cross-training. The increased size of the team made the problems more complex until the imbalance in training was corrected. In fact, while that imbalance was being corrected, the pilot team at 23rd Avenue identified many other aspects of training that would benefit employees taking part in other pilots.

Team-building training at both the Squaw Peak and 23rd Avenue plants included what is known as the Forming, Storming, Norming, and Performing phases. During the Forming phase, the team is assembled and defines its goals and direction. In the Storming phase, the team defines and sets the rules of the decision-making process and negotiates each person's roles and responsibilities. This phase helps resolve any personality conflicts that arise from differing viewpoints. The size of a team has a direct bearing on how long it takes to go through the Storming phase. The larger the team, the harder it is to reconcile personality differences and the differing viewpoints. During Norming, the team has come to an agreement about how to work with each other. The personality conflicts disappear and cooperation becomes routine.

The last phase, Performing, is when the team works as a single entity to accomplish their objective and goals.

After the pilot was completed at the 23rd Avenue plant in December 1999 and deemed a success, the analysis showed that, because of the team size, it was much harder for the team to go through the first three phases. Even before the 23rd Avenue pilot was finished, the reengineering team implemented a new and improved comprehensive training program—about 160 hours in length—for the next pilot, conducted at the Deer Valley WTP. Thus the Deer Valley staff benefited from the trials and errors of others.

Employees on the pilot teams are aware that challenges are opportunities to improve not only their part of the program, but the department's program as a whole. It is that particular shift in thinking that has made the reengineering program work as well as it has. That shift in thinking has occurred, in part, through the team-building or soft-skills training that was provided prior to the start of each pilot.

That training at both the Squaw Peak and 23rd Avenue pilot projects was considered from the beginning of the reengineering program to be a major reengineering "enabler." Enablers—which are tools that help make the reengineering program a success—must be devised to fit needs at both the operational level and organizational level.

At the operational level, technology enablers help make pilot staff more effective as O&M technicians. Several of these enablers were first used on the 23rd Avenue pilot. The first technology enabler implemented was an alpha pager system that alerts the O&M pilot staff of system alarms that need attention. Historically, plant employees had to access a computer program to run a check of the entire wastewater treatment system at the plant to find out if there was an equipment malfunction occurring. With the paging alarm system, the plant equipment tells the computer what is wrong, the computer, in turn, sends a page to all the O&M technicians on duty, and they can respond to the specific issue immediately. Not only is this a good example of applying technology to become more effective, but it also is a good example of how the reengineering program has inspired employees throughout the department to think "outside the box."

Another enabler instituted at the 23rd Avenue WWTP, and subsequently used throughout the water production and wastewater treatment divisions, is the Computerized Maintenance Management System (CMMS). This system is being used to generate and manage the maintenance work at each plant. It is also a tool that supports the department's new maintenance strategy—Program Driven Maintenance. Before the CMMS, maintenance was performed only reactively. Staff would wait for something to break down or wear out before any maintenance was performed. Now, with the CMMS, equipment histories and maintenance records can be stored in a database along with repair cycle information. The CMMS kicks out a work order so that predictive as well as preventative maintenance and repairs can be made before critical components of the operation break down. Simply stated, the philosophy of the CMMS is to maintain equipment so it does not break.

Along with the CMMS, a new centralized maintenance group has emerged to support the plants' O&M teams by tackling specialist-type work throughout several divisions. Planner/schedulers at the plants work with planner/schedulers in the field to provide assignments for the centralized maintenance teams. Those same planner/schedulers also use the

CMMS to coordinate plant level needs with the centralized maintenance group when the work load and difficulty is beyond the capabilities of the O&M Technicians. Under this new maintenance strategy, the centralized maintenance group handles about 20% of the maintenance tasks in the department, while the O&M Technicians handle the remaining 80%.

At the organizational level, reengineering teams and management created enablers as incentives for employees to enter the new work culture of teamwork required by the new combined O&M job duties. The first enabler created was Pilot Incentive Pay—pay that is given to employees on the pilot team after the pilot is completed, if the pilot program is deemed a success.

Another enabler, Multi-Skilled Pay, which is still under development, is linked to the creation of the new O&M Technician classification. Multi-Skilled Pay is an employee-driven compensation plan. All of the O&M Technicians who enter the plan will receive additional pay when they acquire additional skill sets. The Multi-Skilled Pay plan uses a 10-block matrix that is comprised of three levels consisting of three blocks each, all leading up to a specialist level position at the top (the 10th block). There is one block for each type of skill that makes up the O&M Technician classification. The three skills are Electrical and Instrumentation (E/I), Maintenance (M), and Operations (O). The basics of the plan are simple: employees will use their initiative to progress through the matrix and acquire more skills (an employee will have to complete a set of three blocks horizontally before going to the next level). Following that progression from skill block to skill block will make an employee more multiskilled while they earn additional compensation.

While enablers are helping the reengineering program become highly effective, WSD management has been implementing a new philosophy in regards to the design and staffing of future projects and merging the differences between water plant and wastewater treatment plant operations. One example of this philosophy is the city's future Cave Creek Water Reclamation Plant. When originally planned, the plant was to have a staff of 33. Now, due to advances in automation and applied reengineering principles, the plant will have between three and five employees. Additionally, in the past, this reclamation plant would have been operated by the department's Wastewater Treatment Division. With the plant only a little more than a mile from one of the department's water production facilities, the Water Production Division instead will conduct operations.

Other projects, such as the solids-handling facilities being built at Phoenix's water treatment plants as part of the new requirements mandated by the U.S. Environmental Protection Agency, are designed so that they can be operated without adding additional staff. In addition to these projects, the department has embarked upon a five-year, $1.5 billion Capital Improvement Program. The program includes new infrastructure to cope with the city's growth (averaging more than 6,000 new customers per year), improving existing infrastructure and replacing old infrastructure. The WSD reengineering program is saving money by keeping labor costs in check.

One major area of the WSD that is impacted by the growth Phoenix is experiencing is the Customer Service Division. The division, which handles nearly 350,000 accounts, is using the reengineering program to find ways to improve the service level customers receive despite adding more customers each year. For example, the number of water meters being read has increased from about 283,000 in June 1990 to 345,000 in June 2000, while the number of meter readers has been reduced.

The reengineering strategy in the Customer Service Division is in line with the efforts of other divisions in the department, but a different approach is being employed. With the help of a consultant, the division is doing an Activity Based Costing analysis (ABC) and then benchmarking the business practices of the division against best practices being performed in other water utilities and other related industries. After identifying areas that needed improvement, management and labor formed employee-driven teams to look for solutions to the operational weaknesses. Also, as a result of the ABC and the benchmarking, the teams have identified several types of technology that will enable additional growth to occur without stretching manpower and sacrificing customer service.

There is no doubt that servicing Phoenix residents with such high standards is very important to its elected officials. The high level of accountability that Phoenix has set for itself is evidenced by the awards it has won, such as the Carl Bertelsmann "Best Run City in the World" award and *Governing Magazine's* "Best-Managed City in America" award.

The WSD reengineering program also fits in every way with the City's Vision and Values (see Table 18-1). High standards and innovation underscore those vision and values, and what the WSD has set out to achieve with its goal to be "Best of Class" is highly reflective of that. To help quantify reengineering achievements, in 1999, management asked the City Auditor to independently audit both the cost savings and cost avoidance. That audit showed $63.1 million would be saved over a six-year period, including final implementation. Phase III may create an additional annual sav-

Table 18.1　City of Phoenix Vision and Values

- We Are Dedicated to Serving Our Customers
- We Work as a Team
- We Each Do All We Can
- We Learn, Change, and Improve
- We Focus on Results
- We Work with Integrity
- We Make Phoenix Better

ings of $3 to $5 million, for a total combined savings and cost avoidance of nearly $70 million over a six-year period.

With that type of achievement, the WSD is very competitive with what privatizers say they can do. But the department did not stop at just being competitive. In regards to the reengineering program he initiated, Michael Gritzuk, our Director of the WSD, says,

> Phoenix's goals have been far superior to the private sector's, whose main goal is to be cost-competitive and make a profit. WSD's reengineering program is a far superior initiative. . . . We are doing much more than required by law to ensure our water meets or surpasses health and safety standards. We are doing more than required to protect the environment. We are surpassing industry standards for service. And we are doing all this while keeping costs for water and sewer service among the lowest in the country.

From a political standpoint, there are many advantages to keeping the WSD "public," but the most important advantage is that we remain accountable to the citizens and we continue to exercise direct control over our superior service levels. If a privatizer came in (with profit as the primary goal), our customers' desires for product, service, and environmental quality might not be so actively addressed.

In conclusion, the WSD reengineering program has proven that very large public utilities can successfully reengineer themselves, not only to save a significant amount of money, but also to guarantee that the city in which they serve, in this case Phoenix, will be able to give its citizens the best product and the best service anywhere. Or, as Mike Gritzuk says, "The Water Services Department is recognized as 'Best of Class'."

Part VI

A View of the Past and the Future

Chapter 19

London's Private Water Supply, 1582–1902

Nicola Tynan

Private water companies provided Londoners with piped water long before the 1989 privatization of water and sewerage in England and Wales.
London's first commercial water company, the London Bridge Water Works
Company, was established in 1582. Private companies owned and operated
most of London's waterworks for the next 400 years. New companies entered the market and private water supply continued to expand until
1902, when Parliament voted in favor of amalgamating the companies' assets and creating the Metropolitan Water Board (MWB).

The development of London's water supply provides an example of
largely unregulated private provision in an industry commonly believed
to require government ownership or regulation. Private companies built,
owned, and operated reservoirs, pumps, and pipe networks, and they supplied water to homes, small businesses, and public taps. The companies invested in technological innovations to improve service (Dickinson, 1954);
they also increased the volume of water delivered and offered customers
new supply options. Despite the switch to public ownership in 1902, much

*Nicola Tynan is Assistant Professor of Economics, Dickinson College. She previously was an econo-
mist with the World Bank and a development officer for the Institute of Economic Affairs in London.*

Reinventing Water and Wastewater Systems: Global Lessons for Improving Water Management,
edited by Paul Seidenstat, David Haarmeyer, and Simon Hakim.
0-471-06422-X   Copyright © 2002 by John Wiley & Sons, Inc.

historical evidence suggests that private water supply helped make the average London resident better off than contemporaries living in other cities.

During the period of private water provision, the population of central London grew almost 22-fold, from 200,000 residents in 1600 to 4,309,000 in 1899 (Finlay, 1981; Booth, 1969). Along with its population growth, London expanded geographically. For the first few decades of commercial network delivery, London was almost synonymous with the City (a square mile covering the same ground as the Roman walled city); by 1900, private water companies had extended service beyond central London on both sides of the Thames. In 1900, private companies supplied over 6.5 million consumers and an area of almost 620 square miles, including the 121 square miles of central London.

By the nineteenth century, London's extensive water system helped make the metropolis "one of the best housed and healthiest [cities in Europe], with a death rate lower than the birth rate by about 1800, at a time when most European cities were 'devourers of men'" (Daunton, 1995, p. 255). In 1850, the Board of Health report to Parliament stated that 94% of London residents received a pipe supply from one of the private companies. At least 82% received delivery to their home. The remaining 12% obtained their water from a company standpipe outside their house.

London's private companies are also owed some credit for bringing the benefits of piped water to other cities, through the export of technical knowledge, direct investment, and managerial skill. In Breslau, Germany, that city's first high-pressure waterworks opened in 1827. The waterworks made use of steam engines and pumping technology developed by the same engineers employed by London's water companies. Similar waterworks opened in Hamburg in 1833 and in Magdeburg in 1844. English investors, using capital from London, established Berlin's water company in 1853 (Brown, 1988; Kraemer, 1991). In Brazil, English entrepreneurs raised capital in London to build and operate waterworks in Rio de Janeiro, São Paolo, Santos, and Recife (Graham, 1972).

This chapter presents a chronological history of the evolution of London's water industry during the seventeenth, eighteenth, and nineteenth centuries. Section 1 looks at the early years of commercial supply from 1582 to the mid-1600s. Section 2 starts with the period of rising demand for water following the great fire of London in 1666 and ends just before another period of rising demand due to the rediscovery of the flush toilet in the late 1700s. The final section shows how a succession of technological innovations during the late 1700s and early 1800s encouraged entry and competition, and changed the service provided by London's water companies. It includes a brief discussion of the move from cesspits to sewerage and the rise of government regulation during the nineteenth century.

1. THE BEGINNINGS OF COMMERCIAL SUPPLY

Domestic piped water started as a commercial venture in sixteenth century London, when Peter Morris established the city's first private water company in 1582. Morris designed and built a tide-driven waterwheel to raise water from the River Thames and deliver it to the fishmongers living and working in Old Fish Street. At this time, most industrial consumers, particularly the large number of local brewers, obtained their own water from private springs, wells, or pipes from a local river. Morris responded to demand for large quantities of water from a small, concentrated, and organized group of consumers—the fishmongers—for whom the combined benefits of large-scale investment outweighed the cost. Morris received payments from the Fishmongers' Company, one of London's largest guilds (Robins, 1946).

Once established, Morris's Water Works not only supplied members of the Fishmongers' Company, but also provided the local community with an abundant supply of water for its public conduit. More importantly, Morris introduced the pipe delivery of water to individual households on a commercial basis. Morris made his distribution network technically possible and cost effective by introducing to England the use of closed pipes made from elm timber (Dickinson, 1954).

At this time, many vestries and parishes provided local communities with a communal domestic supply through springs, wells, and conduits. (Conduits were open or closed lead pipes carrying water from streams and tributaries beyond the City to small reservoirs in residential areas, relying on gravity to move the water.) Any household could take water for domestic use, and local residents either collected the water themselves for free or paid one of the competing water carriers for delivery. Many wealthy Londoners made donations to build and maintain these public conduits (Stow, 1971), and local taxation provided additional funds. Smaller commercial enterprises, frequently run from the owner's home, also relied on the communal supply and purchased water from the vestry.

The vestry system faced a number of problems, however. Water in public conduits was frequently polluted as people dumped human, animal, and other wastes into the city's rivers. Public conduits suffered from overuse because individual consumers had no incentive to reduce their own consumption during periods of high demand. Illegal use of public water supplies made shortages worse; some brewers took water without paying, and local residents illegally laid private pipes to supply their homes. By the end of the sixteenth century, many conduits had run dry or decayed.

When Peter Morris approached the City Corporation (London's local government) for permission to build his waterwheel under one of the arches of London Bridge, the Corporation welcomed the possibility of supplementing its public supply. The City Corporation gave Morris a 500-year lease on one London Bridge arch and a £1,000 loan to purchase materials and labor. This long planning horizon and the loan made the City Corporation's commitment to his project credible and encouraged Morris to undertake his investment. Morris later incorporated his waterworks as the London Bridge Water Works Company.

Entrepreneurial engineers and investors soon followed Morris's example and established pumps to raise water at other places along the Thames. (See Table 19.1 for a list of known water companies in London.) In 1594, Bevis Bulmer received a 500-year lease to establish a pump to supply homes to a small area west of the City. Bevis Bulmer was London's first mining engineer, and his pump was known as the Broken Wharf Water Works. Bulmer's pump continued in operation until the early 1700s, when it could no longer compete with cheaper suppliers using the latest mining technology. Broken Wharf assets were purchased by the London Bridge Water Works Company.

The City Corporation awarded at least two or three additional leases after Peter Morris's. The Corporation had little reason to prevent entrepreneurs from establishing new pumps and delivery networks because it received a set annual income from each lease and new leases brought additional income. The Corporation even gave Bevis Bulmer a £1,000 loan to help finance his investment and help him to enter the market. Later, during the seventeenth century, political uncertainty may have constrained some projectors. In 1664, Queen Henrietta Maria (Charles II's mother) found the pump in the grounds of Somerset House an unsightly view from her new mansion. The King ordered Sir Edward Ford to shut down his pump, overriding Ford's lease with the City Corporation and halting his supply to houses in the west of London and in Westminster. The New River Company purchased Ford's remaining waterworks in 1667.

Pumps were not the only technology being considered for bringing water to London. Engineers were investigating the possibilities of constructing canals using gravity to deliver large quantities of water to the City. Financing required for such a large-scale project far exceeded what was involved in constructing a pump on the Thames, however. Any businessman willing to lead such an undertaking required secure property rights to make sinking his capital worthwhile. This security was made available via rights and obligations already granted by Parliament to the City Corporation. An Act of Parliament, passed in 1605, had given the City Corporation the power to build a river from Middlesex or Hertfordshire to Islington on the northeast side of London with the purpose of supplying water to local residents. The Corporation found the project too risky to undertake itself, but willingly transferred its powers to the businessman, Sir Hugh Myddelton. Initially financing the project using his own capital and money borrowed from 28 fellow businessmen (Adventurers), Myddelton started building a river from Hertfordshire in May 1609. He completed and opened the New River on September 29, 1613.

Halfway through the project, Myddelton claimed that his money had run out and he turned to the King for support. James I agreed to provide additional funding from the Treasury in return for a half-interest in the company. Myddelton's claim to impecuniosity appears incredible, given his highly profitable goldsmith business and his access to loans from wealthy family and friends. At least one commentator (Scott, 1968) has refused to take Myddelton at his word and provided a better explanation of his willingness to share his project with the King. Myddelton faced opposition to the New River from Peter Morris and some holdout landowners and, in borrowing from the King, Myddelton intended to give the King a financial interest in the River. By bringing the King's interests in line with his own, Myddelton helped to overcome opposition to the river and to guarantee his right to a monopoly over large-scale water supply for the foreseeable future. With other entrepreneurs applying for similar patents, Myddelton needed such a guarantee to maintain the value of his investment. To help persuade the King to join him, Myddelton also agreed to provide water free to the poor in St Johns and Aldergate Streets. In 1619, Myddelton split his moiety of the New River into 36 shares with full managerial control, and the New River Company was incorporated.

Within two years of its 1613 opening, the New River Company had 384 customers. By 1619, the Company had well over 1,000. These probably represented the wealthier residents of London. The cost to an individual household of connecting to the water mains, providing a stopcock (an external valve to prevent water overflowing), and negotiating a water supply contract would have presented a barrier to many families. Each water con-

Table 19.1 Private Water Companies in London, 1582–1902

Company	Established or Incorporated	Closed or Purchased	Details of Closure / Purchase
London Bridge Water Works Company	1582 (1701)	1822	Sold to New River Company. Network south of the river became part of Southwark Water Works.
Broken Wharf Waterworks (City Conduits)	1594	1703	Works taken over by London Bridge Company.
New River Company	1613 (1619)	1902	Purchased by Metropolitan Water Board.
Somerset House Waterworks	1655	1667	Works taken over by New River Company.
Shadwell Waterworks	1669 (1680–1681)	1807	Land and assets bought by London Dock Company. Waterworks sold to East London Company.
Millbank Waterworks	1675	1727	Rights and assets transferred to the Chelsea Company.
York Buildings Company	1676 (1691)	1818 / 1829	Assets leased to New River Company in 1818. Officially disolved in 1829.
Hampstead Waterworks	1692	1856	Bought by the New River Company.
Marchant's Waterworks	1695	After 1741	Not known.
Ravensbourne Waterworks	1699 (1701)	1810	Assets absorbed by Kent Water Works.
Bank End	1720	1771	Became Old Borough Water Works (?).
Chelsea Waterworks	1722 (1721)	1902	Purchased by Metropolitan Water Board.

Company			
(Old) Borough Water Works	1771	1820	Taken over by Southwark Company.
West Ham Waterworks	1743	1807	Bought by London Dock Company & sold to East London Company.
Lambeth Waterworks	1785	1902	Purchased by Metropolitan Water Board.
South London Waterworks Company (renamed Vauxhall)	1805	1845	Merged with Southwark to become Southwark and Vauxhall Company.
West Middlesex Waterworks Company	1806	1902	Purchased by Metropolitan Water Board.
East London Waterworks	1807	1902	Purchased by Metropolitan Water Board.
Kent Water Works	1810	1902	Purchased by Metropolitan Water Board.
Pocock's Waterworks Company	1810	1815	Not known. New River Company obtained customers.
Grand Junction Waterworks Company	1811	1902	Purchased by Metropolitan Water Board.
Southwark Water Works	1822	1845	Merged with Vauxhall to become Southwark and Vauxhall Company.
Southwark and Vauxhall Water- Works Company	1845	1902	Purchased by Metropolitan Water Board.
Richmond Water Company	N/A	1861	Taken over by Southwark and Vauxhall Company.
Bush Hill Waterworks	1875	1887	Works taken over by New River Company.
North Middlesex/Colney Hatch Waterworks	1867	1871	Taken over by New River Company.

Sources: Dickinson, 1954; Matthews, 1835.

nection required a long-term agreement between the company and customer at a time when goods generally changed hands through spot-market transactions. To overcome this novelty of long-term contracting, the New River Company based its water supply contract on an already well-known, but complex, model—the property-lease contract.

Myddelton's arrangement with the City allowed him to charge any price for pipe water as agreed in individual contracts with each customer. For the first few decades of piped delivery, the New River and other companies charged each household a different rate, though many tenants living in the same street paid similar amounts. The price that a household would have to pay for water carrier delivery placed some upper bound on the price of piped supply.

Peter Morris's introduction of for-profit private water supply sent a signal to other investors that individuals were willing to pay for piped delivery to their homes. Seeing this as a profitable investment opportunity, Hugh Myddelton embraced piped water delivery on a large scale. During the second half of the seventeenth century, Myddelton's belief that the demand for piped water would increase was proven correct.

2. INCREASING DEMAND: 1666–1800

The Great Fire of 1666 damaged many public conduits and rapidly increased the domestic demand for piped water. This rising demand for water, combined with the growing acceptance of joint-stock companies and a boom in the City of London, attracted new investment into the water industry. A number of entrepreneurs petitioned the King and Parliament for permission to establish water companies for supplying the newly populated areas of London. Many succeeded. Between 1666 and 1725, at least eight new water companies started to deliver piped water to households living south of the Thames and in the north and west of London. (Table 19.2 gives regions supplied by each company.)

The New River Company also benefited from the increase in demand after the great fire. Although it faced competition from new suppliers, the New River Company maintained its majority share of London's piped water market and supplied about 65% of customers throughout the eighteenth century (Dickinson, 1954, p. 57). After the Great Fire, the New River Company took advantage of its competitor's fire-damaged waterwheel to take over some of the London Bridge Company's customers. On many London streets, neighbors received water from different sources, as the New River and London Bridge companies competed for customers. Surviving lists of the streets supplied by each company reveal the existence of overlapping networks.

Each new water company received a patent and private or public act from Parliament granting it permission to operate. These patents and acts addressed firm-specific issues—such as the location of the company's pump and the type of pump the company would use to raise water—and gave the inventor of a particular technology property rights in his invention. The acts did not impose economic or industry-wide regulation, or give companies exclusive rights to deliver water. Each of the acts did, however, follow similar formulas and specified a company's rights to take water from particular sources. The acts conferred the right of the company to purchase new land for waterworks and gave companies the right to fine vandals. Each act gave the company a right to enter its customers' homes to confirm they had a working stopcock and clean cistern (vessel for storing water). Acts of incorporation also specified the maximum equity capital each company could raise; companies had to reapply to raise more. The acts did not specify the rates that each company should charge, but left them free to make individual contracts with customers. Some of the companies' acts did, however, include pledges to supply water for particular customers, such as hospitals.

In 1676, King Charles II granted a patent to two entrepreneurs, Ralph Bucknell and Ralph Wayne, to build a waterworks on grounds near the Thames. Bucknell and Wayne supplied the residents of St. James' Fields and Piccadilly, an area to the west of the City. The York Buildings Company, incorporated in 1691, soon supplied water to about 2,700 households. The Company brought direct competition to the New River and London Bridge companies. This competition induced Morris's descendants to sell his London Bridge lease to Richard Soames, and Soames responded to the York Buildings' competition by expanding production. He installed improved waterwheels under the bridge's third and fourth arches, and offered to supply water to any four households living within 60 meters of an existing water pipe.

During the eighteenth century, London's water companies continued to invest in research and new technology to help them improve their service. Promoters employed by the York Buildings Company, for example, transferred innovations from the mining industry to help provide cheaper and more abundant water. Miners in the north of England had developed coal-powered engines to help them raise water. The York Buildings Company

Table 19.2   Water Companies Supply Regions

Company	Established or Incorporated	Closed or Purchased	Supply Region
London Bridge Water Works Company	1582 (1701)	1822	Southwark, South London
Broken Wharf Waterworks	1594	1703	Fleet Street area (west of the City)
New River Company	1613 (1619)	1902	Northeast London
Somerset House Waterworks	1655	1667	Near the Strand (west of the City)
Shadwell Waterworks	1669 (1680–1681)	1807	East of the City
Millbank Waterworks	1675	1727	Parish of St. Margaret's, Westminster
York Buildings Company	1676 (1691)	1818 / 1829	City of Westminster, Charing Cross, west of the City
Hampstead Waterworks	1692	1856	Hampstead, Camden Town, Kentish Town, northeast London
Marchant's Waterworks	1695	N/A	Near Charing Cross
Ravensbourne Waterworks	1699 (1701)	1810	Deptford, South London
Bank End	1720	1771	Bankside and Southwark, south London
Chelsea Waterworks	1722 (1721)	1902	Chelsea and Westminster, northwest London
(Old) Borough Water Works	1771	1820	Southwark, south London

West Ham Waterworks	1743	East of the City
Lambeth Waterworks	1785	Lambeth, south London
South London Waterworks Company (renamed Vauxhall)	1805	Vauxhall, south London
West Middlesex Waterworks Company	1806	Hammersmith, northwestern London; western suburbs
East London Waterworks	1807	East of the City
Kent Water Works	1810	Kent (outside London)
Pocock's Waterworks Company	1810	Holloway, north London
Grand Junction Waterworks Company	1811	Northwest London
Southwark Water Works	1822	South London
Southwark and Vauxhall Water Works Company	1845	South London
Richmond Water Company	N/A	Surrey County, southwest of London
Bush Hill Waterworks	1875	Edmonton and Barnett, greater London (north)
North Middlesex/Colney Hatch Waterworks	1867	Middlesex/greater London

Source: Dickinson, 1954

followed suit, and installed a coal-powered waterwheel alongside the Thames, replacing the horse-driven wheel used previously. This innovation allowed the York Buildings Company to pump water higher than the London Bridge Company, with its tide-driven wheels, to the benefit of York Buildings Company's customers.

The New River Company commissioned the top engineers of the day to improve its pumping engine and help it meet an increasing demand. Water companies based in Cornwall, Durham, and other English cities soon imitated the New River Company and employed this new technology for their own pumps. In 1786, the New River Company commissioned and financed an engine to enable the provision of "high" service throughout its distribution area. The London Bridge Company invested in an engine to help power its waterwheels during low tides and dry summers.

After the Chelsea Waterworks Act, in 1721, no new water companies were incorporated until 1785. During the mid-1700s, already-incorporated companies supplied water to their established customers and continued to obtain new ones. The companies invested in new technology to improve their service and to meet additional demand. If a customer or company felt that the other had broken their contract, the complaint would be settled in court. During most of the eighteenth century, evidence of complaints on either side is rare.

Most Londoners used the water supplied to their homes primarily for cooking and washing. Few wealthier householders drank undiluted water; they preferred home brewed or commercially brewed beer, ginger ale, tea or coffee (Hardy, 1984, p. 252). Consumers had much lower expectations of water quality than we have today, partly because the risks of drinking impure water had not been proved. More importantly, the quality of water supplied by the water companies at least equaled the quality of water available to consumers from public sources.

Each water company charged consumers a fixed fee for connection to its mains and an annual water tariff. In return for the connection fee, a company provided connection pipes to the consumers' property, but consumers took responsibility for providing their own internal plumbing. The annual tariff differed across households, but during the 1700s, companies began to introduce standard price schedules. These schedules based each household's tariff on various indexes of probable consumption, or proxies, such as the number of floors in the house, the number of cisterns, or the number of residents. Water companies delivered water intermittently (for two hours every two days, for example) and customers needed to collect water in cisterns to supply their various uses between deliveries. Intermittent delivery probably made the companies' pricing proxies fairly good indicators of actual consumption. Wealthier customers, with larger houses and households (including servants), tended to have bigger cisterns and more of them. Where the size of the cistern determined the tariff, households would install a large cistern only if they would consume large volumes of water between deliveries. Many small businessmen, such as bakers and butchers, worked from home; they also had large cisterns and tended to empty them between deliveries. They were accordingly charged an additional fee. All commercial users were charged according to the type and size of their business. The higher tariffs for greater volumes would have given consumers some incentive to conserve water.

Toward the end of the eighteenth century, water companies started to embrace improvements in pumping technology as a way to offer superior service to their customers. The cost of iron pipes had placed a limit on the extent to which households could receive water above the first floor of their house. As innovations brought down the costs of iron pipes, companies started introduce high-pressure and 24-hour delivery to customers willing to pay for service improvements.

3. TECHNOLOGICAL AND INSTITUTIONAL CHANGE: 1800–1902

At the end of the eighteenth century, the population of London grew significantly, causing a housing boom and a rising demand for water connections. Much new demand came from households located beyond the companies' then-current supply areas. To meet this growing demand, Parliament encouraged new water companies to incorporate and compete with the incumbents. Soon after its incorporation in 1785, the Lambeth Waterworks Company started supplying residents south of the Thames below Westminster. After the turn of the eighteenth century, the Grand Junction, West Middlesex, and East London companies incorporated with a view to supplying customers north of the River. The Southwark and Vauxhall companies joined the Lambeth south of the Thames.

Rising demand for water gave new companies one incentive to enter the market. Innovations in pumping and pipe technology gave them another. The invention in 1790 of a mass production method for iron pipes able to withstand high pressures reduced the cost of high-pressure water delivery. In 1810, the New River Company started to replace the less efficient wooden mains and pipes that connected its existing customers. The company also put iron pipes in new connections, and other established companies followed suit. London's new companies took advantage of the technology to offer high-pressure delivery immediately to all new customers.

Once the costs of constructing a water pump and reservoir had been paid, the new companies were keen to extend connections to households. They started to attract customers away from the old companies by promising a better, cheaper service. The West Middlesex and Grand Junction companies, in particular, sought to take over large parts of the New River, London Bridge, York Buildings, and Chelsea companies' supply areas. The entrants signed up customers at a lower annual rate for water than offered by the old companies, and often waived the connection fee. An Act of 1806 allowed the West Middlesex Company to supply water to Marylebone and a second Act in 1810 allowed the company to raise additional capital to extend its area of service to compete directly with the New River, Chelsea, and York Buildings companies. An Act of 1811 allowed the Grand Junction Company to join the West Middlesex.

The Grand Junction and West Middlesex companies also pledged continuous rather than scheduled, intermittent, supply. The pledges were made voluntarily by the companies in an attempt to attract customers. None of the new companies faced price or quality regulation through their initial acts of incorporation.

As the Grand Junction and West Middlesex cut prices to gain customers, the old companies retaliated. This generated a short period of intense price competition north of the river. In 1810, most households had more than one company offering them connection to a water network, and some households had as many as five companies. Because companies waived the connection fee, customers could switch supplier at low cost. However, with each company cutting prices and connecting customers for free, they were all taking losses. No single company was able to supply the whole of the metropolis but, with three new companies entering the market at the same time, the industry had a larger capacity than required for the current number of customers. Together, the companies would have liked to supply more customers than were willing to pay for a connection, so they competed with each other for those who were willing to pay.

The companies realized that the competition was unsustainable and they needed to do something to prevent further losses. In 1816, the West Middlesex and New River companies applied to Parliament for permission to merge. Parliament refused, however. The following year the companies agreed to divide London among themselves and no longer to compete for the same streets. Under this agreement, each company limited its supply to particular streets and, to enforce the commitment, dug up any pipes it had put down in other companies' streets. The New River Company raised its prices back to precompetition levels, and agreed to allow the West Middlesex Company to take back some of the streets that the New River Company had supplied previously.

The New River Company used a takeover rather than a cooperative agreement to end its competition with the York Buildings Company. As the New River Company expanded its network southwest of the City, it successfully signed up many York Buildings Company customers. In 1818, the York Buildings Company agreed to lease its assets to the New River Company and to stop supplying water. In 1829, an Act of Parliament officially dissolved the York Buildings Company.

During the 1830s, a similar competition for customers took place south of the River Thames between the Southwark, Vauxhall, and Lambeth companies. Companies started to lose money as they reduced prices to compete for customers and developed overlapping networks. In 1842, the companies south of the Thames agreed to cease their competition.

As companies stopped competing and restored prices to precompetition levels, they also introduced new charges for the additional services they now offered. Most companies used either the size of the house or the number of residents to determine each household's basic rate. Lower income households generally paid only this basic rate. Wealthy households now paid not only a higher basic rate, but also additional charges for having installed a flush toilet or extra baths or for receiving 24-hour delivery. Households with two baths paid more for the first than for the second.

By charging additional fees for extra uses of water, a company encouraged customers to reveal their willingness to pay for additional units of water. A household could select its preferred combination of water quantity (using number of uses as a proxy for volume) and tariff.

A customer's risk characteristics were another source of price differences. For example, the risk of nonpayment by poorer consumers living in rented accommodation was much higher than the risk of nonpayment by property owners. Companies used farming out (subcontracting) of rate collection to someone living in the poor neighborhood as a way to increase payment rates by the poor. This reduced the default risk and allowed companies to charge a lower rate for farmed than for nonfarmed consumers.

Customers had enjoyed the low prices under competition and complained when the companies reached agreements and restored prices. Critics of the water companies argued that they had monopolized the market. They pointed out that no company had a legal right to be the only supplier in a particular part of London, but that the companies had made monopoly agreements among themselves and had conspired to raise prices. Critics ignored the improvements in service and new options offered by the companies, such as new connections, "high service," and continuous or more frequent supply.

In response to critics' charge that the companies engaged in monopolistic behavior, Parliament introduced price regulation. First, an Act of 1847 tied companies' basic charges to property values. Then, an Act of 1885 specified that property values were to be those determined by each borough's local government. These acts did not undermine the structure of companies' basic schedules, however. Each company continued to charge consumers additional fees for baths, toilets, and for "high service."

When London experienced its second cholera outbreak in 1849, public attention shifted from the price of water to water quality. Critics argued that private companies had no incentive to provide customers with good quality water. They ignored the fact that the Chelsea Company had developed and employed a sand filtration technique long before the benefits of water filtration were understood. They ignored other companies' investments in screening and moving their intake up river in order to improve water quality. Rather than praising those companies who had invested in quality improvements and supporting regulation only to encourage laggard companies to follow suit, most critics focused only on the laggards and condemned private supply outright.

Recent research has shown that the quality of water in the Thames started to deteriorate in 1815—the year in which wealthy Holborn households were able to flush waste from the newly purchased toilets into public rainwater sewers. Since public sewers deposited wastewater in the River Thames, some of the water companies faced pollution of their source of raw water. As companies supplied more households with piped water, the demand for flush toilets increased. An Act of 1847 made household sewer connections compulsory in an effort to rid the city of cesspits. This encouraged a further increase in the number of households with toilets, and further polluted some companies' water source.

At the time of the cholera outbreak in 1849, scientists had not yet proven that cholera was a waterborne disease. The comparison of cholera rates in areas supplied by different water companies help prove the theory. The

1849 cholera outbreak hit households in some companies' supply areas far more severely than others. Because those areas with lower rates of cholera mortality were those where households received filtered water drawn from further upriver, John Snow was able to show that cholera could be transferred between people by sewage-polluted water. Critics ignored Snow's research in the attempts to regulate the water companies.

By the late 1800s, advocates of public ownership switched the focus of their complaints to the companies' failure to supply a sufficient volume of water. Water shortages during particularly hot summers provided ammunition for the companies' critics. In response to shortages, companies invested in additional reservoirs, bought water in bulk from competitors, and started to invest in network interconnection. The companies also considered reducing demand by charging higher fees for additional supply during the summer, but Parliament prevented them from doing so (Richards and Payne, 1899).

In 1895, the newly established London County Council (LCC) unsuccessfully put forward eight bills to allow them to compulsorily purchase London's water companies. In 1897, the LCC tried again. It failed a second time. By this time, the debate had became highly politicized (Mukhopadhyay, 1981). The LCC put forward a third bill in 1902 with strong support from members of the opposition Liberal party. Conservatives put forward an alternative bill; it accepted the idea of public ownership, but proposed a new Board with representation from London's many local authorities. After many concessions, the Government's bill passed, and the water companies' assets were purchased under the Metropolis Water Act of 1902.

4. CONCLUSION

The history of water supply in London suggests that exclusivity need not be a prerequisite for private water provision and that competing private suppliers have incentives to improve service without regulation. Private companies supplied Londoners with pipe water for over 400 years with little government restriction on entry. Companies invested in service and quality innovations and increased volume and the number of household connections.

During the first 200 years, private companies faced no government regulation. Private water companies operated largely as local monopolies, using differential pricing to maximize the return to their shareholders, but they did face some competition. Neighboring companies offered an alternative source of pipe delivery to households on the boundary of a company's supply area, while water-carriers and private wells offered households other delivery options. Competition with water carriers and potential competition from other companies kept down the price that any one company could charge consumers. Most of the time, however, switching costs prevented excessive competition between companies.

In the face of potential competition, companies had to respond to consumer demands to make profits. By the nineteenth century, companies were investing in technology to improve their service by offering 24-hour delivery and high-pressure supply. Innovations in pumping and pipe technology encouraged entry by entrepreneurs seeking profitable new ventures. Competition brought benefits to consumers in the form of rapid network extension and more frequent delivery, but falling prices wiped out profits for the companies. Consumer complaints about subsequent price increases led Parliament to start regulating the industry.

The quality of raw water in the River Thames deteriorated between 1815 and the 1850s as more London households installed flush toilets. Companies responded by investing in sand filtration and by moving their intakes above London's sewers. Parliament responded to the deterioration of water quality by introducing quality regulation. Critics of private supply argued that regulation was insufficient, and the debate shifted in favor of public ownership. In 1902, London followed the international trend toward public operation of waterworks and authorized the takeover of the companies' assets by a new Metropolitan Water Board (MWB).

5. POSTSCRIPT

As early as 1908, the Metropolitan Water Board's performance revealed that claims of lower costs and better service under public ownership were too optimistic. The MWB had an annual deficit each year between 1908 and 1922, due to the costs of new investment. In 1908, the MWB started building the first of two new reservoirs for which the East London Company had received Parliamentary approval in 1900. The reservoir opened in 1913. Work on the second did not start until 1935. Despite the deficit, the evidence does not suggest customers were the beneficiaries in the form of lower prices. The maximum rate levied by the private water companies in 1902 was 7.5% of the rateable value of a house. Under the MWB, the maximum rate was reduced to 5% in 1907, but by 1922 Parliament gave the Board authority to raise water rates to 10% of property value (Dickinson, 1954, p. 126).

REFERENCES

Booth, C. (1969). *Life and Labour of the People in London, First Series: Poverty II Streets and Population Classified.* New York: Augustus M. Kelley. Original edition: 1902.

Brown, J. C. (1988). "Coping with Crisis? The Diffusion of Waterworks in Late Nineteenth-Century German Towns," *Journal of Economic History,* Vol. 48, No. 2.

Daunton, J. M. (1995). *Progress and Poverty: An Economic and Social History of Britain 1700–1850.* Oxford: Oxford University Press.

Dickinson, H. W. (1954). *Water Supply of Greater London.* London: Courier Press.

Finlay, R. (1981). *Population and Metropolis: The Demography of London 1580–1650.* Cambridge: Cambridge University Press.

Graham, R. (1972). *Britain and the Onset of Modernization in Brazil 1850–1914.* Cambridge: Cambridge University Press.

Hardy, A. (1984). "Water and the Search for Public Health in London in the Eighteenth and Nineteenth Centuries," *Medical History,* Vol. 28, pp. 250–282.

Kraemer, R. A. (1991). *Development of the Water Industry in Germany and the United Kingdom.* Berlin: FFU, Manuscript 91-7.

Matthews, William (1835). *Hydraulia: An Historical and Descriptive Account of the Water Works of London.* London: Simpkin, Marshall, and Company.

Mukhopadhyay, A. K. (1981). *Politics of Water Supply: The Case of Victorian London.* Calcutta: World Private Press.

Richards, H. C., and W. H. C. Payne (1899). *London Water Supply.* London: P. S. King & Son.

Robins, F. W. (1946). *The Story of Water Supply.* London: Oxford University Press.

Scott, W. R. (1968). *The Constitution and Finance of English Joint Stock Companies to 1720,* Vol. 3. Gloucester, MA: Peter Smith.

Stow, J. (1971). *Survey of London.* Oxford: Clarendon Press. Original edition: 1603.

ADDITIONAL READINGS

Clifford, F. *A History of Private Bill Legislation,* 2 vols. London: Butterworths, 1885.

Rudden, B. *The New River: A Legal History.* Oxford: Clarendon Press, 1985.

Chapter 20

Deregulated Private Water Supply

A Policy Option for Developing Countries

Penelope J. Brook and Tyler Cowen

1. THE PRIVATIZATION ALTERNATIVE

Many citizens in developing and transition economies are excluded from enjoying safe and reliable water supply. In many cities, 30–60% of the population has no formal water hookup at all, but rather must resort to wells, buckets, supply by tanker trucks, and physical transport of water through human labor and beasts of burden.

A few simple facts illustrate the serious nature of this problem. In Jakarta, Indonesia, 75% of the population has no formal connection; in Maputo, Mozambique, 65%. In Madras, India, the percentage served is around 50%; and even in the relatively prosperous Philippines, 29% of the citizenry

Reprinted from *Cato Journal,* Vol. 18, No. 1. Spring-Summer 1995, pp. 21–41. Used with permission.

Penelope J. Brook is Program Manager, Rapid Response Unit, Public Sector Advisory Services of the World Bank. Her work includes designing infrastructure sector reforms to ensure improved access for low-income households and on the design of aid instruments to link aid flows to service delivery. Prior experience included the Investment Banking Division of CS First Boston New Zealand. Tyler Cowen is Professor of Economics, George Mason University and the Center for the Study of Public Choice, and Director of James Buchanan Center and Mercatus Center.

of Manila has no connection. When individuals must resort to nonpiped water sources, prices are often at least 10 to 20 times higher. In Luanda, Angola, where the price for piped supply is around nine cents per cubic meter, households can pay as much as $16.00 per cubic meter for tanker supply. Table 20.1 portrays some connection rates and price differentials.

The fundamental problem is institutional rather than technological. Tariffs set by governments at levels below cost recovery fail to encourage inclusion. In developing countries, water utilities recover, on average, around 30% of their total costs (World Bank, 1994). As a result, utilities have no incentive to deliver services to large sectors of the population, almost always low-income households. Incentives for research and development are similarly weak, given that the price-controlled monopolist cannot capture the full benefits of a new product idea.

The human costs of these institutional arrangements have been very high. According to one estimate (the World Health Organization, cited in Cooper, 1997), contaminated drinking water accounts for 80% of disease in India, including a sizable share of the 500,000 Indian children who die each year from diarrhea. Around the world, diarrheal diseases kill more than 3 million people annually, and cause approximately 900 million episodes of illness (World Bank, 1992).

We propose that unregulated privatization be considered as one means of limiting these tragedies. To date, the world has experimented with four kinds of institutional regimes: outright public provision of water (common throughout the world), government-supported natural monopoly with reg-

Table 20.1   Prices for Piped and Vendor Supplies of Water in Selected Developing Countries

City and Year	Percent Without Piped Water	Price for Vendors (US cents/m^3)	Price for Connections (US cents/m^3)
Bandung, Indonesia (1991)	61	9.9	616
Jakarta, Indonesia (1991)	75	17.2	185
Manila, the Philippines (1992)	29	10.5	187
Karachi, Pakistan (1992)	17	7.5	175
Ho Chi Minh City, Vietnam (1991)	35	7.6	151

Source: Asian Development Bank (1993).

ulated price (the English model), government-supported natural monopoly with regulated rate of return (the American model), or a government-controlled franchise, lease, or concession agreement (the French model and its variants). We add a fifth possibility: complete privatization of water assets and unregulated natural monopoly. This scenario involves no price regulation, no rate-of-return regulation, no residual government ownership of assets, and no surreptitious regulation through antitrust law.

The rationale for unregulated privatization is straightforward. An unregulated private monopoly would have an incentive to bring as many potential buyers into the system as possible, so as to maximize profit. Unregulated private monopolies could thus significantly increase the number of water connections in developing countries. If unregulated privatization could produce hookups for currently neglected low-income customers, the poor would end up with higher real incomes, better water service, more time for other endeavors, and a greater probability of a long life. London water supply in the early eighteenth and nineteenth centuries, which was private and relatively unregulated, had a favorable record for extending the number of connections (see Dickinson, 1954, pp. 102–103).

While standard theory emphasizes the output-restricting nature of monopoly, water utilities will use price discrimination and fixed hookup fees (Oi, 1971) to capture as much profit as possible, thereby increasing supply in the process. Unlike the governmental and regulated alternatives, a private unregulated monopoly also would have strong incentives to hold down costs and supply an optimal quality of product. Our main point is that this monopolistic alternative deserves serious consideration.

In a comparative institutional context, the more heavily regulated alternatives may end up excluding more potential buyers. Developing or transition economies, regardless of their historical background or geographic locale, tend to share common problems with their governments. These governments have relatively low levels of credibility, weak track records, and very short time horizons. The government performs especially poorly as an owner or regulator, partly through lack of experience and partly through improper incentives and corruption.*

In Guinea, for example, progress under a water lease has been hampered by problems in defining and implementing the regulatory function,

---

*See, for example, Zajc (1996) on the incidence of these problems in transition countries and their implications for water privatization.

and by continuing disputes between the government-owned water holding company and the private water operator over who is responsible for failures in service expansions and water loss reductions. While the lease contract has increased the number of connections and improved water flows, progress has fallen short of expectations (Brook Cowen, 1996).

Unregulated privatization has received little serious attention, and has generally been rejected or dismissed, albeit without serious analytical consideration. Few sectors have been classified as market failure more universally than the supply of water to households and residences. Throughout the world, water systems are characterized by extensive government ownership or thorough regulation and control. Water supply typically is regarded as a natural monopoly, and therefore a poor candidate for unregulated market provision. In the technical literature, Guislain and Keff (1996) restrict their discussion of infrastructure privatization to concessions and divestitures accompanied by a regulatory license. Klein (1996) takes as a starting assumption that the weakness of competitive pressures in the water sector creates a need for at least some form of regulation. Breyer (1982, p. 17) considers the idea of a price-discriminating natural monopolist briefly, but does not explore the unregulated alternative in depth. Loeb and Magat (1979) attempt to replicate private price discrimination through regulation; their scheme has the government award monopolists for the consumer surplus they generate. Armstrong et al. (1994) essentially argue that if competition is not possible, regulation will be necessary.

In the remainder of the chapter, we provide an analysis of potential sources of inefficiency in water markets, explain why government ownership and regulation have failed to provide fully satisfactory results, discuss how unregulated privatization might resolve the efficiency problems that plague regulation and government ownership, consider the ability of an unregulated monopolist to price discriminate when selling water to residential users, and focus on some residual problems with the proposed policy option of unregulated private natural monopoly. We find that the case for unregulated privatization is not conclusive, but that the proposal deserves serious consideration.

2. SOURCES OF EFFICIENCY LOSS

The relevant natural monopoly problem comes from the distribution of water rather than from water itself. Water, considered apart from the problem of distribution, satisfies the traditional definition of a private good—nonrivalry in consumption. If one person consumes some water, another person cannot use the same water. While there are public health benefits to a clean water supply, the private benefits of clean water are high as well, giving individuals a strong incentive to pay for water quality. Consistent with the private-good nature of water, we observe the efficient private supply of water in a variety of circumstances. We buy bottled water at the supermarket, and private, for-profit car washes supply water to clean our cars. The private sector has had considerable success in supplying and running wells, at least where wells are a reasonably efficient means of water delivery.

The construction and maintenance of water distribution networks presents the difficult problem, and the potential cause of market imperfection. Once a system of water pipes is built, the owner of the system has a monopoly advantage in the market. If only one set of pipes exists, the owner of those pipes can exercise market power and charge a price for water in excess of its marginal cost. As discussed above, other means of obtaining water, such as water delivered by tanker vendors, typically involve costs from 10 to 20 times higher than buying water through a piping system. Alternatively, we might imagine a system of competing pipes, or competing pipe systems. Such systems did occur in Canada and the United States in the nineteenth century, and still arise occasionally when water of different qualities is being supplied.

For example, in Hong Kong seawater pipes supply flushing water (Klein, 1996). In this case, however, consumers ultimately must finance both piping systems. The high prices needed to recoup the costs of multiple piping systems will imply a restriction of water output, just as the monopoly did. Furthermore, the stability of market equilibrium is problematic when multiple, competing suppliers own networks with high fixed costs (Bittlingmayer, 1982). Competition will tend to force prices back down to marginal cost, but at marginal cost, no supplier can break even and recover the fixed costs spent on constructing the piping system.

Some treatments emphasize market means of overcoming the natural monopoly problem. Under one proposal, water is distributed through a club-owned network, with different suppliers competing against each other to win contracts with consumers. Either the suppliers or the consumers themselves own or control the club. We see merit in this idea, but for the purposes of exposition, we assume that the natural monopoly problem cannot be overcome so easily. The potential ability to make the market competitive, however, would only favor our basic proposal.*

Assuming that competition is not possible, the fundamental problem involves the construction of a distribution network with fixed costs that are high relative to marginal costs of supply. The problem behind the private provision of water thus resembles analogous problems with the sale of cable television services, electricity, and natural gas.

Institutional regimes for water provision face three kinds of efficiency problems: inefficient levels of output, inefficient levels of cost, and inefficient levels of product quality. A non-price-discriminating monopolist, in the absence of regulation, will set price above marginal cost and restrict output, compared to a first-best social optimum. That is, consumers would be willing to pay more for additional units of output than it would cost society to produce them. The monopolist does not expand output, however, because the extra units of output could be sold only by lowering the price for all units, and thus reducing profits.

Regulators have found it difficult to address this problem of monopoly without inducing other distortions. One approach grants private ownership but places a cap on price (the British model for water provision). Placing a cap on price, however, gives the private supplier an incentive to skimp on service and product quality. As with all price controls, the supplier will raise the real price to its desired level by lowering the quality of the product. Not only will quality decline in the short run, but also long-run investments in system maintenance will be suboptimal. As discussed above, this problem is particularly drastic in developing and transition economics.

An alternative method of regulation, common in the United States, uses rate-of-return caps to limit the profits of the private monopolist. In practice, rate-of-return regulation usually involves price caps as well, whether implicitly or explicitly, and in that regard also leads to skimping on service and product quality. Furthermore, rate-of-return regulation brings a new set of distortions in the form of higher costs. As rate-of-return regu-

---

*We also see some problems with the club proposal. Even if many suppliers compete by selling water services through a single pipeline, the fixed costs of the pipeline still must be covered somehow. Competitive pricing allows no means of financing the pipeline and allowing each company to break even. Presumably, some kind of Ramsey pricing is necessary, where inelastic demanders face the highest markups, but this introduces some of the welfare losses of market power. Along other lines, Demsetz (1968) analyzed ex ante competition "for the market." In this proposal, utilities offer competitive bids in communities for water supply; see below for a discussion of franchising.

lation is practiced, firms typically are guaranteed a minimum as well as a maximum rate of return. Without the minimum guarantee, firms would not participate in the arrangement, given that they have sacrificed upside potential for profit. Firms therefore can use cost increases as a justification for price increases; not surprisingly, the incentive to keep down costs is low. The end result is high costs and a lower level of water consumption than is optimal.

Leasing and concession agreements, in their various forms, provide yet another attempt to overcome the basic problem with natural monopoly. These institutional arrangements, however, do not avoid the fundamental problems associated with regulation. Leasing and concession agreements typically regulate prices and rates of return to various degrees, either implicitly or explicitly. In this regard, they involve the welfare losses associated with price and rate-of-return regulation. Leasing and concessions may provide for a looser or more informal kind of regulation, given the ongoing relationships between the water company and the relevant government, but in the final analysis either the supplier is free to adjust its prices or it is not. If prices can be set freely, we return to de facto unregulated natural monopoly (of course, this may be an advantage of concessions, as we discuss below). If prices and rates of return are not free to adjust, we return to the distortions of regulation and the weak incentives to expand the number of system hookups. Particular problems arise where concession contracts mandate expansions into low-income areas, while also mandating "life-line" (below-cost) tariffs for low-income consumers.

Leasing and concession agreements involve further distortions through the government's role as residual asset owner. As the leasing or concession agreement nears an end, the private concessionaire has an incentive to cease maintenance, or even strip the water assets. Leasing works poorly when the company faces a short time horizon. The government can alleviate these problems by promising a forthcoming renewal of the lease, or by offering comprehensive provisions for compensation upon contract termination, but if these promises are credible, leasing and concessions do not differ greatly from asset privatization with regulation. If the promise is not credible, we return to poor incentives for maintenance.

A fourth proposal involves outright government ownership of water assets and full governmental control. The record of governmental provision in this sector, however, is extremely poor. In developing countries, where government ownership has been the norm, tariffs are routinely set well below cost recovery levels, routinely less than half of supplied water is ac-

Table 20.2 Examples of Private Sector Arrangements in Water and Sanitation

Contractual Type	Water	Sanitation	Water and Sanitation
Management Contract	Colombia Gaza Malaysia Turkey	United States	Puerto Rico Trinidad and Tobago
Lease	Guinea Italy Senegal Spain		Czech Republic France Poland
Build-Operate- Transfer	Australia China Malaysia Thailand	Chile Mexico New Zealand	
Concession	Cote d'Ivoire Macao Spain	Malaysia	Argentina France Philippines
Divestiture	England and Wales		England and Wales

Source: World Bank, 1997.

tually paid for, and large segments of the population go without formal services (World Bank, 1994). For political reasons, governments have weak incentives to reduce costs, price water at marginal cost, maintain water systems, introduce innovations, and cut staffing to efficient levels. Not surprisingly, countries around the world are moving away from the government ownership option, and embracing various forms of private sector participation (Rivera, 1996). Table 20.2 shows examples of private sector contracts that are now in place.

3. SKETCH OF AN UNREGULATED NATURAL MONOPOLY FOR WATER

Consider a scenario where a government allows complete private sector ownership of all water system assets, including the impounding of bulk water, water treatment, and distribution. The private sector would own all water system assets (which may or may not be vertically integrated) just as the private sector owns the assets in the automobile industry or the computer industry. Furthermore, suppose that the owner of the water assets could set prices and quantities without regulatory interference. Water suppliers and customers would rely solely on contract to set the terms and conditions of water delivery, and the courts would agree to uphold any contracts that are written. The absence of regulation, as defined in this paper, also implies a credible laissez-faire antitrust policy with regard to pricing and output decisions. If water companies set their prices with an eye to avoiding charges of "anticompetitive behavior," or "price gouging," we would return to an implicit form of price regulation. The laissez-faire antitrust policy also would allow complete freedom of merger and cooperative relations across differing firms.

Our use of the term "unregulated" refers to the absence of a set of government regulations found in today's regimes—specifically, restrictions on asset ownership, pricing, service delivery, and so on, and exclusivity arrangements. Under laissez-faire, the provision of services is regulated by market forces and economic incentives. In this sense, our analysis compares one kind of regulation to another, rather than comparing regulation to an unregulated state of affairs. Furthermore, all regimes possess an implicit form of government regulation through ex post liability law. In the scenario we consider, private water suppliers would remain liable to lawsuits for breach of contract, fraud, or provision of water of dangerous quality. Nevertheless, we continue to use the word "unregulated" for purposes of expositional simplicity and for lack of a more accurate descriptive term.

The forces for natural monopoly within a single geographic area would be strong in an unregulated environment. Experience suggests economies of scale in the operation of distribution networks for populations of at least 50,000–100,000 people. Economies of scale in system management as a whole are more extensive (evidence from Britain indicates that managerial economies of scale are exhausted at populations of 500,000–1 million). The natural monopoly may be limited at certain margins, such as when industrial users develop their own wells. At the residential level, some households may find it more profitable to dig wells, or to collect and store rainwater for at least some uses. For the typical residential user, however, we envisage a situation where water can be obtained at lowest cost from a single dominant supplier within that geographic region.

We expect suppliers to offer standard packages to their consumers. If an individual is building a house, the water supplier will offer to outfit the house with pipes for some fixed sum, perhaps based on the value of the house and the neighborhood. Where real estate developers are responsible for installing household connections, they routinely use this approach. If a house is already in place and already possesses a hookup (perhaps as a legacy from a previous, regulated regime), the water supplier will offer so many units of water at a given price, so many more units at another price, and so on. Households will either accept or reject these offers, depending on the promised bundle of price and service.

The water company has strong incentives to set initial offers that will be accepted. The company will try to capture as much surplus from each household as possible, but the company also wishes to ensure that each household signs up to purchase water. Given the initial assumption of natural monopoly, the company can serve subsequent households at relatively low marginal cost. Note that in the polar case where the company has perfect knowledge of household demands, and can precommit to a series of price offers, a "first-best" result will obtain. The water company will extract all of the consumer surplus associated with water purchases. We do not present the first-best as an attainable real world outcome; the relevant comparison is between imperfect markets and imperfect government regulations. Nonetheless, presentation of the first-best ideal illustrates some basic incentives behind unregulated monopoly and also serves as a foil, by contrast allowing us to see ways in which unregulated monopoly falls short of an ideal outcome.

The relevant consumer surplus can be extracted in either of two ways. The company may charge a fixed fee for a hookup, and then sell remaining water units at marginal cost over some specified period of time. Both the fixed fee and the subsequent per unit prices would be determined by initial contract; Oi (1971) has analyzed the efficiency of this arrangement. Alternatively, if the hookup is already in place, or if it is too costly to bargain over the hookup fee, companies will simply supply the hookup and then sell water at some price above its marginal cost of production.

This situation, if it can obtain, solves all three of the efficiency problems discussed above. First, the supplier will produce a socially optimal amount of output. For any unit whose value exceeds its marginal cost, the supplier will produce it and offer it on the market. With perfect price discrimination, a supplier never increases profit by withholding output from the market. Second, the supplier has first-best incentives to engage in cost reduction. Any reduction in cost translates into a one-to-one increase in profits. Suppliers therefore will reduce their costs to the point where the social benefits of cost reduction equal the social costs. Such costs cannot be socialized but rather eat directly into profits. Third, a perfect price-discriminating monopolist has first-best incentives with regard to product quality. The supplier captures all of the consumer surplus in the form of profits. That same supplier will therefore offer the product qualities that maximize consumer surplus, net of the cost of production.

The ability of a natural monopolist to perfectly price discriminate may be problematic, under a variety of assumptions. For that reason, the first-best results may not strictly hold. Nonetheless an unregulated, privatized natural monopoly obtains first-best results under the basic assumption that the water company succeeds in maximizing profit. Even in a second-best setting, the monopolist may produce a greater quantity and quality of water outputs than do today's highly regulated alternatives. In most developing and transition economies, the key problem is to get users some minimal amount of clean water, not to satisfy all optimality conditions.

Note that the potential efficiency of price discrimination also indicates why leasing and concession agreements, and divestitures with a license, may sometimes result in first-best or near first-best outcomes. If the company holding the concession has sufficiently cozy relations with the host government, that company may be given latitude to replicate the efficient price-discriminating natural monopoly outcome. Quantity and quality decisions will again be optimal, if the "unregulated" outcome can be obtained under the guise of regulation. In practice, however, governments frequently impose uniform tariff rules, or otherwise restrict price discrimination by regulated private water companies.*

4. THE FEASIBILITY OF PRICE DISCRIMINATION FOR WATER

Price discrimination is most feasible when four primary conditions hold. First, the seller must hold some degree of market power. Second, the product cannot be vulnerable to low-cost resale from low-price buyers to high-price buyers. Third, the seller must be able to make good estimates of buyer demands. Fourth, the supplier must be able to commit to initial price offers. Each of these assumptions characterizes the water market to some degree.

We take the presence of market power as given, and as following from the natural monopoly assumption. If somehow no market power were pre-

---

*Philips (1983) provides an overview and survey of the economics of price discrimination.

sent, price discrimination would be impossible, but a regime of unregulated private water supply would in any case prove effective.

The absence of cheap resale from low-cost to high-cost buyers also follows from the natural monopoly assumption. By construction of the example, it is much cheaper to sell the water through a system of pipes than through bottles, wells, and buckets. Even if some resale were possible, however, market demands would shift without changing the fundamental nature of the problem. Assume, for instance, that, in the absence of resale, low-valuation buyers would be charged $20 and high-valuation buyers would be charged $100. Now consider resale, which is profitable at any price above $80 to the high-valuation buyers. The high-valuation buyers will refuse to pay more than $80, and the price-discriminating monopolist must lower prices accordingly, presumably to just below $80. Even at this lower price, an optimal quantity of output is still produced, and the monopolist still has full incentives to economize on costs at the margin. Optimal quality cannot be guaranteed, since the monopolist cannot necessarily reap the full benefits of a quality improvement (higher prices for quality improvements may be undercut, implying that the innovator cannot reap all of the new consumer surplus that is produced), but some incentives for quality improvement remain nonetheless.

The third and perhaps most problematic condition for effective price discrimination is whether the seller can predict the market demands of the buyers. The water supplier will estimate two differing features of water demand: how much a given buyer values having any water connection at all, and how much a given buyer values subsequent units of water. We envisage a market where the water supplier sets price by examining the previous use patterns of the water buyer, the value of the water buyer's property, and the wealth of the neighborhood. In wealthier areas, the supplier may consider the number of bathrooms in the house, whether the water buyer has a lawn, and other pieces of ancillary information, such as the water buyer's age, job, or credit record.* Insofar as water demand is closely correlated with observable characteristics of the property and the buyer, effective price discrimination will be relatively easy. The water supplier will run information on the buyer and the property through its "pricing office," which will respond with a suggested price offer, both for initial service and for successive units of water use. In Los Angeles, for example, the

---

*Since US public utilities routinely run credit checks, this need not involve a significant loss of privacy.

water department has the capacity to customize base tariffs across consumers, according to such factors as lot size, temperature zone, and size of household (Mann, 1996).

Price discrimination will inevitably be imperfect in practice. Prices sometimes will be set too high, thereby excluding buyers from either participation in the piping network or from the purchase of additional units, even when the social benefits of added output would exceed the social costs. While some inefficient exclusion will occur, water supply may well be higher than under most current regimes in developing economies.

Even when suppliers make pricing mistakes, they need not exclude buyers altogether. To the extent that monopoly power is considerable, price will exceed marginal cost by a large amount, and the profits of water sales will be large. Each excluded buyer represents a chunk of foregone profit. Consider the position of a water company that believes that a given buyer values regular water use at, say, $1,000, and where the company can produce those same water services at a cost of $300. If the company knows that the buyer's valuation is in the neighborhood of $1,000, but the company is not sure about the exact valuation, the company will more likely price the services too low rather than too high. If the company charges $1,001, it loses $700 of potential profit. The expected return to guessing low will tend to exceed the expected return to guessing high. The microeconomic intuition here is simple: individuals or institutions that face a good chance of capturing a significant prize will behave cautiously when they are within range of winning the prize. For similar reasons, monopolistic firms in other contexts will choose high levels of product safety, reliable service, and easy access to their product, all in the desire to protect their monopoly profit (Klein and Leffler, 1981).

The excluded buyers will tend to be those whose valuations do not much exceed the marginal cost of producing water services. If the marginal cost of production is $300, and the buyer values service at $320, the firm has less marginal profit to lose by trying to squeeze out all of the buyer surplus. Some of these buyers may end up excluded, since the firm will sometimes guess incorrectly and offer a take-it-or-leave-it price above $320. Even when exclusion results, however, the welfare costs of this exclusion tend to be relatively low. In the example, the buyer valued the product only slightly more than its marginal cost of production. When expected profit, and expected social surplus, are low, fewer resources will be spent trying to capture that profit and some potential gains from trade may be foregone.

If such resulting instances of exclusion prove unacceptable, perhaps for reasons of fairness or equity, a government may decide to intervene in the market and require service to low-income buyers at prices they can afford. In this case, our proposal would cease to be purely unregulated, and would involve the costs of price controls, at least for some buyers or some neighborhoods. This outcome, however, represents a worst-case scenario for our proposal, which still appears to provide superior overall performance, compared to a regime with full regulation across all contracts and all buyers.

Many cases of harmful exclusion will come in the form of overpriced marginal units, rather than overpriced fees for basic hookups. Companies often will choose price discrimination in the form of a schedule, where the prices for water services vary with the quantity consumed. Assume that a buyer values the first unit of water services at $100, the second unit of services at $60, and the third unit at $30. The company will try to offer a price schedule that matches these demands exactly, but if the company calculates demand incorrectly, it may offer, for instance, a schedule of $100–$60–$40, thus excluding the buyer from the third unit of water services. The buyer will take shorter showers than would be socially optimal, but some amount of safe water will still be supplied.

Fragmentary data and lack of experience with unregulated privatization prevent us from offering an empirical assessment of the relative magnitudes of these exclusion costs across institutional regimes. Nonetheless, we see no prima facie case for dismissing the unregulated alternative. The unregulated monopoly has a continual incentive to reduce exclusion problems, whereas the regulated monopoly does not, and may even have an incentive to increase costs, and therefore prices, such as under rate-of-return regulation.

The foregoing discussion has assumed that water companies make single, take-it-or-leave-it offers, which customers must either reject or accept. The analysis becomes more complex if the company must engage in bargaining with its customers.

Bargaining with customers may have either positive or negative effects on welfare, compared to the take-it-or-leave-it alternative. When bargaining is present, some of the initially excluded customers may receive price reductions until they are no longer excluded. Low-valuation buyers face a lesser danger of complete exclusion. On the negative side, consumers may hold out for excessively low prices, if they cannot observe the marginal cost of the firm. If the marginal cost is $30 and an individual values the service at $40, the individual may nonetheless hold out for a price of $20, in the mistaken belief that marginal cost is $19. Since customers probably cannot observe the marginal cost of the firm with great ease, the potential for such losses exists. Furthermore, some quantity of real resources will be consumed in the bargaining process. Customers may delay buying hook-ups or may try to masquerade as low-valuation buyers, for instance, or the company may invest in signaling its resoluteness as a bargainer. All of these real resource investments are made for the purpose of receiving transfers, and thus violate first-best efficiency.

We expect that bargaining costs will be a significant issue only for very large users, such as large businesses or perhaps condominial developments that buy their water services collectively. We envisage the water company as being able to commit to a price offer to individual users, rather than having to bargain on a house-to-house basis. Most unregulated large-scale suppliers of household services offer their wares on precisely such terms. If a city has only a single newspaper, for instance, that newspaper may be sold at a price above marginal cost. Yet the newspaper company does not bargain with each household, but rather can precommit to a given schedule of prices, and then sell papers to interested subscribers. We expect a similar practice to develop with water. Bargaining over prices is most likely when the purchase is occasional, rather than repeated, and when the item has significant value, such as an automobile, a home, or an expensive painting. Even in these cases, such as with automobiles, bargaining is often largely a ritual of convergence on a publicly available "book price."

Those institutions that can bargain with the water company, such as large businesses or developments, will consume some resources in the form of bargaining costs. Longer-term rent-seeking costs may arise as well. Individuals will be more likely to live in large condominial developments, for instance, if such decisions hold the promise of reducing their water bill. Residential decisions will be made inefficiently, as the search for transfers from the water company will lead to too many cooperative developments and too few stand-alone houses. In these regards, an unregulated privatized monopoly will again fall short of a first-best optimum.

5. FURTHER ISSUES

We see three other potential problems with unregulated privatized monopolies in the water sector: equity and distributional objectives, rent-seeking costs, and the imperfect ability of governments to precommit to a laissez-faire regime. We consider each problem in turn, and how privatization might be structured to overcome the relevant objections.

5.1 Equity

Commentators often find the distributional implications of perfect price discrimination to be disagreeable. If the water company succeeds in price-discriminating, it will capture all of the produced social surplus for itself, and leave consumers with very little benefit. We do not regard this as a decisive objection to unregulated privatization for two reasons. First, it is possible to structure privatization in such a way as to prevent wealth transfers away from consumers. Second, water policy may be an inefficient means of realizing distributional objectives.

If the distributional implications of price discrimination were objectionable, the income transfer could be reversed by giving water customers an equity stake in the water company itself. The government could privatize water company assets using a Czech-style voucher plan, and send the vouchers to potential water customers. High company profits would then imply high values for the shares, thus reversing the initial transfer of income or social surplus. As long as the company continued to maximize profit, an efficient quantity and quality of water would be produced, without objectionable distributional consequences.*

The water market could even be used to redistribute income toward the poor if the government distributed especially high numbers of shares to the poor (we are not necessarily recommending this policy, however). Even if a foreign company were supplying water, the government could require that company to set up a local subsidiary, and the government could then purchase shares in that subsidiary for its poor. Alternatively, the government could demand that the company distribute such shares for free, as part of the payment for being allowed to market water in the country. The government also could charge the foreign company an entry fee, up to the size of the expected profits (adjusting for risk), and rebate these funds to disadvantaged groups. Even in the absence of rebates or voucher-style privatization, the distributional consequences of unregulated privatization are unlikely to be strongly negative, and may even be positive.

To the extent that clean, potable water brings external benefits, the com-

---

*The firm may deviate from profit maximization if enough of the shareholders are customers as well. The customers, if they have enough voting power, may eschew direct profit maximization and instruct the company to mimic the price and quantities of a perfectly competitive firm. Even in this (unlikely) case, however, the water monopolist will produce a first-best outcome.

munity will gain under price discrimination, even if the monopolist water company extracts the full consumer surplus for each individual. Each individual would fail to reap surplus from his or her water purchase decision, but the community as a whole would receive the external benefits of the additional supply. The widespread provision of clean water would help break the well-known cycle of disease, poverty, and poor sanitation that plagues so many parts of the world. From the community's point of view, the potential status of clean water as a good with positive externalities strengthens the case for unregulated natural monopoly. If water is a public good, from the community's point of view, it becomes less important how much consumer surplus is retained by buyers, and more important to increase the absolute number of hookups as rapidly as possible.

Developed countries also are unlikely to experience significant distributional problems with unregulated natural monopoly. Households currently purchasing water from tankers are likely to face lower per unit prices once they receive a piped connection, even with price discrimination. A government also could offset any undesired distributional consequences of its water policy by changing tax rates or by using the numerous other policy instruments that influence the distribution of wealth.

Using water policy to implement distributional objectives has had an undistinguished track record. Governments often have required water companies to set price below marginal cost to achieve distributional objectives. Fortunately, such practices are now almost universally discredited, even though they continue in practice. Using pricing to achieve distributional objectives has caused many water utilities to be insolvent, and has brought unfavorable long-run distributional consequences as well, again as discussed above. For the same reasons that we reject the distributional argument for pricing below marginal cost, we do not accept the distributional critique of unregulated privatized monopoly.

5.2   Rent-Seeking

A regime of unregulated privatized monopoly may involve significant rent-seeking costs if firms can compete for that monopoly position. In traditional rent-seeking models, the resources expended on capturing a monopoly position are exactly equal to the monopoly profits at stake. If the water company would earn an expected $500 million in profits (in present value), companies would be willing to invest up to $500 million dollars to earn that position (Tullock, 1967). The more successfully a monopolist could price discriminate, the greater the corresponding rent-seeking costs. Unregulated privatized monopoly could cease to serve as a first-best optimum. We see rent-seeking costs as a potential problem for unregulated privatized monopoly. Nonetheless, the transition to privatization could be structured to keep rent-seeking costs to a minimum. The theory of rent-seeking implies only that the would-be monopolist will pay a sum equal to the available rents; this sum may take the form of a transfer rather than the consumption of real resources. Assume, for instance, that the government is selling or auctioning off existing water assets to private companies. The winning company will be willing to bid a sum up to the expected profit, adjusted for risk. So if expected profits are $500 million, companies will bid some sum just short of this amount (again adjusting for risk), and transfer the funds to the government. Rent-seeking takes the form of a pure cash transfer and consumes no real resources. In fact, the transferred funds could be used to satisfy distributional objectives, such as cash rebates to low-income water customers, as discussed above.

Rent-seeking for monopoly positions will consume real resources only when cash transfers are not available. We can imagine water companies that court the local politicians, engage in expensive advertising campaigns, and send costly signals of their trustworthiness. In all these cases, the search for a monopoly position will lead to real resource consumption, and in fact, we do observe all of these phenomena in the contracting process. Nonetheless, both the government and the water company will attempt to replace costly signals and investments with pure cash transfers, simply because the latter are both cheaper and of greater value to the recipient. Rent-seeking costs also can be limited by noncompetitive procurement practices. If one company stands in a favored position to win a given contract, that company need not invest large sums of real resources to capture the subsequent rents. When rent-seeking costs are potentially high, governments may obtain superior results by limiting entry into the profitable activity. The winning company will still be able to serve the entire market, and other companies will be dissuaded from investing resources to capture that position. In sum, we see rent-seeking costs as a potential problem for unregulated privatization, but not necessarily a decisive problem. A comparative analysis also must consider the rent-seeking costs involved with various forms of government ownership and regulation. These costs may be quite high, given the profits at stake.

5.3 Government Precommitment

The imperfect ability of governments to precommit provides perhaps the most serious problem for the unregulated privatization of water. By construction of our policy proposal, water companies and customers are free to set whatever prices and quantities they can agree to. The analysis so far has simply assumed that governments would honor and enforce these contracts with credibility. In reality, governments often do a poor job of enforcing contracts. Many governments are too incompetent to enforce contracts efficiently, or political pressures intervene and the government deliberately voids or rewrites certain contracts. Even in developed countries, governmental interference into the contracting process is common. In the context of an unregulated water market, we can imagine the government rewriting a contract where buyers promise to pay high prices in return for an expansion of capacity or additional hookups. Once the hookups have been made, political pressures might induce the government to regulate or cap prices. Knowing this in advance, the water company might be reluctant to conclude certain kinds of contracts with potential water buyers. In particular, they will be reluctant to conclude contracts that require them to sink significant amounts of capital. (The water sector typically is the most capital-intensive of the infrastructure sectors.) The absence of government credibility will limit the gains from trade.

To a considerable degree, imperfect government credibility simply mimics or recreates the costs of regulation. The costs of forthcoming regulation resemble the costs of having regulation now. In this regard, the initially unregulated alternative should not produce inferior performance, compared to regulation. In some cases, however, the initial absence of regulation may create more risks for companies than if regulation were already in place. If a water company knows that future regulation is forthcoming in any case, the company may prefer to know the nature and extent of regulation up front. Transactions costs may be lower if regulation is present from the onset. Although we regard this problem as a serious one, we do not see regulatory risk as a decisive argument against unregulated privatization. First, an initially unregulated system will not necessarily imply more regulatory risk than a system with initial regulation. Even when initial regulation is present, the water company and its customers always face the risk of additional regulation. A noncredible government cannot make policy risk disappear, or even diminish, by instituting regulations today. In fact, the appearance of regulation may be a signal that more regulation is forthcoming in the future. Typically, we expect greater credibility from the governments that are willing to experiment with market solutions, even if those governments cannot recommit in absolute terms. Today's world exhibits a significant positive correlation between a government's willingness to allow the private sector to operate and the credibility of that government. Starting with a laissez-faire experiment may increase rather than decrease a government's ability, as it has in Singapore, New Zealand, Chile, and other countries in a variety of (nonwater) contexts. Experimenting with unregulated privatization thus might lower regulatory risk, rather than increase it.

The regulatory risk argument is overstated. We could, for similar reasons, argue that the government should regulate every economic sector immediately, to reduce the uncertainty about subsequent regulation. Yet successful economies do not typically approach regulation in this fashion. Rather, a responsible government first attempts to discover what a good policy might be, and then implements that policy. It should not shy away from good policies for fear that the policy might later be abandoned.

Furthermore, a policy "proposal" is precisely that—a proposal about what would work, not a prediction about what will be adopted. Governments might be unwilling to embrace credible commitments to favorable policies, but policy analysts nonetheless should continue to hold such commitments as an ideal or aspiration (Philbrook, 1953). Credibility is, in part, a function of what a government, its citizens, and its advisors believe. By attempting to persuade and to change beliefs about what will work, policy analysts themselves manufacture credibility for policies. To argue that a policy will not have credibility is to assume what is at stake in the policy debate itself.

6. CONCLUSION

The need for water policy reform is pressing, given the stakes in terms of economic development and human health. The lack or very high cost of access by the poor to safe sources of water has devastating social and economic consequences. We have considered unregulated, privatized monopoly as a potential policy improvement. Under some conditions, this policy can approximate a first-best solution across the quantity and quality of output. While we do not expect this first-best result to hold, a laissez-faire approach to water supply may nonetheless result in a significant increase in the number of water hookups. Given the number of individuals who have no access to clean, safe water, this factor should weigh heavily in our evaluation of the policy. The unregulated natural monopoly will bring problems of partial exclusion, bargaining costs, rent-seeking costs, and imperfect government credibility, but in comparative terms we do not see a knockdown argument against unregulated private provision in this context. Unregulated privatization should join the roster of plausible policy alternatives for the water sector.

REFERENCES

Armstrong, M., S. Cowan, and J. Vickers (1994). *Regulatory Reform: Economic Analysis and British Experience.* Cambridge, MA: MIT Press.

Asian Development Bank (1993). *Water Utilities Data Book.* Manila: Asian Development Bank.

Bittlingmayer, G. (1982). "Decreasing Average Cost and Competition: A New Look at the Addyston Pipe Case," *Journal of Law and Economics,* Vol. 25 (October), pp. 201–229.

Breyer, S. (1982). *Regulation and its Reform.* Cambridge, MA: Harvard University Press.

Brook Cowen, P. (1996). "The Guinea Water Lease—Five Years On: Lessons in Private Sector Participation," World Bank *Viewpoint Note* No. 78 (May).

Cooper, K. J. (1997). "Battling Waterborne Ills in a Sea of 950 Million," *Washington Post,* February 17, pp. A27, A30.

Demsetz, H. (1968). "Why Regulate Utilities?" *Journal of Law and Economics,* Vol. 11 (April), pp. 55–66.

Dickinson, H. W. (1954). *Water Supply of Greater London.* London: Newcomen Society.

Guislain, P., and M Keff (1996). "Concessions: The Way to Privatize Infrastructure Sector Monopolies," *Public Policy for the Private Sector: Infrastructure* (Special Ed.) (June), pp. 21–24.

Klein, B., and K. Leffler (1981). "The Role of Market Forces in Assuring Contractual Performance," *Journal of Political Economy,* Vol. 89 (August), pp. 615–641.

Klein, M. (1996). "Economic Regulation of Water Companies," Policy Research Working Paper No. 1649, World Bank, Washington, DC.

Loeb, M., and W. A. Magat (1979). "A Decentralized Method for Utility Regulation," *Journal of Law and Economics,* Vol. 22 (October), pp. 399–407.

Mann, P. (1996). Paper presented at the World Bank Seminar, Washington, DC, December 17.

Oi, W. Y. (1971). "A Disneyland Dilemma: Two-Part Tariffs for a Mickey Mouse Monopoly," *Quarterly Journal of Economics,* Vol. 85 (February), pp. 77–90.

Philbrook, C. (1953). "'Realism' in Policy Espousal," *American Economic Review,* Vol. 43 (December), pp. 846–859.

Philips, L. (1983). *The Economics of Price Discrimination.* Cambridge, UK: Cambridge University Press.

Rivera, D. (1996). *Private Sector Participation in the Water Supply and Wastewater Section.* Washington, DC: World Bank.

Tullock, G. (1967). "The Welfare Cost of Tariffs, Monopoly, and Theft," *Western Economic Journal,* Vol. 5 (June), pp. 224–232.

World Bank (1992). *World Development Report 1992: Development and the Environment.* New York: Oxford University Press.

World Bank (1994). *World Development Report 1994: Infrastructure for Development.* New York: Oxford University Press.

World Bank (1997). *Toolkits for Private Participation in Water and Sanitation.* Washington DC: World Bank.

Zajc, K. (1996). "Private Sector Participation Options in the Water Sector in Transition Economies," Ph.D. dissertation, George Mason University, Fairfax, VA.

Chapter 21

The Future

Structural Change and Competition in the UK Water Industry

Rebecca Reehal

If stock market valuations are systematically lower than regulatory asset base valuations for reasons associated with the regulatory regime . . .
this could be addressed by reconsidering the regulatory issues involved,
irrespective of ownership structure.

—Ofgem, 2000a.

The privatized water industry in England and Wales is at a crossroads.
The second price review since privatization in 1989 has radically transformed prospects for all stakeholders. The regulatory regime is mature,
most companies until recently have remained vertically integrated, debt
is getting increasingly expensive, and yet the industry is required to finance a £15.6 billion capital program over the next five years.

Substantial share price underperformance and sector underperfomance
relative to index-linked bonds following the most recent regulatory price

*Rebecca Reehal is Senior Analyst, European Utilities Research, with Lehman Brothers. In the most
recent Reuters survey, she was the top-ranked water analyst by UK finance directors. Formerly, she
was a consultant with National Economic Research Associates.*

Reinventing Water and Wastewater Systems: Global Lessons for Improving Water Management,
edited by Paul Seidenstat, David Haarmeyer, and Simon Hakim.
0-471-06422-X   Copyright © 2002 by John Wiley & Sons, Inc.

review announcement have been clear measures of the increased perception of regulatory risk and reduced financial flexibility within the industry. Ten years on from privatization, a key issue being faced by the privatized water companies is how to unlock value in the regulated UK water business and improve capital structure to reflect the stability of the underlying business, in order to provide flexibility for expansion. Determining the optimal capital commitment in non-UK regulated activities is becoming increasingly difficult in the absence of this change.

What key factors are triggering the change that is reshaping the existing regulatory and legislative framework in the UK water industry? Within this umbrella, what has been the industry response to change to date? What has been the regulatory response? And what types of reforms are best aligned to the industry characteristics and most likely to maximize value for all stakeholders in the longer term? In this chapter, we briefly review these issues, and in doing so, highlight the changing shape of the model of water privatization in England and Wales.

1. WHAT FACTORS ARE TRIGGERING CHANGE?

Two factors are triggering change within the industry: a mature and tightening regulatory regime; and as a logical consequence of this, liberalization.

1.1 The First Trigger for Change—A Mature and Tightening Regulatory Regime

The water companies in England and Wales operate under an incentive-based system of price control. Prices for the regulated water companies have been based on maximum price changes above or below the level of inflation, known as K factors. Calculation of the K factor takes into account such factors as quality enhancements and an efficiency element based on the expectation of future cost trends and efficiency savings. The essence of incentive regulation is that, because prices are set in real terms for a fixed period of time (currently five years), companies have a strong incentive to seek efficiency gains. By allowing cost savings to be used in the short term to reward management effort and shareholder risk-taking, the regulators try to ensure that customers will receive better service at a lower price. Decoupling the link between prices and costs for five years seeks to avoid the deadening effect of cost-plus forms of regulation (e.g., the "gold plating" of the asset base).

The main mechanism to date for passing gains back to customers has been the Periodic Review, which in effect allows the passing back of extraordinary levels of profit, via, among other mechanisms, a price cut. Some companies have also voluntarily reduced prices between reviews or given out customer rebates. The uncertainty that arises from having to set prices in advance has been mitigated by allowing the price cap to change in accordance with changes in prespecified costs. A "Relevant Change of Circumstance" can trigger an "Interim Determination" of the K factor. Since privatization in 1989, there have been two price reviews, the second of which resulted in new prices for each of the water companies in England and Wales, for five years, from April 1, 2000.

The method used for deriving K factors is illustrated in Figure 21.1, but essentially companies can create value by outperforming the regulatory deal. This occurs through beating the cost-efficiency targets built into the pricing deals, or through beating the allowed cost of capital.

In setting the level of K for each company, the economic regulator, the Office of Water (Ofwat), therefore, takes account of three key variables:

1. The allowable weighted average cost of capital (WACC)
2. Allowable operating costs (based on targets set by the regulator for achieving cost savings)
3. The capital value on which the allowable return is allowed.

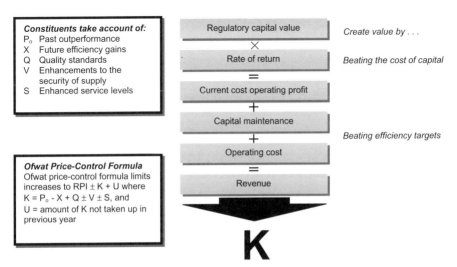

Figure 21.1   Building up to price limits.

The "regulatory capital value" is a current cost measure based on the initial market value (the average over the first 200 days of trading postprivatization) and inflated by new capital expenditure net of current cost depreciation.

The second price review led to an average initial price reduction in 2000–2001 (Po cut) of 12.3% to be sustained for the following two years, with an average rise of 2.6% in total over the following two years (see Figure 21.2). Average annual reductions in prices over five years amount to 2.1%. In their business plan submissions, companies had sought average price increases of 3.8% above inflation.

We explain in the following subsections why the mature regulatory regime is making it tougher for companies to outperform their regulatory contracts going forward. In essence, erosion of the "financial cushioning" that has existed since privatization is beginning to highlight some of the fundamental long-term weaknesses and inconsistencies inherent in the current regime.

1.1.1   The Allowable Weighted Average Cost of Capital Provides Little Headroom

The weighted average cost of capital (WACC) identifies the return required by the providers of capital (debt and equity) in order to fund the capital of the business. To date, any differences between Ofwat's view of

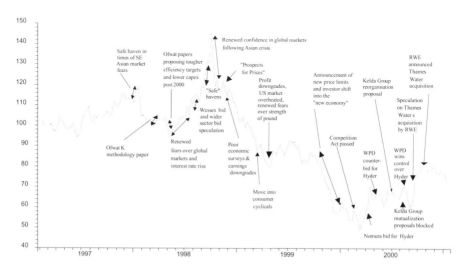

High: 145.62, 10/5/98. Low: 49.36, 3/13/00. Last: 71.99, 1/29/01.

Figure 21.2   UK water price index. *Source:* Datastream.

the cost of capital and that of investors has been "swamped" by the ability of companies to generate additional cost savings. Given the limited scope for outperformance of cost-efficiency targets (discussed below), the available headroom between allowed and achieved returns is now even more crucial. The key variables are summarized in Table 21.1. The allowable WACC range of 4.25–5.25% posttax real on new investment has severely reduced headroom in comparison to the glidepath to the 5–6% posttax real assumption deemed appropriate at the first price review.

An additional shortfall is that the WACC calculation assumes a spot rate cost of debt, but some companies have existing fixed rate exposures above this rate that pushes the average cost of debt beyond the assumption. While Ofwat has allowed a specific company premium of up to 0.4% on the cost of existing capital to take account of the cost of embedded debt (deemed to be 7.9% across the industry on average), the overall approach to setting the cost of debt introduces a high degree of regulatory risk. This is because, under the current pricing framework, although prices are set for five years, the allowed return is calculated using a real interest rate applying at the time of the price review. Companies will only benefit, therefore, if real interest rates are below those assumed by Ofwat. In addition, given that prices are linked to inflation, changes to inflation should, in theory, have no impact on a shareholder's real return. However, Ofwat's methodology means that under a rising inflationary environment, companies benefit, since the real value of debt would have reduced. Conversely, under a falling inflationary environment, companies that have locked into long-term debt may lose out.

The current treatment of debt in the WACC calculation, alongside relatively short price review periods (five years) in relation to the long-term financing requirements, is clearly destabilizing.

Table 21.1 Components of the Werghbed Average Cost of Capital (WACC) at the 1999 Price Review

Risk-free rate	2.5–3.0%
Debt premium	1.5–2.0%
Posttax cost of debt	2.8–3.5%
Equity risk premium	3.0–4.0%
Adjusted equity beta factor	0.7–0.8
Posttax cost of equity	4.6–6.2%
Gearing	Around 50%

Source: Ofwat (1999b), p. 130.

1.1.2   Built-In Operating Expenditure Savings of 2.4 – 3.1%, on Average, Provide Less Scope for Outperformance

Ofwat's price limits for the next five years assume operating expenditure savings on average of 2.4% per annum for the water service and 3.1% for the sewerage service. Assumed savings on capital maintenance amount to 10% for the water service and 12% for the sewerage service. For capital enhancement expenditure, Ofwat assumed that all companies could make a minimum efficiency saving of 2.1% per annum for the five years 2000–2005. Judgments were also made on the scope for the least efficient companies to catch up with the most efficient ones. Operating expenditure efficiency targets proposed by companies in their strategic business plans were more likely closer to 1–2%, although few companies revealed these publicly.

A mature regulatory framework and the capital-intensive nature of the industry make these targets extremely challenging. The latter means that only a relatively small proportion of costs (between 25% and 50%) are likely to be controllable. Moreover, as companies become more committed to maintaining the network (i.e., spending on capital maintenance vs. enhancement), their ability to modify expenditure phasing in response to market movements (e.g., changes in interest rates) is becoming more constrained. The regulatory deal also substantially reduces the built-in incentives for outperformance through cost-efficiency savings through the introduction of a rolling five-year clawback mechanism for past operating expenditure and capital expenditure efficiency. This shortens the period for which companies and shareholders can benefit from past cost savings, over and above those already assumed by the regulator. See Table 21.2.

1.1.3   Continuing High Investment Burdens

The structure of the value chain in the water industry means that the capital costs of building and maintaining assets dominate. The water industry is a long-term highly capital-intensive industry with high investment relative to turnover. The 26 water and sewerage companies and water-only companies have aggregate annual revenue of almost £7 billion. In aggregate, they have a regulatory capital value of £26 billion and invest over £3 billion per annum. Since privatization, there has been capital investment of over £33 billion.

A £15.6 billion investment program across the industry is built into price limits over the next five years. Around £7 billion of this total is re-

Table 21.2   Capital and Operating Expenditure Included in the Latest Price Determinations (Projections of Expenditure, 2000–2005)

	Water (£ million)	Sewerage (£ million)	Total (£ million)
Capital Expenditure (Five-Year Total)[a]			
Base service:			
Infrastructure renewals expenditure	1,260	890	2,150
Noninfrastructure capital maintenance	2,130	2,130	4,260
Enhanced service levels	1	137	138
Supply/demand balance[b]	1,129	556	1,685
Quality enhancements	2,260	5,120	7,380
Total	6,780	8.833	15,613
£ per property per year	57	78	135
Operating Expenditure (Annual Average)			
Base service	1,395	917	2,312
Enhanced service levels	0	0	0
Supply/demand balance	23	24	47
Quality enhancements	33	156	189
Total	1,451	1,097	2,548
£ per property per year	61	48	109

Source: Ofwat, 1999b, p. 74.

[a]Capital expenditure is gross of grants and contributions for new development.

[b]Only £315 million of this impacts on the price limits. The remaining expenditure is broadly self-financing.

lated to quality enhancements and £6.4 billion to capital maintenance. As such, operating costs account for only 40% of revenue, and capital charges for a further 30%. The return on capital accounts for the remaining 30% of revenue. Just less than two-thirds of cost is therefore accounted for by capital investment.

The trend for the industry is a growing capital base, broadly similar infrastructure renewal expenditure, and a declining operating expenditure profile. Prices will therefore be influenced more by the cost of building and maintaining the assets than by the cost of day-to-day operations or raw materials.

The scenario that companies risk with declining profits and a squeeze on infrastructure maintenance charges is one where cash flows are just in balance if the network is being maintained. This leaves little scope for further improvements to meet new obligations and, all else being equal, might risk the creation of incentives to run down the asset base in the long term. Increasingly, companies are being required to spend capital expenditure, which is not depreciable. While historic cost depreciation in relation to the regulatory asset base remains pretty stable over the next five years, capital expenditure in relation to the regulatory asset base and current cost depreciation in relation to the regulatory asset base are on a declining trend.

1.1.4 Increasing Regulatory Uncertainty Between Formal Price Reviews

Price limits can be adjusted between formal reviews if, for specific reasons set out in their licences, a company experiences a material increase in costs, or loss in revenue, and seeks an interim determination. Ofwat assesses the net additional costs or revenue loss (arising from "notified items" or new water quality and environmental obligations) against a specified materiality threshold. Three new notified items relating to the effects on debt recovery of banning disconnections, uptake of optional metering, and administrative costs of protecting vulnerable groups were introduced in the November 1999 price determination. The increasing emphasis on relevant change of circumstances to trigger interim determination of the K factors represents yet a further move toward rate-of-return regulation. Redetermining prices on an annual basis is inconsistent with the long-term nature of the industry, increases uncertainty for shareholders, and is likely to encourage inefficiency and overinvestment.

1.2   The Second Trigger for Change—Liberalization

Competition in the water industry in England and Wales has been possible under the existing regulatory framework for some time. This is primarily through "inset appointments," where incumbent operators are replaced by new entrants supplying water on specific sites, and through cross-border supplies, where customers are supplied by a company from a neighbouring region. These types of "within market" competition have, until recently, been slow to develop. We explain in the following subsections why the pace of liberalization is likely to quicken and, within this context, we consider the options for change that are best aligned to industry characteristics.

1.2.1 New Legislation Has Transformed the Outlook for Competition

Under the European Union (EU) Treaty (Articles 81 and 82), companies were already forbidden from entering into anticompetitive agreements or from abusing a dominant position. The 1998 Competition Act, which came into force on March 1, 2000, changes the way in which these Treaty provisions will be enforced. It introduces two prohibitions:

1. *The Chapter 1 Prohibition.* This is a prohibition of agreements between undertakings that have the object or effect of preventing, restricting, or distorting competition in the United Kingdom and that may affect trade in the United Kingdom.
2. *The Chapter 2 Prohibition.* This is a prohibition of conduct by one or more undertakings, that amounts to the abuse of a dominant position, in a market in the United Kingdom, that may affect trade within the United Kingdom.

In respect of water companies, the Competition Act grants Ofwat the powers to investigate complaints, and to impose financial penalties of up to 10% of turnover if a company's actions are found to be anticompetitive. There are a large number of areas where powers could be applied to enforce the prohibitions. The Competition Act and company licences can now be used to investigate pricing conduct, and whether the prices charged to different classes of customer are appropriate. The Competition Act also makes the development of third-party access to some assets inevitable, providing a catalyst for widescale competition through shared networks from now onward. In future, refusal of access, or the imposition of unreasonable price or nonprice terms for access (unless objectively justifiable, on, e.g., quality grounds) could be regarded by the regulator as abuse of a dominant position. Outsourcing arrangements, such as billing, engineering design, and meter reading, could also be scrutinized via the Competition Act. The regulator has also made it clear that an abstraction licensee's conduct falls within the scope of the Act. Examples of potential abuse could include a reliance on protected abstraction rights or denial of access to a resource or a refusal to supply a resource to a new entrant.

1.2.2   Third-Party Access Is Developing

Third-party access (TPA), or common carriage, is prevalent in many industries, for example, in the oil industry, where pipelines may be used by a number of independent companies. In utility industries, deregulation usually requires third-party access to a network, where users pay for the use of another company's assets in exchange for a fee. In the United Kingdom, TPA is now common in electricity, gas, rail, and telecommunications networks. TPA in water networks has been less common to date, but following the introduction of the Competition Act, all companies now have in place a set of access codes.

In order to make TPA work, it will be necessary for certain access rules to be defined to ensure that water quality can be maintained, as well as principles defining how access charges will be determined. Rules of access for the water industry would be analogous to the Grid Code in the electricity industry. The Grid Code specifies *inter alia* minimum technical specifications for installations, rules for the sharing of information and plans between the transmission company and other system users, and certain standards to which the network operator should adhere. Specific issues that need to be resolved in a water market are:

1. *Access to What?* The definition of what assets are parts of the network can be blurred. Water companies have storage reservoirs and third parties may wish to use these. But are they part of the network? Ofwat has made its position on this clear:

> . . . (The decision) will depend in part on whether access to the network facilities is essential to competition. A facility is essential if access to it is necessary in order to compete in a related market, and duplication is impossible or extremely difficult owing to physical, geographic or legal constraints (or is highly undesirable for reasons of public policy). In deciding whether a facility is essential, I will consider whether it is needed to compete in the relevant market, and the ease and cost of its duplication. To favor new entrants by interpreting the concept of essential facilities too

widely would damage long-term incentives to invest or to develop competing facilities that might be more efficient. I will be careful to strike a balance, such that market competition and common carriage is facilitated, but not forced or distorted.

—Ofwat, 1999a

And more recently:

. . . My view is that many of the incumbents' capital assets are likely to be essential facilities. If they refuse access (or set terms that are unreasonable), the Director may be called upon to decide whether the incumbent's behaviour is illegal.

—Ofwat, 2000

Separate charges for a use of a number of essential facilities would therefore be needed in an approach analogous to the gas industry's method, where storage charges are levied separately from transportation fees.

2. *How Should Services Be Safeguarded?* Competition will need to take account of supplier of last resort. As such, at least for a while, compliance with legal duties to ensure certain aspects of service provision will continue to be underwritten by the existing undertaker to guarantee the supply of water and sewerage services to a domestic customer within a certain geographical area.

3. *How Should Supplies Be Secured?* New entrants will need to ensure that inputs to the network (e.g., flow rates) are matched by customer demands. It might also be necessary, where the new entrant fails to make enough water available to supply its customers, for the incumbent to receive adequate payment for being left automatically to supply the entrant's customers via some form of financial security. This type of mechanism would be similar to the balancing mechanisms that are used in the electricity markets in the Nordic region, Germany, and Spain, and in the UK gas market.

With a proper consideration of technical and legal issues, TPA in the water industry could be made to work. Moreover, the fact that the legislation is already in force means that companies have no choice but to make it work. The real issue is how and where in the value chain should it be accommodated? And is it sensible to accommodate it within the existing industry structure? Going forward, there are two main ways in which TPA could be accommodated within the industry.

Companies could remain vertically integrated and TPA be accommodated as a deviation from this. This is the approach that has been adopted within the industry to date, with minimal impact. It requires minimal change to business process, but the associated transaction costs are potentially high. With this approach, it is difficult to demonstrate nondiscrimination as the terms and conditions to related companies are not transparent.

The monopolistic network business could be separated out and incorporated within a separate company, as is the case in more liberalized electricity markets such as the United Kingdom, Spain, and the Nordic regions, and in the UK gas industry. This would involve having TPA with separate businesses for resources, production, water distribution, and supply. The water production division of an incumbent water business would need a contract with the "network" business for carriage of its water. A set of charges would be established, and it would be clear that the incumbent's production division was treated in exactly the same way as those of new entrants.

The Competition Act 1998 imposes enormous financial risk on companies that cannot demonstrate that they are acting consistently with the law. Formal separation with transparent contracts would protect the company from the risk of investigation, and so would appear to be a logical step. It would also place the responsibility for issues such as quality and liability in the hands of a single identifiable organization. Such restructuring would require changes to primary legislation (e.g., for licensing different activities separately) and has not attracted much interest from policy makers to date. In the following couple of sections, we consider the response of the industry to change. We then go on to explain why separation would appear to be the most sensible way forward for the industry, both from a competitive standpoint and from the need to ensure that the optimal risk-reward structure is in place.

2. THE COMPANY RESPONSE

The changes outlined in the previous section has, over the last year or so, led to an unprecedented amount of internal restructuring within companies and to new ideas for contracting out operations; more recently, it has triggered a host of financial restructuring initiatives aimed primarily at lowering the cost of finance. We provide below brief background on the main three regulatory test cases that have been tabled, each of which provides some guidance on current thinking.

2.1 The First Test Case—Mutualization

In June 2000, Kelda, following a strategic review, announced its plans to create a community-owned corporation (Registered Community Asset Mutual, or RCAM), which would negotiate on an arms' length basis to acquire the shares and assets of Yorkshire Water Services for a total consideration that, at the time of the original announcement, Kelda believed to be close to the Regulatory Asset Base. The RCAM was to be 100% debt-funded, with any surpluses being returned to customers via improved tariffs or higher levels of investment. It would have 160 staff and an independent board. Following the sale, the RCAM would put all contestable operational activities to competitive tendering. Kelda expected to be awarded all initial operating contracts to ensure continuity of service, after which it intended to become a dedicated water and waste management facilities management company and will be rated accordingly.

Therefore, in their initial form, the proposals implied that a large degree of risk would reside with debt holders and customers:

1. All regulatory risk would be transferred to the RCAM, but Kelda and the RCAM would share outperformance benefits.
2. All responsibility for contracting out new capital expenditure enhancement would reside with the RCAM.
3. Most of the costs associated with the transaction were to be borne by the RCAM.

And yet, at the same time, Kelda maintained that it expected the RCAM to buy Yorkshire Water's assets for close to the Regulatory Asset Value less debt (around £1.1 billion). While Kelda proposed the establishment of a number of financial reserves to finance exceptional operating costs and hedge against changes in debt costs, it was never quite clear as to where this money would come from, both initially and on a continual basis.

It was not surprising that the proposals were blocked subsequently by Ofwat in a Stock Exchange announcement on July 25, 2000. Ofwat highlighted several key factors it felt had not been adequately addressed: a failure to set out clearly how customers would benefit from a change in ownership; a failure properly to inform Yorkshire Water's customers about the proposals and consult with them, a failure to ensure that the Drinking

Water Inspectorate and Environment Agency would be able to enforce the required quality standards; and a failure to demonstrate clear independence between the proposed mutual and Kelda.

2.2 The Second Test Case—Procurement Rules

Western Power Distribution (WPD), under the terms of its offer for Hyder in 2000, intended to retain the core assets and responsibilities under Welsh Water's Instrument of Appointment, but to contract out the day-to-day running of Welsh Water's operations to United Utilities. On completion of the offer, WPD intended to set up a subsidiary, UUCo, and transfer various Hyder assets and employees into UUCo. WPD would then procure that Welsh Water would enter into an operations and maintenance (O&M) agreement with UUCo, and then sell UUCo subsequently to United Utilities.

Subsequently, Severn Trent challenged the seven-year £800 million O&M agreement in the High Court on the basis that it infringed public procurement rules. The judgement, delivered in October 2000, ruled in Severn Trent's favor on three counts: (1) that it had brought the action promptly; (2) that an injunction would be the more appropriate remedy; (3) that the O&M agreement was covered by the procurement rules regardless of the facts that it did not involve a procurement of services, that it was incidental to the WPD offer, and that, because UUCo would start life as a WPD subsidiary, it would benefit from an exemption in the procurement rules for intragroup services contracts.

A fourth issue remains outstanding, whether the services under the O&M agreement are characterized under the procurement rules as "Part A" services (which require an open tender procedure to be followed) or "Part B" services (which do not). This was to have been considered at a subsequent hearing, but was rendered academic by the decision of WPD and United Utilities not to proceed with their arrangements.

One option for awarding a services contract without the need to put it out to tender is to do so in a way that will allow it to take advantage of the intragroup services exemption. This exemption excludes from the scope of the procurement rules the award of contracts to "affiliated undertakings" (i.e., group companies). This would imply that care must be taken to structure any arrangements so that they either rely on existing group companies to take on service contracts or, where it is necessary to set up a new company, that such companies are allowed some form of independent existence before they are transferred out of the group. Arrangements that rely purely on hiving off operations into a new company and putting in place a contract that can then be taken on by a predetermined purchaser are likely to breach the procurement rules.

There are still a number of hazy areas surrounding this issue that might be seized upon by a company in a restructuring initiative in order to avoid putting operations out to competitive tender.

2.3 The Third Test Case — Ring-Fenced Company Limited by Guarantee

On November 3, 2000, Glas Cymru, a new independent company, offered to acquire Welsh Water from Western Power Distribution (WPD) for £1.8 billion, representing around 95% of Welsh Water's Regulatory Asset Base (RAB). Under the proposals, Welsh Water would outsource, under competitive contract, around 80% of the day-to-day operations of the assets and the provision of customer services. Glas is a company limited by guarantee, owned and controlled by 50 members that will not receive dividends or have any other financial interest in the company. As such, it pledged to reinvest financial surpluses for the benefits of customers. Glas's aim is to establish itself as a provider of low risk and essential public service, thereby allowing Welsh Water to raise new capital more cheaply.

Glas claimed that reduced investment risk came from having no ordinary shareholders, strong reserves, and a competitive procurement plan. Of key importance was its pledge not to diversify into other activities in order to remove diversification risk.

In addition to tackling head on a number of the regulatory hurdles raised at the time of Kelda's mutualization proposals (management accountability, no transfer of liabilities to customers, independence of the board from WPD, operation under strong commercial incentives via a deferred consideration, competitive procurement of 80% of services accompanied by a detailed procurement plan, and wide consultation with interested parties), Glas's proposals were fundamentally different in the implicit size and allocation of risk between stakeholders.

To start with, they were based explicitly on a sub-RAB transfer value, which under the initial proposals equated to 95% of the RAB equating to reserves of £100 million. The board in its initial proposals stated that these would be built up to £300 million by 2005. In addition, the proposals sought licence modifications to the mechanisms for setting price limits between periodic reviews in its licence.

3. THE REGULATORY RESPONSE

The regulatory response to the financial reengineering proposals tabled has resulted in a public debate on new forms of structural ownership that might be suited to the water industry in England and Wales. Ofwat published its response to the radical restructuring proposals tabled by Glas Cymru [referred to as the Position Paper (Ofwat, 2001) from here onward] on January 31, 2001, giving the green light for a debt-financed not-for-profit company, Glas Cymru, to acquire Welsh Water from its owners, Western Power Distribution. The key elements emerging from this ruling that are likely to influence the viability of any future financial restructuring in the water industry are highlighted below.

3.1   Glas Deemed to Be a Specific Case

Ofwat made it clear in the Position Paper that this case should not be considered as a template for others. Specific circumstances cited included:

> The broad support of the democratically accountable body of the Nafw [National Assembly for Wales]; the wish by the current owner, WPD, not to be involved in the water sector; the significant discount in the purchase price to (RCV); and the clear independence of the purchaser and seller.
> —Ofwat, 2001, p. 4

Moreover, Ofwat stated that it:

> Would itself be concerned if a significant number of companies were to seek to follow, before the model of a wholly debt-financed, outsourced company had been tested . . . A model which appears suitable in the particular circumstances of Dwr Cymru would not necessarily be appropriate for the English water industry as a whole, where the equity shareholder model has performed well . . . however Ofwat will continue to review with care and fairness any proposals put to it.
> —Ofwat, 2001, p. 11

3.2   Level Playing Field Toward Regulation

Welsh Water will continue as the appointee, with all of the obligations of a water and sewerage undertaker. Ultimately, all responsibilities for service, safety, and quality standards remain with Welsh Water, regardless of its intention to outsource most of the operations, and strengthening of internal controls and licence modifications are being imposed to underline this liability.

Ofwat made it clear that there must be a level playing field for all in the industry. As such, there can be no guarantee that contract prices will be reflected in price limits that will continue to be reset every five years using the industry-wide methodology. A fair amount of risk resides with Glas, which will be influenced by the ability of management to hedge adequately against risk in the outsourcing contracts.

In addition, Ofwat proposed two licence changes to ensure the availability of information for regulatory and market purposes. First, a licence modification requires Welsh Water to publish information relating to its financial and trading position and the results of its operations in line with listing rules for a company with ordinary shares listed on the Stock Exchange. Second, a licence modification requires Welsh Water to maintain the listing of a financial instrument (either a bond and related coupon or a preference share).

3.3 Attractive Financing Terms Secured Through Tighter Ring-Fencing

Ofwat pointed to the fact that the elimination of risk from diversification into nonregulated activities is significant in terms of securing the right financing terms. Subsequently, it proposed a new licence condition to secure that this restriction apply to Glas and all its subsidiaries. Other license modifications aimed at tightening the financial ring-fencing include prohibiting the raising of any finance for the regulated business on terms which include cross-default covenants and the requirement that Welsh Water must at all times maintain an investment grade corporate credit rating. Glas secured commitments from its financiers, including MBIA (a US-based credit insurance company), which will be insuring a large proportion of the bond issue, and most of Welsh Water's existing lenders.

3.4 Large Reserves Required to Absorb Cost Shocks

The intention is for Glas to initially have committed credit facilities of around £150 million, implying a purchase price at 92% of RAB (estimated at £1.8 billion). These are estimated to grow, via the addition of trading surpluses, to £350 million by 2004–2005.

3.5   Safeguards to Share Benefits with Customers

Glas believe their proposals will enable them to reduce their financing costs (which account for around one-third of Welsh Water's total expenditure) by around 25%. In order to safeguard the sharing of these savings with customers, Ofwat requires Glas to make a public policy statement on customer rebates from 2003–2004 onward, "analogous to a statement of dividend policy," to provide reassurance that these benefits will flow to customers as soon as possible.

3.6   Adequate Efficiency Incentives in the Absence of an Equity Buffer

Ofwat made it clear that it continues to think that incentives for making efficiency savings are likely to be stronger under equity financing, highlighting the point that "lenders are likely to focus largely on protection of their interest payments and capital rather than maximisation of efficiency" (Ofwat, 2001, p. 7). Key factors of the deal that it thinks will offset these concerns are the linking of management remuneration to performance both in terms of delivering customer rebates and improving its relative efficiency rankings, the existence of a majority of nonexecutive directors, and the involvement of MBIA, which is to ensure a large proportion of Glas' bond issue. On the latter, it sees benefits from MBIA providing an external monitoring role.

3.7   Procurement Rules

Severn Trent's High Court challenge helped to establish some clear ground rules in the area of outsourcing and the ruling by Ofwat on Glas crystallized the need for full competitive tendering. The Glas ruling highlighted the fact that, under such a deal, Ofwat attaches particular importance to the fact that the operations and customer service contracts will be competitively tendered from the outset. Ofwat is also requiring a licence condition requiring Welsh Water to submit a procurement plan.

As added protection over control of operations, a licence modification will also be made prohibiting Welsh Water from delegating responsibility for the proper performance of its statutory functions to any other party.

3.8   Member's Powers

Ofwat expects Glas to make clear the principles governing the appointment process and to negotiate the recruitment process in an open and transparent manner. Members will have formal power to direct the policies of the Glas board, strengthened by the introduction of a new licence condition that requires Welsh Water to obtain an undertaking from Glas prohibiting it from making changes to the memorandum and articles of association without Ofwat's consent.

3.9   Maximizing Efficiency for Stakeholders in the Longer Term

The debate on new forms of ownership in the water sector in England and Wales has reinforced the need for some change going forward, although financial reengineering proposals in themselves do not go far enough in lowering overall risk in the industry. The Glas proposals go some way to highlight the deficiencies that need to be addressed and highlight the current lack of visibility of the protections currently offered in the licences, but ultimately there is a risk that the asset company may be left with too much risk and, in the longer term, customers may be required to bear too much risk. As Ofgem, the electricity and gas regulator, indicated in its response to Ofwat's consultation on the Glas proposals:

> The additional returns which companies are seeking can only come from a redistribution of risk. Unless the overall level of risk can be reduced, companies can only gain at the expense of customers (or possibly other companies). Moreover, it is not clear that the equity buffer proposed in the form of reserves will be sufficient in all future periods, hence posing an additional source of risk for customers, accentuated even further by the proposed Licence modifications . . . in Ofgem's view serious concerns would remain if similar proposals were put forward in the gas and electricity sectors.
>
> —Ofgem, 2000b, Par. 2.4

While financial engineering of this type provides some temporary relief, there are wider reforms that could potentially lead to a more transparent allocation of overall risk and retain the benefits of having an equity buffer.

What is the optimal way forward for the water industry in England and Wales? At the core of the way forward, for the industry, is the need to develop new structures and ways of working that allow businesses to be run more efficiently. This is tied inextricably to getting the competitive framework in the industry right, and in doing so, having regard to the different risk-reward structures and types of competition that are appropriate for different parts of the value chain.

Despite the Glas ruling, many questions surrounding optimal risk allocation in the industry remain unanswered. Does it need to be lowered overall? Or just better redistributed between different parts of the value chain? Or do just the existing protections need to be made more visible?

The logical way forward toward addressing these issues and bringing about longer-term efficiencies, while maintaining the benefits of having an equity buffer, point to:

1. Unbundling to better align risks and rewards and limit what is regulated to the monopoly elements while maintaining the benefits of retail price index plus an adjustment factor (RPI–X) regulation for these.
2. Further ring-fencing safeguards for equity holders—no more perhaps than just a longer stable price review for the monopoly part of the value chain that needs to be regulated. Maybe associated with this, a ring-fenced minimum dividend guarantee backed up by an insurance policy.

4. THE LIMITATIONS OF CURRENT VERTICAL INTEGRATION

As detailed at the beginning of this paper, water companies in England and Wales have traditionally been viewed as vertically integrated and regulated on that basis, that is, at price reviews they are set an overall price limit (and inherent cost of capital) that applies to all parts of the value chain (abstracting water through to distributing it). In reality, they perform a number of different activities relating to water production, water treatment, wastewater treatment, and network management and customer services. Experience in other countries such as France and Spain has demonstrated that vertical integration of water and sewerage companies is not essential for adequate service provision. But there is a clear division of activities:

1. Trading activities, where companies can distinguish themselves on price. Such activities could include water resources abstraction and permit handling.
2. Production-related services, where companies can distinguish themselves on price. Services provided in this segment include treatment and storage on the clean water side and sewage treatment and disposal on the dirty water side.
3. Network or infrastructure-related services, where the ability for a new entrant to undercut the incumbent on price may be limited. Infrastructure services on the clean water side cover the transportation of treated water by large trunk mains (bulk transport) and the distribution of treated water across the local network to smaller customers (local distribution). On the dirty water side of the business, activities include bulk transport of sewerage and the local distribution of sewerage.
4. Retail services, where companies can distinguish themselves on standards, level of service, branding, and reputation.

It makes sense to consider each of these activities differently within utilities, because each has very different associated skills, and water companies are no exception. For true comparative advantage to be exploited, the structure of the industry must allow companies to focus on their specific skills in specific markets such as running and maintaining network assets, financing assets, commodity trading, managing and negotiating concessions, packaging related environmental services products together, handling relationships with local municipalities, or marketing and customer service in utility-related businesses.

Only through separation can companies improve efficiency and create a strategy based on their skill base. The absence of vertical disaggregation limits flexibility for merging specific functions (i.e., parts of the value chain) where there are genuine benefits from doing so. How does it make sense to unbundle water activities?

Any future unbundling should be related to consideration of two key factors. First, the different forms of competition that are workable at different parts of the value chain ("within market" or "for the market") are linked to asset characteristics. Second, the risk-reward structures that should be associated with different parts of the value chain are very different, given the very different financing risks. It makes sense, therefore, to permit more stable and long-term returns for the asset-intensive part of the value chain, consistent with the need to raise finance over 20–30 years.

This would not necessitate a move from incentive regulation, nor from the equity-based model. Rather, it would:

1. Increase transparency for capital markets and facilitate a better understanding of the current cost value of the asset base.
2. Provide scope for setting more stable returns where they are needed without compromising the shorter-term interventionist regulation more suited to less capital-constrained parts of the business.
3. Provide companies with flexibility to exit those activities that are not central to their skill set (for example, sell off their supply customers, as is the case in the electricity sector). Such transactions are needed to place a value on parts on the business and customers that are not being valued at all.
4. Allow shareholders to choose to invest in a type of company that matches a portfolio's requirements. Those requiring income and low risk should be able to invest in a long-term asset company with a guaranteed long-term return with any outperformance driven by an ability to achieve lower cost financing and through built-in incentives for commissioning new asset building efficiently.

What is the sensible way forward? Where do we go from here in order to improve efficiency in a vertically integrated privatized water industry operating in a mature regulatory regime? There would appear to be a few logical steps.

4.1   Step 1

Get the structure right: decide on the monopoly assets that need to be price-regulated. The current cost characteristics of these assets in the water industry mean that "within market" competition for this part of the value chain will be less effective than competition for the market. Fundamentally, this is because the substantial discount to asset value at privatization means that the regulatory asset value is way below replacement value. Water prices are way below long-run marginal costs, so that the return that companies earn on their assets is way below that which a new entrant would demand. Therefore, while new legislation will undoubtedly make entry into the water market easier, it is unlikely, in itself, to lead to a substantial increase in competition for network-related activities.

4.2   Step 2

For the monopoly (i.e., network) assets that need to be price regulated, make the length of the price review period consistent with the financing requirements. Water companies are currently raising debt for 20–30 years, so there seems to be little rationale for not lengthening the review period to at least 10 years, maybe even longer. At privatization, price review periods were initially for 10 years, callable after 5 years. A reduction in the overall cost of debt through greater stability and transparency would benefit all stakeholders.

4.3   Step 3

For supply activities (treatment, customer services, etc), facilitating within-market competition makes sense as an incentive to improve efficiency. Over time, it becomes questionable as to whether these activities should be price regulated at all. Unlike the network activities, it makes sense to deaverage prices in order to avoid cherry-picking. As in the electricity sector, it might make sense to apply a short-term second price limit for "supply" activities for a limited period until within-market competition develops fully.

4.4   Step 4

Eventually, put a system in place to encourage multiple suppliers. In the gas industry, this was achieved through auctioning storage capacity. In water, ultimately, it might involve auctioning abstraction licenses.

5. CONCLUSION

A mature regulatory environment and the inevitable move toward liberalization are prompting significant change in the water industry in England and Wales. Recent financial restructuring proposals in the industry involving changes to ownership structures bring to the fore the crucial issue of appropriate safeguards to lower and redistribute risk between stakeholders. Ultimately, any new ownership structures will need to coexist with any competitive reform. Getting the overall risk-reward structure and competitive framework in the industry right probably points to unbundling of the true monopoly assets. See Figure 21.3.

The Water Bill, currently in its draft form, may provide the next impetus toward achieving this goal. In its draft form, among other things, it proposes to:

1. Bring the duties of the water regulator in line with those of the electricity and gas regulators by including an additional primary duty to "promote competition." This compares with current legislation, where the director-general of Water Services has only a secondary duty to "facilitate" effective competition. The two primary duties are the existing ones, ensuring that the functions of an undertaker are properly carried out and that undertakers can finance the proper carrying out of those functions.

2. Simplify the process for water abstraction licence applications and modifications. It also proposes that the duration of licence and the compensation regime surrounding licence revocation should be changed. It proposes that new licences will be time-limited and lower the period of nonuse after which a licence can be revoked from seven years now to four years. Additional changes would need to be put in place for the proper implementation of licence trading.

Ultimately, policy makers face two choices for maintaining an equity-financed, incentive-based model. Either the framework must change to bring about greater visibility and stability for the capital-intensive parts

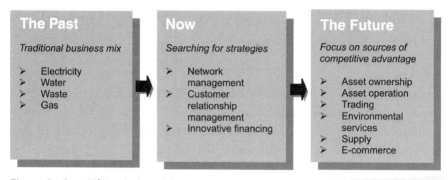

Figure 21.3   Utilities in transition.

of the value chain, or higher prices must be allowed through a higher cost of capital. It makes good sense for all stakeholders for decision-makers to opt for the former.

REFERENCES

Ofgem (2000a). "New Ownership Structures in the Water Industry: A Response to the Director General of Water Services Consultation Paper," Birmingham, UK, July.

Ofgem (2001b). "The Proposed Restructuring of Dwr Cymru Cyfyngedig by Glas Cymru Cyfyngedig: A Response to the Director General of Water Services Consultation Paper," Birmingham, UK, December 22.

Ofwat (1999a). MD 154, "Development of Common Carriage," Birmingham, UK, November 12.

Ofwat (1999b). "Final Determination Future Water and Sewerage Charges 2000–05," Birmingham, UK, November.

Ofwat (2000). MD 161, "Maintaining Serviceability to Customers," Birmingham, UK, April 12.

Ofwat (2001). "Position Paper: The Proposed Acquisition of Dwr Cymru Cyfyngedig by Glas Cymru Cyfyngedig," Birmingham, UK, January 31.

Chapter 22

Creating Liquidity in an Illiquid Market

How the Internet Can Facilitate the Development of Efficient Water Markets

Patrick J. Meyers

One of the most critical issues facing cities and municipalities today is the answer to the question: How do we increase our water supplies to allow growth to continue? In the United States, this issue is particularly important in Western states, although the recent water "wars" between Alabama and Georgia illustrate that Eastern states are not immune (ABC, 2000). To underscore the importance of this concern, the former Governor of California, Pete Wilson, has stated that he believes that the looming water crisis may make the current California energy crisis pale in comparison (Wilson, 2001). An open and transparent on-line market for water is one of the most powerful tools that federal, state, and local water professionals and regulators should employ to help avoid this future crisis, and improve their current situations. A well-functioning water market will

Patrick J. Myers is Director with Resource Conservation and Management Consultants, Houston, Texas, specializing in consulting services to public and private water companies and other utilities. Formerly with Enron's water utility, Azuria, he developed the first online water exchange, waterzwater.com.

Reinventing Water and Wastewater Systems: Global Lessons for Improving Water Management, edited by Paul Seidenstat, David Haarmeyer, and Simon Hakim.

provide substantial near-term benefits to agricultural, commercial, and residential water consumers.

The scarcity of reliable water supplies constrains economic growth, and it is the job of water authorities to quench the growing thirst of agricultural, industrial, and municipal users. For example, the Texas Water Development Board's recent $16.9 billion plan to meet the water demands of a burgeoning population—expected to double by 2050—calls for the construction of 7 major and 10 minor new reservoirs, as well as hundreds of miles of new conveyance systems—yet fails to adequately supply 8.5% of the water consumer groups studied. The report shows that the current solution set is not adequate to solve 100% of the demand. Therefore, other solutions must be examined, and one of most powerful and successful methods to allocate a valuable resource is a well-structured marketplace. A marketplace allows the resource to find its highest and most valuable use, and helps to avoid or minimize wasteful practices.

The terms "water market" and "Internet-based on-line water market" will herein refer to distinct institutions for facilitating the exchange of water. The terms may be interchanged in many instances, although it is a key assumption in this chapter that an Internet marketplace contains certain efficiencies of scale and scope that make it the preferred choice. Currently, there are many places around the world, including the United States, Chile, Australia, and Mexico, where traditional water markets exist. The US markets are located primarily in rural and agricultural regions of California, Colorado, and Texas. Although not a prerequisite, locations with active marketplaces are likely to experience higher adoption rates, as participants are already comfortable with the process of water trading.

The more structured markets usually operate in local coffee shops, using a simple bulletin board where buyers and sellers post contact information on index cards.* Less formal markets rely on word-of-mouth communication and established relationships. These markets create value for participants by reducing transaction and search costs. In fact, if the marketplace did not exist, the transaction costs might be large enough to prevent a transaction from occurring in the first place. The fact that these markets emerged spontaneously is evidence that these markets create

---

*The Westlands Irrigation district in California has experimented with an on-line exchange. The marketplace uses outdated technology and functions merely as an on-line bulletin board, where sellers create a listing and interested buyers must contact the sellers directly. In essence, the system acts as an on-line "classified advertisement" for water.

value for their participants. However, these micromarkets remain inefficient because the transferability of water is geographically specific and, until recently, there was no efficient method to centralize the information.*

Now, however, the emergence of the World Wide Web and the development of trading software allows for the efficient aggregation of micromarkets into regionally managed marketplaces. Once a regional market infrastructure is established, the incremental cost of adding additional marketplaces is very low. An Internet marketplace will improve upon the traditional marketplace efficiencies by reducing transaction costs further by increasing the number of participants, by decreasing the time required to communicate with potential counterparties, and by improving transparency. Once a market develops liquidity, the transferability issue is addressed, as participants attempt to arbitrage pricing differences between regions. The differences in prices encourage free-market "for-profit" participation to find, and build, efficient methods for transporting water. The benefits this "for-profit" motivation brings to the market is enormous, as it reduces the need for the government to finance and construct projects. A well-functioning water market could thereby delay or obviate the need for costly state and federal projects, as the newly informed private sector endeavored to undertake more water projects. As the governments funding requirements are reduced, those funds can be applied to other projects where the benefits may not be as clearly defined, such as environmental and recreational projects, or a completely different class of projects, such as educational improvements.

A marketplace also provides the economic incentives needed to help reduce wasteful practices. In many regions, low-value, high-water-demand crops are grown with irrigated water. Water users system-wide will benefit when a farmer is provided the incentive to view water as a cost of production. The farmer may make the economic decision to "harvest" water in dry years instead of growing a crop. This incentive occurs naturally when buyers and sellers view real-time, transparent pricing information, allowing them to more accurately value their assets, in this case water, and make informed decisions.

---

*This is one of the major distinctions that separates water from other commodities like gas or power. The infrastructure does not exist by which to move large quantities of water over large distances, and thereby arbitrage regional differences, thus forcing the markets to function as microgeographical markets. There are very few locations that could act as regional "hubs" that would allow for the creation of a trading environment that resembles, for example, the Henry Hub for gas and the California/Oregon border for electricity.

A factor that has limited the creation of water markets in the past has been the lack of clear and readily available information concerning the value of water and water resources. This lack of information has created suspicion among potential market participants that they may not be selling or buying their water at the "best price." As a result, market volume is thin, liquidity is limited, uncertainty continues, and a scarce resource is not allocated where it creates the greatest value for society. An Internet marketplace will help alleviate some of these limiting factors.

1. INTERNET MARKETPLACES

One does not have to look far to find stories about how the Internet is changing the way business is done today and will be done in the future. In certain areas, particularly marketplaces, the Internet will serve as an important tool to help more efficiently allocate and distribute resources and information. One of the key questions is "What can the Internet do that other current technology cannot?"

What separates the Internet from other current technologies, that is, from faxes, phones, and mail, is its ability to allow multiple users access to large volumes of information, helping them make quick, well-informed decisions. The Internet economizes information, enabling an open system for users to access, analyze, and act upon market information in a convenient time and place. This increased access lowers the transaction costs associated with buyers and sellers finding each other, and allows for an easy comparison among various alternatives. Most of the business-to-business (B2B)* sites generally involve "eHubs" (Sawhney and Kaplan, 1999), third-party Internet-based intermediaries that focus on a specific industry or commodity, that host electronic marketplaces, and that enable transactions among participants. In accordance with Metcalfe's law,[†] the value created through an eHub grows exponentially with the number of users. Examples of marketplace eHubs are: Alta Energy, Paper Exchange, and Plastics Net. Aggregation mechanisms present in eHubs are most appropriate and beneficial when the current market framework is characterized by high transaction (search) costs, transactions are negotiated one-off, and the product is fairly uniform or commodity-like.

---

*Or government-to-business (G2B), or government-to-consumer (G2C).

[†]Commonly referred to as a Metcalfe network. In a normal group, without a "hub," all points must be connected to every other point to exchange information. In a Metcalfe network, only one connection is needed, to the hub, to connect to every user.

The current water market is one in which aggregation mechanisms will be very valuable. For example, as mentioned above, compared to a highly transparent and liquid market, the current water market may be characterized as inefficient due to the fact that buyers and sellers may spend a great deal of time locating potential trading partners; in other words, transaction search costs can be high. A buyer, usually a municipality or a private entity, that wishes to purchase a large volume of water must first find and then aggregate many potential sources. The buyer usually conducts many parallel negotiations with potential counterparties on a one-off basis, adding additional transaction costs.

A more efficient approach would be for the buyer to create, or participate in, an on-line marketplace, with a standard purchase contract and pricing structure, and notify all water owners in the region of the auction. Potential sellers could examine the contracts on line, and determine if the terms were acceptable. The sellers will also be assured that they received a fair price, as all transactions conducted are completed at the market-clearing price. The buyer will benefit through a reduction in negotiations with counterparties, and a more transparent marketplace that assures their constituents, ratepayers, that all transactions were conducted in an open marketplace, and not behind closed doors.

There are several keys to creating a successful eHub, but the most important is the ability to supply liquidity. Liquidity can be created either by aggregating all the suppliers into one location (thus generating a market where buyers are at an advantage), by focusing all the buyers in one location (thus generating a market where sellers are at an advantage), or by creating a neutral location (where neither buyers nor sellers are at an advantage). In most cases, successful eHubs require a neutral site. For example, most commodity markets are neutral in nature, and thus buyers and sellers are assured that the transaction they completed was "at-the market," and their search cost is reduced. Examples of this type of market structure are eBay and the NASDAQ Stock Exchange. An example of a marketplace where sellers are at an advantage is in highly engineered or patented products. For example, in the prescription drug market, the buyer must pay the price the pharmaceutical manufacturer lists as the sales price.

Other factors that can affect the success of an eHub include managing the cost and speed of aggregating sellers and acquiring customers, dealing with inertia of ingrained business practices, managing credit risk, choosing the appropriate software platform, and investing the appropriate resources. Additional factors that will determine success are the depth of industry expertise, degree of market fragmentation and inefficiency, and an understanding of how markets should, and do, function. eHubs and marketplaces have the promise to provide valuable services in many industries and sectors. The water market is one in which a neutral eHub could provide significant benefits to water consumers.

2. INTERNET MARKET-BASED SOLUTIONS FOR WATER

To move to an on-line water market, the current structure of water "markets" will need to be modified and new rules and policies instituted that will aid both in determining title and ownership of rights and in building the economic incentives needed to form a neutral and transparent marketplace. There are several areas and types of markets for which an Internet solution will provide the best alternative. Such a market would be open and available to all, and it would allow pricing information to be quickly disseminated into the marketplace. It would also allow users to determine the economic benefit and cost of proposed environmental policies and accurately value their current assets.

There are several steps that governments would need to take in order to facilitate the creation of an on-line water market. One of the basic policies that should be enacted is defining water as a property right and establishing clear, unambiguous title to a discrete quantity of water. The issue of how water is linked to land will also need to be clarified. For example, in many areas in Texas, the current users of groundwater have unlimited access to the water beneath their land, leaving open the potential for significant overdrafting. Few problems have arisen to date, primarily due to the fact that most of these areas are rural, and the users withdraw only a portion of the water in the aquifers. However, unlimited access creates a problem when cities begin to search for new water supplies. They often acquire land and install massive pumping stations. Over time, aquifers may be dewatered, unless there is a system to control and monitor how much water can be withdrawn. The current system basically allows the user with the largest pump to withdraw as much water as possible, regardless of the impact on neighboring property owners. This effectively takes water from landowners without their knowledge.* An example of this can be seen

---

*A good source for how the oil industry solved this same dilemma can be found in Libecap and Smith (2001).

in certain areas in West and North Texas, where municipalities and private entities are acquiring agricultural land to build pumping stations and "ship" the groundwater to their customers.*

One alternative is to publish data on each aquifer to reflect the sustainable yield that precludes dewatering of the aquifer. Based upon this sustainable yield, landowners over the aquifer should be assigned a right to use the groundwater based on a per-acre basis.[†] This process has been put in place on several aquifers in Southern California, where landowners have been assigned, or adjudicated, an annual right to pump a specified volume of water. It is also important that water rights are defined as separable from the land and that data are easily accessible in county or state offices to determine the current status of the land and water on a piece of property. This requirement will aid the marketplace in the creation of liquidity, as potential buyers could accurately ascertain what a seller owns. Once these rights have been assigned and recorded, an Internet marketplace would allow landowners to buy, sell, or lease rights to meet their particular needs. It would also allow cities to either lease or acquire water rights in order to meet their projected demands. A good model to follow would be the current mineral rights law; groundwater rights could be defined in a similar manner. The same structure should also be applied to surface water rights, reservoir ownership, and conveyance systems. Utah, Colorado, and Texas are currently altering their respective water laws to provide some of these benefits.

The water rights market will require one critical component before it can begin to function properly. That component is the creation of a database that contains a clear line of title from ownership records. The most analogous approach is that used by real estate title companies, since water rights would relate to land and should be overlaid on the current title and registration of land. It may be advantageous for the "water title" policy to be included in all land transactions to assure that buyers and sellers are perfectly clear on what is being exchanged in every transaction. Utah and

---

*If an on-line water market existed, the municipalities would gain efficiencies based on a reduction of search costs. However, the municipality might lose its ability to profit from the lack of transparency, as market participants, that is, sellers, would have access to pricing information that would enable them to alter their asking price to reflect market conditions. In this case, the reduction of transaction costs and increased liquidity should provide more value than that lost to buyers due to price ambiguity.

[†]For example, if it is determined that the sustainable yield of the aquifer is one acre-foot per year per acre, a land owner would have the right to withdraw one acre-foot of water from the aquifer per year for every acre of land owned.

Colorado are both in the final stages of requiring that some real estate transactions contain a water title policy rider.

Once the water has been defined as a clear property right, separable from the overlying land, and the right is insurable under typical title insurance products, the water market participants, that is, government, municipal, environmental, and agricultural entities, should be encouraged by the states to move to the Internet.* Entities interested in determining the value of their rights would examine historical data and determine a fair market price. By giving landowners the ability to value their water economically, the system will allow all market participants to have an incentive to allow users who have assigned the highest value for the resource to use it, thus benefiting all participants. As described above, an eHub will add significant value to this market by aggregating buyers and sellers, and insuring that information is available to all and that water is valued to reflect its true opportunity cost—the value of the second-best use.

There are three different products that an eHub should trade that will benefit water users. A water lease would allow water to be sold (leased) on a short-term basis from one user to another. Leases in the traditional water market are typically one year, but they can be structured as a series of one-year leases to increase the duration of the contract. This is typically the most active market where "coffee shop" water markets exist, as both the buyer and seller are more comfortable entering into shorter-term contracts, since the true value of the water may not be realized in the initial stages of a market due to a lack of liquidity. As liquidity improves, buyers and sellers gain confidence in the ability of the market to accurately reflect the true economic value of water. Hundreds of one-year leases are conducted annually in the Lower Rio Grande Valley, the Eastern Slope of Colorado, and the Central Valley of California (*Water Strategist,* 2000, 2001). The Water Master† for the Lower Rio Grande is currently experimenting with an on-line marketplace.

Secondly, market participants could sell the right to extract water, that is, a water right. This involves a permanent sale and allows the buyer the legal right to extract a certain volume of water on an annual basis. These

---

*The state and federal governments need to be involved in this process due to the fact that they have regulatory review over water transactions in many areas. Reducing the regulatory "transaction cost" would help the marketplace create further value for market participants.

†The Water Master, a part of the Texas National Resources Conservation Commission, allocates, monitors, and controls raw water resources in the Rio Grande Valley.

rights allow an owner to either use the water or sell water in the lease market.

The third product involves options on the ownership of rights, which requires the sale of the water right to the option purchaser at a specific price during a prespecified period of time at the buyer's discretion.

These three products together will provide the financial underpinnings for significant follow-on benefits in the form of water infrastructure financing. The combination of these three products will generate the information and comfort level necessary for banks to finance private or public water projects.

The key insight from this analysis is that, once markets are allowed to form, users begin to get a better idea of the "true" economic value of the traded asset, that is, water, by comparing it to the resource's alternative uses. As liquidity improves and transactions become more transparent, other market institutions will emerge to broaden and deepen the market. For example, accurate price signals will facilitate the efficient evaluation and financing of water infrastructure projects such as conveyance, storage, and conservation.*

When a water market begins to function, agricultural users, for example, will value the water that is being applied to their land and the economic benefit they receive from it. Once growers know the real "cost" of water, they may continue using the water as it is currently used, or may "harvest" it. The additional information that a functioning market provides the grower will also provide incentives to invest in conservation measures if the conserved water can be sold. This set of follow-on benefits has a positive impact beyond growers. By harnessing the market, governments will not have to rely on costly and heavy-handed mechanisms to achieve conservation. Water consumers will benefit from the "new" water created by the growers' conservation efforts, and taxpayers will benefit by having to fund fewer water projects.

In the short term, conserved water can be delivered more cheaply than water obtained by developing new conveyance infrastructure. Conservation delays the need for local and state governments (or private entities)

---

*In order to perform this analysis, it is assumed that the current user of the water, a grower, would own any water that has been "saved" by investments the grower has undertaken. If a grower installs a drip irrigation system, the difference between the prior usage and new usage could be sold and benefit the grower. The current laws may need to be modified in order for the grower to retain this right.

to invest and carry large amounts of debt. A portion of the cost savings may be made available to the landowners through conservation and reuse programs that pay for capital improvements, that is, for more efficient irrigation methods, in exchange for a portion of the water conserved. With all the incentives aligned, significant amounts of "new," more efficiently allocated water is likely to be available locally.

The functioning of an Internet-based water market will dramatically benefit the regional water planning process by providing a framework that puts a clear value on alternative uses of water and water projects. According to the regional water plans of some districts, "new" water supplies to service future needs are expected to come from conservation and reuse programs. In some of the regions, the volume of water that is to be "created" through conservation programs is up to 30% of the "new" water (Texas Water Development Board, 1999). By allowing a clear and transparent marketplace for water to exist, the incentives for conservation are built into the system, and the goals of conservation programs will be reached with less effort on the part of the regulators, while at the same time, minimizing the cost to taxpayers.

3. APPLICATIONS OF ON-LINE WATER TRADING

The following three cases are examples of how a functioning on-line water market could be applied to different geographical market situations. The first case shows the potential positive impacts that a functioning water market may have in aiding the implementation of the Texas Water Development Plan. The second case shows how an on-line water market could be applied to help the San Antonio Water System solve its drought-year constraints. This example is easily applicable to any regional problem where water demand fluctuates widely. Finally, the third case shows how the CALFED Environmental Water Account could be implemented using an on-line marketplace.

3.1   Case I: Texas Water Development Board

The recently published Texas Water Development Board Plan (2001) calls for the construction of $16.9 billion of infrastructure projects in Texas, most of it financed over the next two decades by state taxpayers. Many of these projects are intended to move groundwater from rural areas to urban areas, sometimes over large distances at great expense. However, before these projects are able to obtain financing in the public and private capital markets, two requirements must be met. First, the project must obtain a reliable water supply that is large enough to justify the capital-intensive construction.* Second, the project operator must be able to sell long-term supply contracts to urban centers (i.e., there must be demand for the water). This is where a functioning on-line water market would allow the potential pipeline builder the opportunity to either purchase water rights, sign a series of multiyear leases, or option water rights prior to obtaining financing for a project. By defining supply and demand, these mechanisms provide the financial underpinning for the project. Without a functioning water market, these mechanisms could not exist, and it is difficult to compare the costs of constructing competing projects.

A water eHub also allows project owners to sell, through an auction or exchange, their "capacity" in a conveyance or storage system. Operators and contractors may sell/auction "unused" capacity. For example, the natural gas industry uses this approach to allocate space or "capacity" within its pipelines. Currently, gas pipelines are required to provide open access to pipelines at competitive prices. This will provide an efficient mechanism allocating and transporting water on an as-needed basis.

As a tertiary effect, these changes will enable the state to stretch its investment dollars. In the above example, a private entity may be willing to undertake some projects, thus allowing the state to invest in projects where the economic impacts are not directly attributable to one set of users (e.g., environmental and recreational). If the conservation programs are successful, potential project construction may be delayed or cancelled as market participants find private financing, thus allowing Texas to use public funds to invest elsewhere. Functioning water markets will enable legislators and the public to more accurately value projects that are primarily environmental in nature.

For example, consider the case where an environmental study shows that in order to maintain the aquatic life in a river, a certain minimum annual flow must be maintained. An economic analysis based upon the prevailing water market could demonstrate exactly what the economic cost would be to maintain the water flows. This analysis could be conducted wherever water markets exist, and prior to a policy being adopted, its true economic cost could be determined. Further analysis based upon both the

---

*Measured in the tens of thousands of acre-feet.

environmental and financial impact of a policy would better educate the public and legislature, and aid in policy decisions.

These impacts will be substantial, and will only be achievable if well-functioning on-line water markets are allowed to operate. A liquid market in water will increase the likelihood that a project may be financed at the private level, and therefore will increase the state's ability to fund a different class of projects, that is, environmental, recreational, or educational.

In conclusion, this case illustrates how a well-functioning water market would enable the residents of Texas to enact the policies of the Texas Water Development Board's Plan. The market could reduce the obligation of the State to finance projects, as private enterprise would be more able to construct and finance water projects on a "for-profit" basis. A water market would also provide the incentives needed to achieve conservation goals, by encouraging current users to view water as an economic good, and invest in conservation methods.

3.2   Case II: San Antonio Water System

Like most of the Western United States, San Antonio's water demands fluctuate dramatically on an annual basis. As shown in Figure 22.1, between 1980 and 1998, San Antonio had a minimum annual usage of roughly 155,000 acre-feet, and a maximum usage of 190,000 acre-feet, or an increase of approximately 22% over the minimum requirements.

Under the current market framework, the San Antonio Water System (SAWS), as with most municipalities, is required to obtain water supplies for peak years, that is, 1984 and 1989 in the figure, to insure reliability in

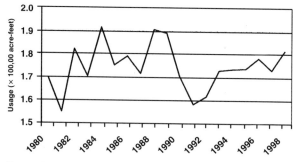

Figure 22.1   Variability of San Antonio water use.
Source: Texas Water Development Board, 1999.

case of drought. However, in nondrought years, such as 1981 and 1992, the additional water supplies needed to supply peak demand in drought years is not used. This infrequent use dramatically increases the "true cost" of water, as the marginal cost of obtaining water that is utilized in only one out of every five years is high. However, by providing more options to customize water transactions, a well-functioning and transparent on-line water market will allow SAWS to obtain water at a much lower cost. There are several different ways in which a marketplace may lower the total, delivered cost of water for SAWS.

To purchase an Edwards Aquifer water right, the primary source of water in San Antonio, SAWS would have needed to pay $1,600 per acre-foot in 2000 (*Water Strategist,* 2000), which would amount to $163 per acre-foot per year amortized at 8% over a 20-year life. Currently, this is the only option available to SAWS. However, if this water is used only 20% of the time over 20 years, users would pay the equivalent of $820 per acre-foot, assuming the water is used in years 3, 8, 13, and 18 at an 8% discount rate.*

In 2000, the average cost of a one-year municipal water lease in the Edwards Aquifer was $80 per acre-foot (*Water Strategist,* 2001). Currently, SAWS is unable to reliably use this source for water, primarily because the market is extremely illiquid, however, if SAWS were able to enter the market on an as-needed basis, they could acquire water at substantial savings to their ratepayers. The rate for water in drought years would need to increase by a factor of 10[†] to make the acquisition of the water right make economic sense.[‡] This is despite the fact that annual water leases are unlikely to increase on this magnitude, primarily due to the fact that the value of crops grown in the area is unlikely to increase as they are primarily lower valued crops that are also grown in other areas (i.e., corn). The analysis also ignores the positive economic impact of the farmer being able to grow a crop in four out of five years, whereas if he sells his water rights his land must be fallowed and no crops will be grown. Therefore,

---

*For this calculation, it was assumed that the water was needed in years 3, 8, 13, and 18, and the cost of the water right in year 0 was $1,600. The cost of the water was obtained by discounting the four future payments at 8% until the present value was $1,600. Thus, a cash flow of $820 in years 3, 8, 13, and 18, discounted at an annual rate of return of 8%, would generate a present value of $1,600.

[†]As per the prior calculation, the price of the water would have to trade at $820, or ten times the current lease rate of $80.

[‡]There is also another possibility. If a functioning market existed, SAWS could sell water in the four out of five years that the water was not needed. This should provide SAWS the comfort of knowing they could always provide water to ratepayers, while at the same time earning a "return" on an asset that is currently unused.

given the above example, SAWS' (and by extension, their ratepayers') and the farmer's economic positions are improved.

Finally, there is an additional method that SAWS may use to obtain its "peaking" water. SAWS could approach water rights holders in the Edwards and offer to pay them a small up-front annual fee, an option payment, in exchange for the right to potentially call the water at a later point during the water year. The options could be structured to have several different "strike" prices and dates, and allow SAWS to claim the water during the dry summer months on short notice. This would allow SAWS to further reduce the cost of obtaining peaking water, as water is paid for only when it is most critically needed, at a prearranged price that may be below the current market price in dry periods. The farmer's economic position is improved, as he receives added income in every year where the option is not exercised, and has the option to dry farm (rely on rainfall) the crop in the event that the water is called.

In conclusion, each example above shows how an on-line water market may be structured to improve SAWS' flexibility to handle its water acquisitions and manage its supplies. In each of the examples, SAWS' ratepayers benefit through a reduction in the actual cost of obtaining water supplies. The costs will be further reduced through a decrease in SAWS' search and negotiations costs, as water contracts in the region become more standardized.

3.3   Case III: CALFED Environmental Water Account

In 2000, the CALFED Bay–Delta Program released their findings relating to California's water needs. "California's Water Future: A Framework for Action" (CALFED, 2001) is an effort to build a framework for managing California's most precious resource: water. The proposal calls for a partnership between California and the federal government to launch the largest and most comprehensive water management program in the world. Taken as a whole, the program would be the largest ecological restoration project, most intensive conservation effort, and one of the largest infrastructural programs (based on cost) undertaken in decades, with a projected cost in excess of $10 billion. The proposal also contains the goal of having a positive impact on the quality of water provided to millions of Californians.

One specific proposal contained in the Framework for Action is the Environmental Water Account (EWA). The goal of this program is to provide increased water supply reliability to water users while, at the same time, assuring the availability of sufficient water to meet fishery protection, restoration, and recovery needs. As a means to achieve this, the Program will provide commitments under the Federal and State Endangered Species Act for the first four years of Stage 1, which will be based on the availability of water from existing regulation, the EWA combined with the Ecosystem Restoration Plan (ERP), and the ability to obtain additional assets should they be necessary.

One of the critical difficulties that CALFED faces is in determining how to begin to implement some of the proposals contained in the program. As discussed in this case, a combination of market-based solutions, in conjunction with the Internet, will help CALFED begin to institute the EWA proposal contained in the Framework for Action in a quick and efficient manner. The EWA program outlined in CALFED's framework is an excellent example of a case where market-based solutions and Internet technology may be used to help implement the suggested policies. There are two different areas where an Internet marketplace may be used to benefit the state, as well as an additional benefit that the EWA may provide to a water marketplace.

The Plan calls for the EWA to be jointly funded by the state and federal governments and will be authorized to acquire, bank, transfer, and borrow water, as well as arranging for its conveyance. The assets will initially be acquired for the EWA by federal and state agencies, and subsequently managed by federal and state fisheries.

The marketplace would be created using a reverse auction* for the purpose of purchasing the water for the EWA. To conduct this auction, the purchasing agencies would select the location of the water that they wish to purchase as well as a dollar amount that the state is willing to spend. This information would be used to create a reverse auction "market" that conforms to the agency's specifications (a separate auction would be set up for every tributary or water source). These reverse auctions would be conducted over a set period of time (several per year or one per year, as de-

---

*A reverse auction is a mechanism where the buyer, in this case the state or other agency, defines the specific requirements that the water must satisfy as well as the maximum price they are willing to pay and sellers compete to set the lowest bids. Using an on-line water market, registered landholders that fit the desired specifications will be assigned a password-protected account and access to the auction. Users would log-in to submit bids (which they may lower over time if new bidders underbid them) and could check whether or not they currently have a winning bid. After a specified time, the auction is closed, and the winners are notified.

fined by the availability of purchasing funds) and water right owners would be informed of the market in the traditional fashion. The advantage of this "dynamic" system over a typical closed-bid process is that the agency is assured of obtaining the lowest price and best value as a result of market participants lowering their bids in competition with other users.

The Framework also states that normal-year supply improvements may not be achieved in all years due to annual hydrologic variability and its impact on carryover storage.* It further states that in dry years there will probably be shortages. An Internet marketplace could be designed to allow current water users to sell or lease their water in dry years by creating a "Dry Year Market" when the environmental conditions demand. The value in using this approach is that users who may see their deliveries decreased, possibly below the annual requirements of their crops, will be allowed to make an economic decision to determine if they could generate higher returns by "harvesting their water," changing crops, or planting a reduced acreage.

An additional approach that may be used by CALFED to help insure reliability would be to allow users to voluntarily purchase water contracts that are interruptible during dry years, a product similar to that used in the electricity and gas markets. In return for accepting interruptible demand, users would pay less for their water. Due to the subordinate claims of these users, senior users could count on water in all but the most catastrophically dry years. In exchange for this right, the senior users would pay a higher fee, in exchange for firm water deliveries, compared to the users who assume subordinate claims. This would allow permanent crops growers, or those wishing to enter this market, to plan ahead, while allowing seasonal and less valuable growers to pay less for their water. Finally, the EWA could be used by the state to help control price volatility and minimize uncertainty in the market during certain stresses, such as natural disasters and drought. The state could add the EWA water to the marketplace, if needed, to help smooth the problems that might exist during a crisis event.

Using the Internet with dynamic pricing would enable the state in each one of these examples to fulfill the requirement stated in the Framework: "Acquisition of fee title to water will be from willing sellers only, and will be used when neither available public water nor partnerships are appropriate or cost-effective to the specific need." CALFED and the EWA program

---

*These years are referred to as Tier 3.

would benefit through the creation of an on-line marketplace by experiencing a reduction in transaction fees, a lowering of the price paid for water, and a simplification of the negotiating and purchase process. CALFED would also increase its flexibility to react to drought conditions by using its water account to dampen volatility and by providing order to the market during stressful events.

4. CONCLUSION

The current situation facing water resource planners in the United States, and much of the world, is forcing society to look at alternative ways to help insure that reliable and cost effective water supplies remain available to all people. As with many economic problems, the constraining factor for water in many cases is not insufficient supplies, but instead, how to establish efficient mechanisms for allocating the resource among competing users. A good leading example of the "allocation problem" can be seen in California.

Depending upon which report you read, California has another crisis looming that may make the electricity "shortages" experienced by the state in 2000 and 2001 pale by comparison. Reports by various state, federal and environmental agencies expect municipal and industrial water supplies will be several million acre-feet short over the next couple of decades. However, what is often not discussed is how the current supplies are currently allocated throughout the state. Roughly 80% of the state's water is used in agriculture, with the remaining 20% used by municipal and industrial users. If only 5% of the water currently going to agriculture can be reallocated to municipal and industrial uses, the state goes from a net shortfall to a net surplus. What is required for this to occur is an effective way to allow water consumers in agriculture to transfer water to municipal and industrial users in a win-win fashion.

One of the tools that will aid this transfer is a well-functioning on-line water market. An Internet-based market for water will enable this resource to be valued based on economic considerations. Current users of water, primarily agriculture, will be able to value water as an economic input into production. When water becomes more valuable than the output, that is, alfalfa, cotton, or rice, the user can decide to "harvest" water instead of the agricultural commodity. When this occurs, water will be allocated to those who value it most, and supply will be in balance with demand.

The problem facing California is not unique, nor is its ratio of agricultural use of water to municipal and industrial uses. As stated above, current Internet technologies and market-based solutions may be applied to solve the challenges facing water users to create transparent and well-functioning markets. The market functions should be organized to maximize the benefits to all market participants. Similar to the development of other markets, with a clear and transparent market in place, market participants and the general economy will experience important follow-on benefits in the form of resources and incentives to build necessary infrastructure projects.

REFERENCES

ABC World News Tonight (2000). "Water Wars," March 14.

CALFED Bay Delta Program (2001). "California's Water Future: A Framework for Action," Sacramento, CA, February.

Libecap, G., and J. Smith (2001). "The Economic Evolution of Petroleum Property Rights in the United States," Paper prepared for the Conference on the Evolution of Property Rights, Northwestern University School of Law, Evanston, IL, April 20–22.

Sawhney, Mohanbir, and Steven Kaplan (1999). "The Emerging Landscape of Business to Business E-Commerce," *Business 2.0 Magazine,* September.

Texas Water Development Board (1999). "Water for Texas," Austin, TX, August.

Texas Water Development Board (2001). "Water for Texas: Summary of Regional Water Plans," Austin, TX, February.

Water Strategist (2000). P.O. Box 963, Claremont, CA 91711, February.

Water Strategist (2001). P.O. Box 963, Claremont, CA 91711, February.

Wilson, Pete (2001). Keynote speech given at the UCLA Anderson Forecast Conference, Los Angeles, June 28.

Chapter 23

Project Financing of Water Desalination Plants

A Worldwide Trend

James S. Harris

1. INTRODUCTION

Historically, the supply of water to a population has been viewed as a social responsibility. Such a view regarding water supply services has left little room for the private sector. As a result, water management has lacked many of the benefits that the private sector can bring to industry. Generally, ministries, municipalities, or public boards have managed water services along with other public responsibilities, resulting in little or no investment planning and, consequently, a less than optimal system. Furthermore, the public sector, as compared with the private sector, lacked technical and managerial skills needed to improve services and reduce costs (Lokiec and Kronenberg, 2001).

Added to this is the fact that a significant part of the world's total population does not have proper access to safe drinking water. Therefore,

James S. Harris is a Project Finance Partner with the international law firm Lovells based in Singapore. He has advised governments, corporations, and investors in the Asia-Pacific region on infrastructure projects.

Reinventing Water and Wastewater Systems: Global Lessons for Improving Water Management, edited by Paul Seidenstat, David Haarmeyer, and Simon Hakim.
0-471-06422-X   Copyright © 2002 by John Wiley & Sons, Inc.

growing water scarcity and inefficient public sector operations have led to an increasing movement toward private sector involvement in the supply of water to the public. At the same time, technologies capable of converting saltwater into potable water have continued to improve, leading to their increasing competitiveness with alternatives such as water tankers and pipelines.

Taken together with all types of worldwide water projects, participation by the private sector in desalination plants has been relatively minimal, due primarily to the relatively high production costs associated with this process historically. However, in recent years, technological advances and increasing competition have pushed down the cost of desalination. One of the key delivery mechanisms the private sector has used efficiently in organizing the finance, design, construction, and operation of desalination projects is project finance. This structure, which has been used throughout time and across all infrastructure sectors, has the merit of ensuring that risks and responsibilities are allocated properly and that a project's cash flows are sufficient to meet financial and operating obligations. To procure the private sector's services, governments have relied on BOOT (build, own, operate, transfer) contractual schemes and variations. Supported by a competitive and open procurement process, this has resulted in steeply falling water production costs. Whereas a few years ago, production costs were over $2 per cubic meter, recent projects today have production costs below $1 per cubic meter.

This chapter examines a number of areas relevant to the project financing of desalination plants, namely:

- Desalination processes
- Desalination advances
- Financing water desalination projects
- Project procurement
- Risk allocation

Before touching on these topics, however, it is useful to provide some of the applicable background relating to desalination.

2. BACKGROUND

It is well known that water covers approximately three-quarters of the Earth's surface. However, it is less commonly known that 94% of the

Earth's water is saltwater from the oceans and that only 6% is fresh. Of this 6%, about 27% is in glaciers and about 72% is underground. According to the World Health Organization (1996, 1998), only water with a dissolved solids (i.e., salt) content below about 1,000 milligrams per liter is considered acceptable for a community water supply. The application of desalination technologies over the past 50 years has permitted cities and industries to develop in many areas of the world where freshwater is lacking but seawater or brackish waters are available. This development has been prominent in parts of the Middle East, North Africa, and some of the islands of the Caribbean (Buros, 2000) (see Figure 23.1).

Key factors to consider in deciding whether or not to develop a desalination plant usually include:

- Water quality;
- Water availability from other sources such as rainfall, underground and surface water resources, and so on;
- The level of water demand and the effects of population growth;
- Social acceptance of the plant, including the environmental aspects (for example, the discharge of waste brine water back into the ocean);
- The use to which the desalted water will be put (industrial purposes have a lower tolerance for total dissolved solids than do human consumption purposes); and

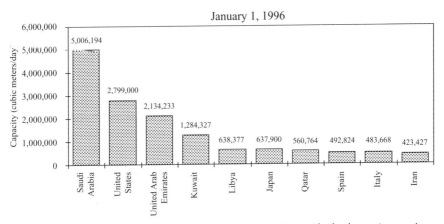

Figure 23.1   Desalination capacity for the ten countries with the largest capacities. *Source:* Gleick, 1999.

- Whether desalted water is the most economic manner in which to obtain potable water.

Often considered an expensive last resort solution for water supply, desalination technology is becoming increasingly affordable in its own right. It is increasingly a viable long-term source of water, providing alternatives to major water transport schemes. The cost difference between desalted water and conventional supplies has narrowed significantly in recent years (Bremere et al., 2001). Current contracted water prices quoted for desalinated seawater range, for example, from $0.46 per cubic meter for the Tampa Bay desalination project, $0.71 per cubic meter for the Trinidad desalination project, to $0.804 per cubic meter for the Larnaca Cyprus desalination project (Lokiec and Kronenberg, 2001).

Furthermore, start-up and operational costs are expected to continue to fall. A cost comparison performed recently in the Middle East showed that desalted seawater was of a comparable price to tanker-imported water and two to three times cheaper than pipeline transported water (Macoun, 2000).

3. DESALINATION PROCESSES

The increasing use of seawater for the provision of potable water has been derived largely from the advances made in the three major desalination processes, explained below.

3.1 The Multieffect Distillation (MED) Process

This is the oldest process for seawater desalination and has been in use in one form or the other since the first half of last century. It relies principally on boiling seawater (by increasing temperature or reducing pressure) through different stages and collecting the vapor created and condensing it into desalted water. Efficient MED plants are based on a large number of stages and benefit from the provision of low-cost steam, usually from the boiler of a power station.

3.2 The Multistage Flash Distillation (MSF) Process

This process was first used by US Navy ships in the early 1950s. It is similar to the MED process in that it relies on distillation. However, the process used causes the water to boil immediately and "flash" into steam. This operation is repeated many times in stages and, typically, an MSF plant can contain 15 to 25 stages, also benefiting from the provision of steam.

3.3 Reverse Osmosis (RO) Desalination

These plants began to appear in the 1970s in the form of small units for island resorts and offshore oil rigs. Reverse osmosis uses the natural tendency of a solvent in a solution to move through a semipermeable membrane into a solution of higher solute concentration (osmosis) in a reverse manner by applying pressure to the higher solute concentration side.

4. DESALINATION ADVANCES

Although RO is the most recent of the technologies used in desalination, current trends indicate that RO has become an increasingly popular technological choice (in preference to MED and MSF) for desalination plants because it has relatively low energy requirements, can operate on standalone basis (RO does not require steam), and can attain high availability with high quality membranes. MED, popular initially, began to encounter problems with scale that precipitates out of saltwater at high temperatures and causes mechanical problems with the plant. The scaling problems of MED thus allowed MSF to become popular, even though MSF consumes the most energy.

Unfortunately, it is not easy to compare the production costs of desalinated water between countries because energy costs, which vary depending on location, are a major component of operating costs. However, the Tampa Bay RO plant, the Trinidad and Tobago RO plant, and the Lanarca (Cyprus) RO plant have all reported a contracted price of below $1 per cubic meter. Furthermore, the introduction of the BOOT structure and project finance has facilitated the accessing of low-cost finance. For example, the Tampa Bay project took advantage of the availability of nonrecourse tax-exempt project bonds. The bonds financed 90% of the project and resulted in a project finance rate of approximately 5.5% (Lokiec and Kronenberg, 2001).

Ultimately, the choice of desalination technology depends on the volume of water required, the quality of the intake water, and other project-specific factors, and these are not consistent between different projects. Given the competitiveness of technologies, when the Singapore government recently sought expressions of interest on a 30-million-gallons-per-day BOOT project, it encouraged the bidders to choose from the range of generally accepted desalination processes. Moreover, they could also include a merchant power plant element, if desired. Singapore may provide insight into future water supply arrangements around the world—especially in those countries that have no groundwater supplies and little surface water.

At present, Singapore purchases water from Malaysia transported by pipeline under long-term contracts. With demand for water continuing to increase and security of supply an issue, however, it is not surprising that Singapore has turned to desalination of seawater.

5. FINANCING WATER DESALINATION PROJECTS

Traditionally, desalination projects were developed, owned, and operated by the public authority requiring the water. This resulted in the relevant authority assuming significant capital and technical risks which, because of its nonexpertise and lack of commercial incentives, it was not in the best position to manage or mitigate. However, more and more public authorities are now passing these risks onto water project developers and private financiers. This group accepts various responsibilities ranging from developing, financing, designing, building and owning, to operating the facilities.

Under a BOOT or similar arrangement, the project sponsors provide project equity (generally about 10–30% of the capital cost of the project) while the remaining 70–90% of the project funds are raised by the sponsors as nonrecourse or limited-recourse debt through project financing. The exact mix of equity to debt will depend on the risk profile of the individual project and will be set to provide all parties with the necessary incentives to see the project completed on time and on budget, and operated efficiently.

The debt funding arrangements in such projects are usually underpinned by the relevant water offtake contract between the relevant project company and the promoter. It is the cash flow of the project, in other words, that provides the revenue stream necessary for the provision of the project debt, because the water offtake contract will normally contain a take-or-pay element obliging the promoter to pay a minimum amount for water, regardless of whether or not it uses it. Typically, the water offtake contract is based on a fixed number of years (often 15–20 years).

The water price in the water offtake contract is normally based on a detailed tariff formula, comprising usually following cost components:

- A capital charge (investment recovery);
- An operation and maintenance charge; and
- Depending on the type of plant, an energy charge.

Figure 23.2 shows a simplistic diagrammatic representation of a typical project finance structure. Desalination plants make good candidates for project financing because of the following characteristics:

- The existence of large fixed assets with a specified construction period;
- The creation of a product (potable water) sold to a monopoly that requires, in most cases, a long-term contractual offtake;
- The presence of private sector investors providing either equipment, EPC (engineering, procurement, construction) expertise, or operational skills; and

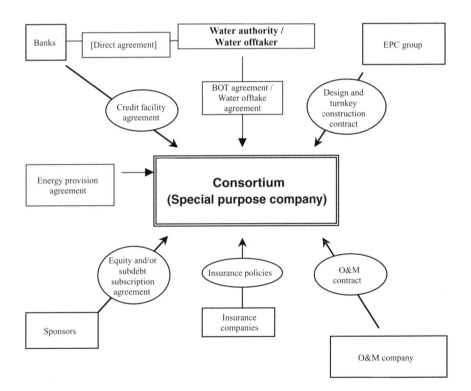

Figure 23.2   Financial structure for a desalination project.

- The traditional public sector interest in relieving capital burden, transferring construction and operational risk to the private sector, and attracting foreign investment and technology.

Typical BOO/BOOT structures have been applied to the financing of desalination plants. For instance in the Lanarca project in Cyprus, a nine-year loan was raised to finance an approximately 50,000-cubic-meters-per-day desalination plant.

Given the risks involved in desalination BOO or BOOT, a typical contractual structure for a project financed desalination project would include:

- A water purchase agreement or concession between the promoter (i.e., government) and the successful consortium (itself usually acting through a project company or special purpose company (SPC) covering the performance and output specifications of the desalination plant, offtake purchase, tariff structure, the targeted commencement date, and so on;
- An engineering, procurement, construction (EPC) contract between the SPC and its contractor(s) for the provision of the completed desalination plant, usually on a turnkey basis covering performance specifications of the desalination plant, targeted completion date, and so on;
- Service contracts between the SPC and service providers covering the operation and maintenance of the desalination plant, and so on;
- Supply contracts;
- Debt and equity funding agreements;
- SPC insurances covering such events as business interruption, and so on; and
- Potentially export credit support may also be involved to support the project borrowings.

6. PROJECT PROCUREMENT

The procurement process that occurs in selecting a suitable sponsor (or consortium of sponsors) to build a desalination plant is typically as follows.

First, a request for prequalification is undertaken to introduce potential investors to the proposed project and to enable them to make an decision as to whether they are interested in participating in the proposed project or not and to allow the promoter to select only the most qualified of interested investors to participate in the remainder of the tender process.

Next, a request for proposals (RFP) stage may then take place, which requires the invited participants to respond with their detailed proposals. The RFP documentation issued by the promoter usually contains detailed provisions as to the project's specific requirements, such as performance specifications, bid forms, drawings and diagrams, technical information, and draft legal documentation. It also advises bidders of the evaluation procedures that will be used. Typical evaluation criteria include price, the financial viability of the SPC, technical viability, the timing of the project, the expected level of employment of local workers, and any qualifications to the RFP requirements.

Evaluation of proposals then follows, where each bid is scrutinized against the RFP requirements and detailed financial, commercial, legal, and technical evaluations are made by the promoter.

Shortlisting and negotiation then take place, where usually the contract is awarded to the most favorably priced and technologically sound bid. Often, however, the top two or three bidders are invited to clarify and negotiate their final proposal, following which a successful bidder is selected and agreements executed by the promoter and SPC. The negotiations will, among other matters, deal with the delicate balance required to be found between the sponsors (who wants an adequate return on their equity investment), the promoter (who wants a guaranteed water supply), and the providers of finance (who want to ensure they have adequate protections, particularly as regards security over the assets of the project).

Given the nature and size of these projects, providing as much information as possible to all bidders at all stages of the bidding process is vitally important. A transparent process increases the credibility of the project and can, in some cases, partially allay sovereign risk concerns. Further, a transparent procurement process is usually more successful in attracting a number of reputable firms as bidders. In turn, this competition leads to significant benefits by delivering low-cost desalinated water.

7. RISK ALLOCATION

A key part of the process of arranging project finance is risk allocation. Before a desalination project can proceed, a prospective sponsor will conduct a detailed analysis of the various risks the project presents and cost the project appropriately in light of the likely allocation of the risks between the parties concerned. The common notion that "Risk should be taken by the party best able to bear it" is the guiding principle of project finance as it explains how projects risks are best mitigated and project costs are kept at a minimum. Risk allocation involves:

- An analysis of rights and obligations under general law, contract, licences, and so on; and
- An evaluation of the probability and effect of risk occurrence in realistic scenarios, in light of:
  - The extent of the project company's resources; and
  - The extent of risk passed through the project company to subcontractors or the sponsors.

The specific risk issues unique to desalination projects include:

- *Raw Water Availability and Quality.* The availability and quality of the raw water supply is a key question. The availability of seawater is never in doubt. However, changes in raw water quality can have severe consequences on the operation of the plants as, for example, the quality of effluent discharges or other pollution in the vicinity of the plant could vary over time. The SPC would not ordinarily be expected to accept this risk. Rather, it would seek to allocate the risk to government by specifying clear water quality standards and default provisions that would operate in its favor if the raw water quality differs to such an extent that it is unable to commercially operate its plant.
- *Role of Energy.* The three main desalination processes all rely on energy, either through electricity alone or electricity and steam. Because energy forms a large proportion of the operating costs of a desalination plant, the ability to secure a reliable energy source at a competitive price is a key parameter in structuring these transactions on a project finance basis. This, in many cases, requires the negotiation of a power purchase agreement (PPA) which is less typical in a traditional water treatment plant. Furthermore, the scope to integrate power/steam generation and desalination means that desalination projects can be, in effect, "desalination plants

producing electricity" or even "IPPs [independent power producer] producing desalinated water" (generally called cogeneration plants), as has been the case in the Middle East. For instance, the Taweelah A2 project (710 megawatts + 50 million gallons per day) was sponsored by the CMS Energy Corporation, best known for its power expertise. This creates a much more complex risk profile: multiple offtakers with different priorities; differing sets of players; and difficult questions related to marginal pricing of electricity, steam, and treated water. Ultimately, the allocation of risks in cogeneration plants will depend on the underlying contracts and the particular characteristics of the plant.

- *Unique Technological Risks.* Finally, technological risks faced by desalination projects are unique to this sector and can be challenging in an industry where plant size has been rapidly increasing and dual purpose plants (water + steam/electricity) or hybrid plants (using a combination of processes) are common. This is something the promoter will usually require the sponsor to bear, often through performance standards relating to the quantity of water to be produced each day, and so on.

8. CONCLUSION

Increasing demand for water, limited fresh surface water and groundwater reserves, and continued technological advancements in desalination methods have driven the trend toward greater use of desalination plants to supply potable water. At the same time, the involvement of the private sector in the provision of water has been introduced through competitive procurement processes that give private sector participants the opportunity and incentive to innovate and provide the best possible water solution for a relevant location.

The involvement of the private sector in desalination projects has been organized through BOOT schemes and facilitated with project finance structures. Such contractual arrangements facilitate efficient risk allocation between project promoters, sponsors, and lenders, and at the same time, provide incentives for the private and public sectors' commitment to a successful outcome. As a result of these mechanisms, competition and innovation in the delivery of desalination projects has increased significantly, driving down the cost of water to the consumers' benefit.

REFERENCES

Bremere, J., M. Kennedy, A. Stikker, and J. Schippers (2001). "How Will Water Scarcity Affect the Growth in the Desalination Market in the Coming 25 Years?" *Desalination,* No. 138.

Buros, O. K. (2000). *The ABCs of Desalting,* 2nd ed. Boston, MA: International Deslination Association.

Gleick, Peter H. (1999). *The World's Water.* Washington, DC: Island Press.

Lokiec, F., and Kronenberg, G. (2001). "Emerging Role of BOOT Desalination Projects," *Desalination,* No. 136, 2001.

Macoun, A. (2000). *Desalination and Water Reuse,* Vol. 10, No. 2.

World Health Organization (1996). *World Health Organization Guidelines for Drinking Water Quality,* 2nd ed., Vol. 2. Geneva, Switzerland: World Health Organization.

World Health Organization (1998). *World Health Organization Guidelines for Drinking Water,* Addendum to Vol. 2. Geneva, Switzerland: World Health Organization.

ADDITIONAL READINGS

Birkett, James D., and Adil Bushnak. "An Introduction to Desalination or How Attention to Detail Can Be as Important as Breakthroughs," Paper presented at the IDA International Conference on Desalination, Singapore, March 21–22, 2000.

Harris, James, and Raymond Bourdeaux. "Project Financing of Water Desalination Plants," *Project Financing International,* Issue 216, May 2, 2001.

Kappaz, Michael H., Loren Rodwin, and Linda Ivanov, Energy, Economics, Privatization and Financing Panel. "Structuring Successful Private Water Projects," International Desalination Association—Asia Conference, Singapore, March 21–22, 2001.

Chapter 24

A Public Sector Alternative for the New Millennium

Michael J. Crean

INTRODUCTION

As the title of this chapter suggests, customers now have alternate means available to them concerning the manner with which their potable water is safely supplied, or their wastewater is returned clean to the environment. As the "natural monopoly" of public water supply and wastewater management gives way to other options of providing this service many jurisdictions are awakening to their rediscovered power to make a choice. Asset sales, contract operations and design-build-operate are just some of the popular alternative approaches now available. Clearly, our industry has been caught up in the competitive frenzy experienced by the electric, gas and telecommunications industries.

Reprinted from *Excellence in Action: Water Utility Management in the 21st Century,* by permission. Copyright © 2001, American Water Works Association.

Michael J. Crean, P.E., served as the Chief Operating Officer for the Washington Suburban Sanitary Commission from 1992 to 2000. He joined the Commission in May 1968 after receiving a Bachelor of Science Degree in Civil Engineering from Villanova University. He served in a variety of engineering and managerial capacities before 1992. He is a registered professional engineer in the State of Maryland and an active member of the American Water Works Association and the American Society of Civil Engineers. Mr. Crean retired from the WSSC in September of 2000. Presently he manages his own consulting practice, which focuses on helping organizations deliver dramatically improved services and value to their customers.

The point here is that the privatization debate is not about demonizing the alternatives, it's about customers exercising their inherent right to make an informed choice. For those utilities fortunate enough to currently enjoy their customers business it is more about earning the right to be the customers continuing choice. If the public sector wants to be the choice of its customers, it must do more than provide excellent quality and service. Customers also want to know that they are paying competitive rates for these services. They want good value.

This chapter is about the journey of one large water and wastewater utility to reacquire the right to be the utility of choice for its customers in the future. It is using the power of its employees to adapt industry best practices, and reengineer its core business processes of treatment, distribution and engineering, to create breakthrough improvement that translates to better value for the customer. I think Southwest Airlines says it best after each flight, "Thank you for choosing Southwest. We know you have a choice in airlines and we will do our best to earn your return business." The Washington Suburban Sanitary Commission (WSSC) also knows the customer has a choice. Therefore, it is our goal to be their choice in the future for water and wastewater services by providing them with an outstanding value, with all of the inherent advantages of public ownership.

BACKGROUND

The WSSC is a large water and wastewater utility created in 1918 by the State of Maryland to serve a 1,000-square-mile Bi-County area just outside Washington, D.C. It has witnessed tremendous growth, especially since the 1960s, reaching its present state of 1.5 million customers and over 450,000 accounts. Two Rivers, the Patuxent and the Potomac, furnish the water supply. There are two water treatment plants, one on each river with a capacity of 72 mgd and 285 mgd, respectively. The average water production is 160 mgd. The distribution system has over 5,000 miles of pipelines. Most of the wastewater, some 130 mgd average, is treated at the Blue Plains Wastewater Treatment Plant in Washington, D.C. The remaining 50 mgd are treated at five wastewater treatment plants located throughout the District and are fed by over 5,000 miles of sewer mains. There are approximately 1,800 employees in the Agency. It has an operating budget of $450M and capital budget of $210M.

Often past success enjoyed by an organization can pose the largest obstacle in moving toward future improvement. The WSSC has been internationally recognized for our award-winning plants, innovative engineering and excellent service. However, our rates for water and sewer service are the highest in the Washington Metropolitan Area and among the highest in the country. Our debt is also very high consuming nearly 40% of the operating budget. It is against this backdrop that the wave of competitiveness washed. As large utilities across the country tested the waters of competition and large savings were being publicized, the public appetite was whetted and things would never be the same in the industry. Almost suddenly, our customers were no longer content with the safe, reliable water and the excellent service. They wanted a better "value" from the Agency that included rate stabilization and debt reduction.

In reaction to our customers' strident demand for better value and in recognition of the change in our industry, we took the first step to make a dramatic improvement in the cost of providing service in early 1997 when we conducted a "competitive assessment." The structured process of improvement that followed was to become known as the Competitiveness Action Program (CAP). Almost a year later in April 1998, the State Legislature passed a bill that required that a Privatization Task Force be created with the purpose of exploring the feasibility of privatizing the WSSC entirely or in part. The Task Force furnished their recommendations (Pirnie, 1999) in October 1999. Briefly, the Task Force recommended that a special Bi-County Committee be established to further evaluate WSSC governance, restructuring and other options for long-term implementation while WSSC pursues further efficiencies through the CAP initiative. These recommendations and other options are currently being debated in the Maryland State Legislature.

COMPELLING NEED TO CHANGE

As customer demands for better value reached the ears of private utilities that were used to operating with a strong financial focus, they began to intensify their presence in the public market place. Concurrently, changes in the tax laws occurred that allowed private companies to navigate around thorny issues such as the operation of facilities financed with tax-exempt bonds. Contract periods that were once limited to five years were suddenly expanded to 10 and then 20 years. This allowed private companies to roll up their annual savings to huge numbers, offsetting initial investments. No longer was it only the case where private companies were called upon to come to the aid of troubled public systems plagued with non-compliances

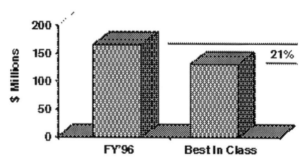

Figure 24.1   Controllable costs.

or huge financial burdens. Now, systems much like our own with excellent records of public service were being reexamined by their political governance in hopes of getting a better deal for their constituents. A case in point is the Milwaukee wastewater system that entered a ten-year contract with United Water at savings of more than $14 million per year.

Given the public sector's inherent advantages of tax exemption, non-profit status and access to low interest bonds the large savings posted by early privatization efforts spoke volumes about the public performance gap. Recognizing this as an excellent way to focus our employees on the need to change we hired a consultant[1] to perform a "competitive assessment." This assessment compared our current organizational business practices with those of the best utilities in the private sector. The result was a "competitive gap" in controllable costs of 21% or about $32M between our FY '96 actual costs and the predicted performance of "best in class." The result is shown in Figure 24.1.

In addition to identifying a significant opportunity to increase customer value, the assessment also outlined "best business practices" in the industry and established a method for measuring performance improvements. This assessment has been the core of our communication with management, employees and stakeholders throughout the improvement process. The important best practices (EMA Services, n.d.) were as follows:

- *Work Force Flexibility.* The largest single dead time factor affecting execution of utility maintenance work is people waiting for other skills. Single-skill work systems artificially separate skills and crafts creating dead time and reducing "wrench on bolt" time. Europeans

---

[1] EMA Services, Inc.

have shown that increasing the range of skills possessed by personnel through cross training increases productivity by up to 40%. The creation of multi-skilled workers reduces the dead time waiting for other personnel.

- *Program-Driven Maintenance.* Many utilities operate in a reactive maintenance mode with the "if it ain't broke, don't fix it" philosophy. The planned maintenance best practice focuses labor resources in planned, preventive and predictive activities, while confining reactive maintenance to a small fraction of all maintenance performed. Materials inventory management is synchronized with planned equipment overhauls.

- *Total Productive Operations.* Traditionally, US utilities have organized around two distinct work groups: operations and maintenance. The TPO best practice eliminates this distinction by changing from a dual work force with all its inherent "hand off" inefficiency, to one of focused operations. Operators no longer "wait and watch" for problems. Instead everyone in a single work force has maintenance assignments to complete, while process- monitoring and control technology alert personnel to process deviation.

- *Technology as Strategy.* In progressive organizations and "best in class," technology is effectively assimilated and strategically employed to boost productivity. They systemize and integrate information so people can share it and reengineer the way they do work. They leverage their scarcest resource, experienced managers, by providing them with meaningful, integrated information tasks.

The results of this competitive analysis were presented to all managers throughout the Operations Branch and many emotional debates took place as people struggled to reconcile the success of the past with this new data about a competitive gap. After a considerable period of "mourning" and denial, the managers finally rolled up their sleeves and moved forward with quiet resignation to undertake the task before them.

COMPETITIVE ACTION PROGRAM

After the completion of the Competitive Assessment, it was necessary to develop an overall plan to close the competitive gap. A Steering Team was formed consisting of key managers and organizational change agents. With the help of a consultant, the Steering Team developed a "Competitive

Action Plan." (EMA Services, 1997). The Plan consisted of three components: a communications part, which established the methodology for communicating the effort to internal and external stakeholders; an involvement process, which designed the approach by which employees would be involved in the reengineering effort; and reengineering team guidelines, which served as the framework for management of the reengineering teams. The program was defined as follows:

- *Goal.* To provide our customer the best possible service at the least possible cost
- *Objective.* To reduce 1996 controllable costs by \$18.4M annually by 2002 without any reductions in quality, service and safety levels
- *Strategy.* To reduce the competitive gap by reengineering critical work processes and adapting best business practices through employee-led teams

There was much corporate-level discussion regarding the issue of how much of the gap to target for reduction. The final decision was to set an aggregate objective for the teams that would reduce the gap to less than 10%. As it turned out the teams would exceed this goal in the end since they focused more on adopting best practices and letting the metrics take care of themselves.

Some of the guiding principles set by the Steering Team for the effort were as follows:

- It was the right thing to do for the customers
- The solutions had to come from the employees or workers
- There would be no reduction in product quality, service levels or safety
- Position reduction would occur through attrition and reassignment
- Management would support the effort with training, incentives and technology
- Communication would be comprehensive, continuous and in person at all levels
- A continuous improvement philosophy would be part of the new culture
- The focus would be on three critical business processes: plant operations; water distribution and wastewater collection system maintenance; and engineering support

- The steering team would be responsible for the vision, urgency, resources and barrier removal
- There would be fewer management levels in the new organization

A three-phase approach was generally employed and consisted of:

Phase I—Preliminary Change Program Design
Phase II—Reengineer Work Practices and Train in New Philosophies
Phase III—Implement New Philosophies and Work Practices

Phase I is the change layout process. This includes establishment of teams of affected employees and incorporates a process to develop the reengineered practices. After the change program is designed, Phase II, which normally takes two years, implements the reengineering of work practices and training. This is referred to as the "transformation phase." During this phase, the employees learn new practices and discover the ineffectiveness of the current practices. Phase III of the change process is implementation of the work practices. The change or implementation phase is one year in length and is a guided implementation. The employees have learned all of the new behaviors in Phase II and are now guided in implementing the new work practices. At the end of Phase III the utility becomes self sufficient, and is committed to the new practices and their continual improvement. While these three phases are sequential, it is important to note that some implementation actually takes place starting in Phase I. In fact, approximately 10% of the total projected annual savings are expected to be realized by the end of Phase I.

The Steering Team, as part of Phase I, went about the task of mobilizing the various work teams in accordance with the guidelines established in the "Plan." The Team decided to focus on three critical areas in the Operations Branch: plant operations; water distribution/wastewater collection system maintenance; and engineering. With heavy input from executive level management, the goal was established to reduce the competitive gap to less than 10% rather than the entire 21%. This equated to a reduction in controllable costs of $18.4M. The plant operations work team, also known as the Total Productive Operations (TPO) Team, would have as its charter to conceptually reengineer and redesign, through the adoption of best practices, the plant operation and maintenance function to generate productivity gains of at least $5.6M or the equivalent of 96 work years (WY) by

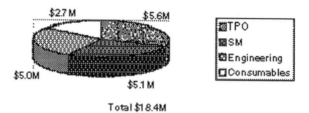

Total $18.4M

Figure 24.2 Gap allocation.

FY '02. The TPO Team had representatives from Wastewater Operations, Water Operations, Biosolids, Project Management, Human Resources, Facilities Maintenance and Engineering. The System Maintenance (SM) Team would also conceptually reengineer and redesign the distribution and collection systems maintenance function to generate productivity gains of at least $5.1M (87 WY) by FY '02. The SM Team had representatives from System Maintenance, Maintenance Reconstruction, Maintenance Services, Supply and Inventory, Budget, Systems Analysis and Engineering. In the engineering area, because the best practices were not as well defined, it was decided to use a different approach. Instead of forming a team to reengineer it was decided to form a team whose purpose would be to determine how much internal engineering capability was necessary to support the core business functions and recommend a plan to the steering team on how best to transition from the existing state to the proposed state. This Team was the Engineering Functional Analysis Team (EFAT). Once their plan was implemented, it would be expected to achieve a $5.0M (85 WY) productivity gain. The last category was consumables, which included such things as utility, equipment and chemical savings for an expected total of $2.7M. Most of this was expected to come from electrical cost savings due to the deregulation of the electrical utilities expected in Maryland to begin in FY '01.

Each Team was required to furnish a report, which outlined findings and recommendations for the reengineered process as well as tools, and resources that would be required. They were each assigned consultant resources to facilitate meetings, think out of the box and provide general team support.[2] The final gap allocation is depicted in Figure 24.2 and was integrated into each team charter.

[2] EMS Services, Inc., for the TPO and SM Teams; Camp Dresser & McKee, Inc., for the EFA and FPD Teams.

WORK TEAM FINDINGS

TPO Team

The Team began their journey by sending a subgroup on a series of site visits to other utilities who were noted for their particular practices in one of the targeted best practice areas such as workforce flexibility, program-driven maintenance, team organizations, use of technology to promote reduced plant staffing, and pay-for-performance systems. They visited Denver Metropolitan Wastewater Reclamation District, Colorado Springs Utility, San Jose Water Company, Union Sanitary District and Fairfield–Suisun Sewer District. Members from the Steering Team as well as the Commissions' Board also attended the site visits. These trips were very successful in lending credibility to the process by allowing team members to discuss in detail and witness firsthand the implementation of best practices with fellow professionals and managers from other utilities. At the conclusion of their visit a report that summarized important observations was prepared by the Team and presented to the Steering Team.

Next, the Team addressed its attention to those practices in the Agency that prevented it from being competitive. The Team documented costs associated with the inefficient existing practices and recommended alternatives for improvement. The TPO Team submitted its "Reinvention Report" to the Steering Team for further action. Many of the items could be implemented quickly from within the Operations Branch. Other items were more complex and cut across the entire organization. For those items that fell into this latter category, the Steering Team assigned a member to collaborate with someone from the affected part of the organization. They also were responsible to champion the issue by developing a plan of action, identifying resources, removing roadblocks, tracking progress and reporting to the Steering Team. Some examples of the type of items contained in the TPO "Reinvention Report" included: replacement of the computerized maintenance management system with an enterprise-wide system; more efficient maintenance of vehicles and equipment to reduce lost time; creation of an empowered process control action team to address many problems associated with current system; and using the facility planning process to review the condition of assets at our plant facilities.

The last and most important deliverable from the TPO Team in Phase I was the "Adopt Best Practices Report," which would present the Team's conceptual redesign of adopted best practices that would meet or exceed the metrics of the charter. After the site visits the Team spent many meetings determining how operation and maintenance forces at the plants were currently spending time. Then they superimposed the best practices that were identified in the Competitive Assessment and illustrated in the site visits. The results were astounding. The original goal of $5.6M was nearly doubled. With the Team's new reengineered process, it had the potential of producing productivity gains through training, incentives and technology that would free up more than 192 WYs (from the FY '96 level) or $11.1M in annual cost savings as shown in Figure 24.3. Progress achieved to date through "quick wins" is also depicted in the graphic. Some of the important changes that would generate these quantum improvements are as follows:

Team-Based Organization—Move away from the hierarchical silo organization that presently separates operation and maintenance groups to a team-based organization focused on total productive operations. These teams would have flexible workers who possess a range of skills that contribute to the overall execution of the plant mission. This concept would be underpinned by a skill-based compensation system as well as a team-based incentive. Develop formal cross-training and certification programs to ensure that all employees have the proper training, and verify that they can perform at the appropriate level.

Base Load Staffing—Import resources for specialty, emergency and non-routine work. Labor necessary to meet demands of emergencies should be imported from other areas within WSSC or from contractor crews. Under this concept 80%–90% of the work would be done by the teams of flexible workers at the plant while the remaining 10%–20% of the work would be imported from a centralized support area. Imports would include high-level experts and specialized workers such as welders, engineers, high-voltage technicians, etc. Reduction or elimination of off-shift staffing and the identification of alternate strategies, such as call back compensation, alternate

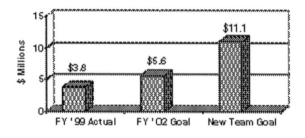

Figure 24.3   TPO controllable cost reduction.

shifts and part-time personnel, for responding to emergencies during these periods.

Program-Driven Maintenance—Develop, implement and train all employees in the proper use of a new computerized maintenance management system (CMMS). This will help avoid the expensive and inefficient activities inherent in the reactive maintenance mode, by substituting planned, predictive and preventive practices. The new system would include work order generation, inventory, maintenance requirements, time recording, and standard procedures for each type of work.

Technology—While WSSC does have advanced process control at its plants, the team discovered that additional tools would be necessary in order to be truly competitive. Some of those that were identified included: remote-control modem access; pagers for alarms; laptop computers for local on-call; cell phones; critical alarms connected to the Supervisory Control And Data Acquisition (SCADA) system, or a set of prioritized alarms; enhanced instrumentation for automatic monitoring and control; and use of PLC to replace older control systems.

Specific Plant Improvements—Finally, the Team made specific recommendations on a variety of plant improvements that would be required at each of the Agency's facilities in order to support the reduced level of staffing, without threatening a reduction in product, service and safety levels.

The TPO Team emphasized that the savings due to reduced staffing could not be realized without undue risk to product, service and safety levels until the best practices were part of the new culture of the organization. This meant that the productivity gains resulting from significant investments in training, incentives and technology had to occur first, or at least in conjunction with the position reduction due to attrition or reassignment. This would be a common theme from all the teams as their fears increased that the political governance would lose sight of the investment and time required for this program to be successful, and focus instead on taking the savings resulting in a downsizing effort with all of its well-documented perils. Thus it was stressed to the Agency's governance that CAP was not about job loss, but about adopting the best business practices in our industry into our culture, and *as a consequence* of this process certain efficiencies manifest themselves such that it is no longer necessary to fill a certain number of positions as they become vacant. *To reduce the staff before the efficiencies associated with incorporating the best practices is to invite a performance failure, which in our industry cannot be tolerated because of our public health mission.*

System Maintenance Team

The SM Team used much the same approach as the TPO Team by beginning with site visits to other utilities known for their best practices in the areas of interest to the team. Utilities visited included the Louisville and Jefferson County Metropolitan Sewer District, Louisville Water Company, City of Houston Public Utilities—Southeast Water Treatment Plant, and City of Houston— Kingwood Estate (operated by Severn Trent). These trips proved to be very beneficial in that team members got an opportunity to interact with their professional counterparts and discuss firsthand the experience of implementing these practices. As with the TPO Team several members of the Steering Team as well as one of the Agency's Commissioners accompanied the Team on the trips.

Next, the Team focused on those in-house practices that they saw as barriers to competitiveness and in desperate need of "reinvention." Suggestions were offered that would address the problems in their report to the Steering Team. Much as it had done for TPO, the Team addressed those items that were in its control with quick wins. Other items that would require interfacing with business units outside of operations were assigned to sponsors who had the responsibility to create a plan of action and seek resolution. Some of the more critical recommendations identified in the "Reinvention Report" were as follows: consolidate the meter services function with the System Maintenance function to address overlaps and better use of human resources; coordinate maintenance of vehicle-mounted equipment with vehicle maintenance; have plumbers set new small meters at time of plumbing inspection under the supervision of code enforcement unit; provide electronic access to Geographic Information System (GIS) for field and office via laptops; make all customer information available to inspectors and crews in the field by granting access to Maintenance Management Information System (MMIS) through mobile computing; and eliminate verbal job dispatching and report taking.

The last deliverable from the SM Team was the "Adopt Best Practices Report." Incredibly, the Team identified the potential to generate productivity gains of such a magnitude that eventually some 187 WYs could be reduced from FY '96 levels for an estimated annual saving of $10.8M annually. This is more than twice the goal that was originally set by the Steering Team as shown in Figure 24.4. Some of the more significant best practice recommendations made by the Team are as follows:

Team-Based Organization—Reorganize the system maintenance zones utilizing work teams, team leaders, and coaches. This would be a major

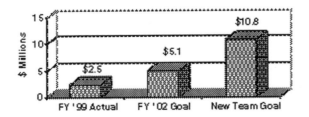

Figure 24.4   SM controllable cost reduction.

departure from current zone organization that uses multiple layers of managers such as section heads, supervisors and crew chiefs. The flexible worker concept would also be employed in three levels encouraging cross training and supported with skill-based compensation. Finally, the team would be supported with a formal training and certification program, as well as a team-based incentive pay system.

Base Load Staffing—Develop a strategy for importing labor in a crisis and staff for the workload that can make full utilization of available resources. Every winter, due to falling temperatures at the water source, there are a large number of water main breaks that must be repaired. One option is to import crews from the mainline section that are normally dedicated to re-placing water mains all year. The final option is to bring in private utility contractors to perform some aspects of the work such as backfilling and temporary paving. This concept also applies to off-shift staffing where crews are kept on duty waiting for emergencies that may or may not require their full capability. As part of this concept, implement flexible crew sizes for performing the work. The crew size should be matched to the type of work being performed.

Technology—Implement a mobile computer system using GIS and MMIS data to provide maximum information to Systems Maintenance field crews. Develop a computer-based system that will provide integration of work orders and materials management to ensure that crews have the right materials to do the job when they leave the depot, and that emergency material needs are met as quickly as possible. Implement an improved, flexible scheduling system to ensure the coordinators have the tools they need to manage the workload in the most efficient manner to properly match the job with the optimum crew size.

The System Maintenance Team determined that it was necessary to conduct a pilot at one of the four depots to prove that the proposed team and organization structure is appropriate, before rolling the concept to the rest of the organization. One depot would be selected for the pilot project and it would be staffed according to the proposed model and number of employees. Since not all of the recommendations will be in place for the pilot project, employees selected to participate in the pilot project should be among the most highly trained, outstanding performers in the Systems Maintenance Division. There will also have to be "work arounds" or human resource substitutes for technology that is not yet in place. The pilot is scheduled to start this spring.

Engineering Functional Analysis Team

As stated earlier this was not a reengineering team like the TPO and SM Teams. The latter Teams had the benefit of well-defined best practices in the area of operations and maintenance of water and wastewater facilities and systems. These were identified in the "Competitive Assessment," and specific work year reductions were identified for each best practice in each area. This was not the case for Engineering. There have been limited examples of privatization in the engineering area and other support functions of utilities. Thus, there have been few acknowledged best practices that can be used as a basis for comparison to current support functions.

Instead of reengineering work practices, the Steering Team created a Team to investigate all of the engineering functions at WSSC from a global perspective. Specifically the Engineering Functional Analysis Team (EFAT) was chartered to: develop a vision of what the engineering functions should look like in the future; determine the internal level of engineering needed to support the core processes of water treatment/distribution and wastewater collection/treatment; and develop a plan for the transition to the future vision.

The Team was formed with representatives from planning, design, operations, maintenance, construction and finance. There was one site visit to Chicago that had recently privatized its engineering function. The Team used extensive benchmarking data from Boston (BWSC), Massachusetts (MWRA), San Francisco, Milwaukee, St. Paul, Philadelphia, and Columbus. These trips helped to create the new model and clearly showed that WSSC's engineering costs expressed as a percent of construction cost were much higher than comparable utilities. The Team also had presentations from well-respected managers of engineering services in the private sector that were useful in understanding these services as a business process.

In July 1999, EFAT presented its "Engineering Services Business Plan"

[4] to the Steering Team. In general, the plan called for WSSC to run its engineering group more like a business and to focus on providing services to support the core mission, rather than an abstract collection of functions. This implied that the old model of providing available engineering talent "on call" was to be replaced by a more business like "billable hours" or client services model. Other important aspects of the new vision were as follows:

- Engineering services would have certain targeted allocations to foster a more business-like approach. Project-related services would enjoy the largest share at 75% while non-project services would be limited to 15%. New services and administrative services would be set at 5% each.
- Based on preliminary benchmark data and metrics the project delivery performance target for total engineering costs was set at 20% of total construction costs. An allocation for internal staff time spent on projects was targeted at 5%.
- There would be less internal engineering staff and fewer specialists. Engineering staff would possess the capability to function in several areas of competence.
- There would be fewer hand-offs in the life of a project and more use of design-build, which tends to blur the distinction between the design and construction phases.

The next step was for the Team to determine the internal level of engineering staff needed in the new model to support the core business processes. Keeping in mind that the preliminary gap originally allocated to Engineering was $5M or 85 WYs the new recommendation was quite remarkable. As shown in Figure 24.5 the Team identified potential savings of $8.7M or about 150 WYs.

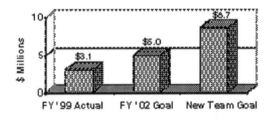

Figure 24.5   Engineering controllable cost reduction.

These new staffing levels were heavily reliant on the model requirements of having most of the engineering services (75%) support ongoing projects and limit internal costs for projects to 5% of the construction cost. Also detailed projections of future project and non-project work were developed and used to generate future staffing levels.

The last step was to propose a transition from the current staffing to the new level. The new model calls for a reduction of 150 WYs from the 1996 staffing baseline. To date, there has been a reduction of approximately 50 work years in the Planning and Design and Construction Bureaus. The staffing targets are deemed as appropriate future goals to be achieved through attrition, reassignments and the implementation of best practices in the engineering area. Specific work-year reductions will be developed in detail by the ongoing and future reengineering teams. The Team recommended the formation of several new reengineering teams as follows:

- Systems infrastructure
- Facility infrastructure
- Marketing opportunities
- Planning functions

The new reengineering team focused on the marketing of engineering services will be very important as a transition tool to fully utilize existing engineering staff "stranded" by the implementation of the business model. Several additional processes were assigned to existing reengineering teams already active in the engineering area. The Development Services (DSP) Team that was focused on the process of extending our service lines to new customers was assigned the on-site review process. The Facility Project Delivery (FPD) Team was assigned major pipelines and the Engineering Support Program approval and delivery process.

ORGANIZATION

At some point, the organizational structure that has supported the old culture must be changed to accommodate the new work processes. The competitive assessment pointed out that the supervisor to worker ratios were not competitive at WSSC where a command and control hierarchical culture prevailed. The vision established by the Steering Team specified that in order to create a more empowered work environment where teams could flourish, management numbers had to be reduced both vertically and horizontally. Since the management ranks experience very slow attrition rates, there was a potential that the old organizational structure could linger too long and suffocate the new reengineering processes.

A decision was made to accelerate management attrition by offering an attractive retirement incentive. For those employees eligible to retire the incentive offered was two years of service credit taken as an annuity or a lump sum. This was coincident with the presentation of a new organization chart that, among other things, drastically reduced the total number of executive-level management positions, and set the stage for the elimination of one layer of management vertically. For those managers who could not take advantage of the retirement incentive, the job abolishment policy was revised to offer a more generous two weeks severance for each year of service.

One disadvantage of the retirement incentive was the impact on the overall employee population. Although the original purpose of the incentive was to reduce management ranks, it was offered to all employees. Of the 486 employees eligible for incentive, 376 employees indicated their intention to take advantage of it. Most of these employees were non-management, making it necessary to fill some of the vacated positions since many of the productivity gains have not yet been realized at the worker level, and service levels cannot be compromised. The advantages, however, far outweighed the disadvantages. For those areas that are in a position to absorb the accelerated attrition, considerable progress will be made toward the overall CAP goals in the next fiscal year. The critics who had once charged that cost reductions were being achieved on the backs of the frontline workers were forever silenced as the number of managers was drastically reduced and a much flatter, more agile organization emerged.

CONCLUSION

Once a team has been given a clear corporate vision, a compelling need to change, a stretch goal, and the freedom to think "out of the box," the synergy that takes over as the group coalesces is truly a phenomenon to behold. In each case, the Work Teams at WSSC "raised the bar" by identifying cost-reduction targets that exceeded those originally established by the Steering Team. In addition to courageous and dedicated team members, the author believes the methodology used by the teams plays an important part. All teams focused on the adoption of best practices and let the metrics take care of themselves. This, combined with the reengineering fundamentals

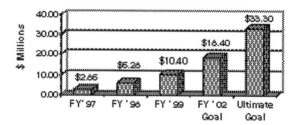

Figure 24.6   Controllable cost reduction.

of starting with a "clean sheet of paper," and allowing for no "sacred cows," pushed the realm of possibility well beyond the original boundaries, as shown in Figure 24.6. This graph also shows the early savings that have accrued due to quick wins, and the focus of the organization on the compelling need to change.

The bottom line however is making sure these improvements translate to value for the customer. As shown in Figure 24.7 there have been no rate increases for the last two years and none proposed for FY '01 budget. This would not have been possible without the savings in payroll cost reduction and consumables due to the Competitive Action Program. To further ensure that the savings accrue to the ratepayer, the General Manager prepared and the Board recently approved a financial plan that committed to no rate increases for the next five years. This will only be possible if the Competitive Action Plan continues to reach its full potential as identified by the Work Teams. Although this puts much pressure on future team performance it also commits the governance to fund the necessary training, incentives and technology needed to successfully implement the best practices at the worker level throughout the organization.

Finally, the process of customers exercising their choice in the water and wastewater industry is a complicated one. In our case, it involves the Mary-

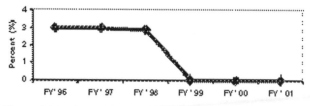

Figure 24.7   Water/sewer rate increases.

land State Legislature, since WSSC was created by the State. At the time this chapter was written, the Legislature was considering several bills involving privatization of WSSC, as well as the findings of the Privatization Task Force. One bill calls for an asset sale, another for contract operations, and finally, one proposes continuing with the Competitiveness Action Program under the current public structure. The latter bill commits the Agency to the FY '02 goal of $18.4M and the ultimate goal of $33.3M in savings from the baseline. As time winds down in the Legislature, the session ends April 10, 2000, the debate is certain to intensify. One thing is certain, WSSC employees have clearly demonstrated their resolve in providing the best possible value for their customers, and in so doing have made a convincing case as the public sector alternative for the next millennium.

ACKNOWLEDGMENTS

This chapter is dedicated to all the men and women who worked long hours under stressful circumstances contributing to the many teams associated with the WSSC Competitive Action Program. In particular, I would like to acknowledge the Operations Branch Steering Team, the System Maintenance Team, the Total Productive Operations Team, the Engineering Functional Analysis Team, the Development Services Team, and the Facility Project Delivery Team.

REFERENCES

Camp Dresser & McKee, Inc. (1999). *WSSC Report on the Development of a Business Plan for WSSC Engineering Services.*
EMA Services, Inc. (n.d.). *WSSC, Operations Branch Competitiveness Assessment, Draft Report.*
EMA Services, Inc. (1997). *WSSC Competitive Action Plan.*
Pirnie, Malcolm (October 1999). *Washington Suburban Sanitary Commission Privatization Study.*

Chapter 25

Global Lessons

Options for Improving Water and Wastewater Systems

Paul Seidenstat

Countries across the globe face the challenge of ensuring that their residents are provided with safe drinking water and nonpolluting wastewater. Water is seen as a special commodity for the sustaining of life, the preservation of health, and the supporting of industrial activity. Viewed as natural monopolies, the water and wastewater industries are capital intensive, have significant sunk costs with highly specialized capital, possess economies of scale, have high costs of entry, and face demand that is not very responsive to price. Technically, the distribution system is the highest cost segment of the production process, and a piped system is the most efficient means of water delivery.

Paul Seidenstat is an associate professor of economics at Temple University. A specialist in the fields of public finance, public management, and water resources, he has written or coedited eight books and several articles on these subjects, including Contracting Out Government Services *(Praeger). Seidenstat and Simon Hakim are coeditors of* America's Water and Wastewater Industries *(Public Utility Reports),* Privatizing the United States Justice System *(McFarland),* Privatizing Correctional Institutions *(Rutgers),* Privatizing Education and Educational Choice *(Praeger), and* Privatizing Transportation Systems *(Praeger). He has conducted funded research projects for the federal government in the fields of manpower and water resources. His current research is in applying economic analysis to the management of public agencies.*

Reinventing Water and Wastewater Systems: Global Lessons for Improving Water Management, edited by Paul Seidenstat, David Haarmeyer, and Simon Hakim.
0-471-06422-X   Copyright © 2002 by John Wiley & Sons, Inc.

To offer a modern piped system requires an organization to plan, finance, design, build, and operate the enterprise. Alternative models can be mobilized to provide and produce water and wastewater services. Elected officials at the national, state, or local level must choose among these alternative models with the goal of a achieving reliable and quality services at a minimum cost. There are two elements to the models, ownership and mode of operations, as seen in Table 25.1.

1. GOVERNMENT-OWNED AND -OPERATED MODEL

In light of the monopolistic market structure with its potential performance imperfections—as compared to a competitive market—of restricted output, high prices, elevated costs, and above-normal profits, controlling the behavior of the monopolist has been treated as a critical policy issue by elected government officials. One policy option is to have public ownership.

In many cases, water or wastewater utilities were not government-initiated. Private companies frequently started piped water systems; eventually, they were taken over by government. Nicola Tynan (Chapter 19) discusses the British history. In the United States, while many public

Table 25.1 Structural Models

Ownership	Mode of Operations
Public	Public
	Executive department
	Department—Enterprise accounting
	Agency—Reports to legislative branch
	Independent agency
	Stock company
	Municipal water district
	Multijurisdictional agency
	Public–Private Partnership
	Outsourcing
	Operations and management contract
	Build, design, operate (or variants)
	Franchise
	Leasing
	Concession
Private	Private, subject to public regulation
	Private, not regulated

water systems began as private, profit-motivated companies in the nineteenth and early twentieth centuries, outbreaks of typhoid and cholera and major fires in the young urban centers led to dissatisfaction and eventual government takeover. Regulators enforcing profit restrictions on private utilities, federal and state government subsidies of publicly owned systems, and the taxation of private investments contributed to the dominance of government-owned water utilities. By the end of the twentieth century, more than 200 communities had shifted from private to public ownership (Westerhoff et al., 1998), and municipally owned water utilities with monopolistic service territories had become the dominant model for service delivery.

The same process toward government ownership also was present in the case of sewer systems. Beito (1993) mentions that private systems were more common in smaller and medium-sized cities, but by the beginning of the twentieth century, almost all cities of more than 30,000 had publicly owned systems.

1.1   Organizational Choices

There are several options available in a government-ownership model. Possibilities include:

1. Government executive department
2. Government executive department subject to "enterprise accounting"
3. Government agency reporting to legislative branch
4. Independent agency
5. Government stock company
6. Municipal water district
7. Multijurisdictional agency

One option is to treat the water utility as a regular government department that not only is subject to common operating rules such as personnel and procurement, but also is treated as a municipal department for budgeting purposes. Water and wastewater may be combined with other infrastructure operations into a broader "public works" department. All revenues and expenditures go into the general fund, and the water department must compete annually for operating and capital funding. The department is subject to the same political and budget pressures as the police, parks, and other typical municipal departments. It is a part of the executive branch and reports to an elected chief executive.

The ordinary departmental budget treatment, however, can lead to budget decisions that obscure the "business nature" of a utility that directly collects revenue for its services. Many governments expect "user fee" services to be completely self-financing. Consequently, many municipal governments in the United States, especially larger ones, establish an "enterprise accounting" system that allocates all relevant revenues and costs to the utility's operations. If properly configured, the accounting device can reveal whether the utility is fiscally viable or has to be subsidized by tax revenues. Also, budget requests, especially for capital funds, can be more easily judged on the basis of their economic feasibility.

In a study for the Los Angeles Department of Water and Power, the Rand Corporation examined five alternative governance models in addition to the typical government department models (Baer et al., 2001). In a number of smaller cities, the municipal utility reports to the City Council (or legislative branch) directly. The City Council sets policy, but the execution is left to the executive director (or chief executive officer) of the utility. The idea is to remove the utility from local politics as much as possible and to give it budget and personnel autonomy. For small utilities, this system may work well.

A variant of the City Council model is to establish an independent city agency that reports to an independent governing board appointed by the mayor and approved by the City Council. The board members serve for fixed and staggered terms to remove them from day-to-day politics. Jacksonville, Florida, and Knoxville, Tennessee, have this arrangement. Again, the board appoints the chief executive officer, who is afforded considerable freedom in running the utility. In this case, the board also sets rates. However, capital outlays typically have to be approved by the city government.

Another option is to have a government-owned stock company with much more freedom to operate. This model is more commonly found in Western Europe. Klaas Schwartz and Maarten Blokland (Chapter 11) explore the Dutch example. Generally, this business structure is independent from direct government controls and can operate more like a private enterprise.

The North American version of "corporatization," primarily used for electric utilities as in Toronto and a few smaller US cities, is where the city establishes a municipal corporation and appoints the board members. As in the Dutch case, significant power is placed into the hands of the appointed executive, but the agency generally is confined to one political jurisdiction.

California has also been at the forefront of using the idea of a municipal utility district. Under the law, voters can establish a separate public agency to operate a public utility. Board members would come from the various geographical parts of the political jurisdiction. Similar to the other structures, the district is managed by an appointed management group with much autonomy. However, this structure allows the utility to float its own bonds for capital improvements.

California law also allows the establishment of a multijurisdictional agency where two or more cities, counties, or public agencies can operate the utility. Initially set up for the electricity market, such a concept could be extended easily to water or wastewater operations.

1.2　Benefits and Costs of Government Model

In many countries, accessibility to clean water has the highest social priority, given the direct linkage between access to water and the quality of water to public health. Thus, it has been argued that government should take the responsibility for this service in order to maximize these positive health externalities. If there is concern about the availability and affordability of safe drinking water to low-income or rural consumers, government could directly offer that service at a subsidized price. Additionally, since water and wastewater systems have been viewed and treated as monopolies, concern about monopoly power and exploitation has led to government itself assuming the role of the monopolist.

Government operations, however, often may not produce water or wastewater services efficiently. The potential inefficiency of government-operated systems flows from the absence of competition,* the difficulty of providing meaningful incentives for managers to minimize costs and provide a high level of service, and the complexity of rewarding employees for maximizing productivity. Lacking a profit-based measure of performance, rewarding managers on the basis of cost, quality of product, or responsiveness of service is problematic.

Often, little discretion is given to the manager in making personnel de-

---

*Often, the prices paid to suppliers and contractors and the salaries paid to managers and workers are in excess of the competitive level. These excess payments are called economic rents. For a discussion of economic rents in government, see Seidenstat (1996).

cisions, including hiring and firing. Also, offering incentives to employees for outstanding performance is constrained in a civil service, typically unionized, environment. Often payrolls may be expanded for patronage or macroeconomic reasons. For example, Mark Dumol and Paul Seidenstat (Chapter 10) document the excess of personnel before the government-owned Manila system was converted to a private concession arrangement. Similarly, in the experience of Great Britain, within a year of privatization the Thames Water's staff was 20% below its peak level (Economist, 1990).

The existence of organizational slack that results in above minimal costs of operations is commonplace in public systems. A study by the Association of Metropolitan Sewerage Agencies and the Metropolitan Water Agencies indicates that operating costs in many public systems can be cut by at least 10%. By applying currently available methods, four public systems were able to cut costs by 20–25%.*

As a government department, additional constraints on managerial decision-making may be imposed on the water or wastewater utility. Budget requests are subject to a political budgeting process. The water department must compete with other government departments for operating or capital funds. Overall, owing to the financial problems of many local governments in the 1970s and early 1980s, there has been a strong effort to strengthen balance sheets and credit ratings by careful management of budgets and debt. The competition for borrowed funds can be especially intense, since municipalities face statutory or constitutional limits on the level of outstanding debt and strive to maintain good credit ratings by limiting debt. Thus, underfunding presents a major obstacle to improving service or product quality. Consequently, maintenance may be shortchanged or plant upgrading or expansion plans may be cut back.

The enterprise or stock corporation variants may mitigate some of these disadvantages. As an enterprise fund, the water utility can be given some autonomy, and could be expected to live within its revenue means with less political interference. However, in practice, in many local governments in the United States, it is still subject to the personnel rules and the budget, especially the capital budget allocation process. Additionally, the borrowings of the corporation may still be counted again any debt limitations that the local government faces.

---

*The four systems were: Fort Wayne, Indiana; Orange County Public Utilities, Florida; Colorado Springs, Colorado; and Houston Public Utilities, Texas (Association of Metropolitan Sewerage Agencies and Association of Metropolitan Water Agencies, 1998).

The corporate stock model, wherein stock is issued and owned by the government, offers even a greater degree of independence. It may be free of government personnel and budgeting rules. It may also be able to reward managers and high productivity employees based on profit and cost considerations and may not be subject to government budget rules. Vivien Foster argues in Chapter 3 that,

> Public sector managers will tend to be influenced by political pressures, although this will depend on the degree of corporatization. Corporatization strengthens the political autonomy of a publicly owned enterprise by making it increasingly self-sufficient financially ... and introducing rules that protect directors and senior managers from being removed on political whim.

However, there are constraints on the level of profits and, thus, it may evolve as a cost-plus monopoly subject to rate-of-return constraints. With government ownership of the stock, the operation still may be subject to political interference.

Another potential limitation of the government operations model is that water and wastewater operations are local government functions. In the United States, municipalities typically have jurisdiction over water and wastewater operations. If there are economies of scale (and scope) in major distribution systems, raw water supply, wastewater treatment, and customer service operations, localized operations may be on too small a scale to achieve these lower costs.

By the 1990s, Latin American countries such as Argentina, Columbia, and Peru saw their highly centralized public water systems become fragmented into hundreds of municipal suppliers. Vivien Foster points out that this decentralization gave rise to a number of problems. Scale economies were lost, managerial and professional expertise was spread very thin, and the financial viability of these very small units was threatened.

Overall, organizational mechanisms such as regional compacts or intergovernment contracts may be used to overcome the scale limitations. However, there are often political obstacles to this form of cooperation.

2. PUBLIC–PRIVATE PARTNERSHIPS

As concern grew in many countries about meeting efficiently the demand for safe and clean drinking water, publicly owned and operated systems began to look at enlisting the help of the private sector in improving the efficiency of their operations. There is a long history of private enterprises supplying various services and designing and building projects, and in some countries, such as France, supplying full contracted operation. Now, private vendors are offering a much more extensive range of services in a much larger number of countries.

2.1   Outsourcing

Today, government water utilities frequently enlist the assistance of the private sector for some routine operations. It is believed that a limited private partnership for one or more specific functions can be cost effective as the private firms compete for the right to provide the service, and the terms (e.g., regarding quality and performance) of the service are regulated by an enforceable contract. Commonly included in these functions are installing and reading water meters, billing and collection, infrastructure maintenance and repairs, and laboratory services. Firms performing these services, especially on a regional or national basis, can take advantage of economies of scale and can better utilize the latest technology.

2.2   Operations and Management Contract

In this arrangement, private firms are contracted to manage the utility. Since managers work for a private company that can increase profits if costs can be constrained, given a fixed monetary contract, they can be rewarded for effective performance. The vendor may assume the risk of ensuring that minimum water quality levels are reached. Moreover, a company experienced in managing water or wastewater systems can derive the managerial and technological benefits that come from operating a number of plants and systems. The private company may not be bound by the government's procurement rules, which can raise the costs of supplies and services purchased by the utility.

To be successful, the public authority has to assure fair bidding and design a clear and comprehensive contract. If the contractor is required to follow the government requirements as to personnel policies and procurement, many of the benefits of private operations may be lost. Rather than focusing on input issues, the contracting government can concentrate on specifying output targets. In this fashion, the contractor is free to utilize the most cost-effective technology and resource utilization to meet the terms of the contract.

2.3   Design, Build, and Operate

When government water utilities decide to expand their facilities, they typically turn to private contractors to design and to build the infrastructure. Usually, a separate contractor is used at each stage of the process. In retrospect, this separation of designing and building usually increases the time to complete the project and can lead to less efficient and more costly projects. Since the design engineers do not have to operate the facility, their interests are not aligned with the operators. Consequently, the design of the project may not have achievement of operational efficiency in mind. Without the experience or knowledge of operations, builders often overlook aspects of the structure that can simplify operations and reduce operating costs.

As Elizabeth S. Kelly, Scott Haskins, and Rodney Eng of Seattle explain in Chapter 12, "While the conventional design-bid-build-operate approach to public works contracting is well understood, and most appropriate for many applications, its shortcomings are numerous and well documented." They go on to pinpoint some of the shortcomings. These include excessive design costs due to "undesirable risk allocation and mixed incentives, and a failure of the designer and builder to collaborate in ways that would lead to reductions in construction costs," problems of low bidding and resultant costly change orders, a greater risk of failure owing to the difficulty in establishing liability among the designer, contractor, and the municipal operator, and several other weaknesses.

Public utilities, including water utilities, have begun to recognize the inefficiencies of using separate contractors. A new organizational model for expanding facilities is now in place that combines all three functions of design, build, and operate.

There are at least two variants of this model. One is a straight design, build, operate model (DBO) in which the winning contractor performs all functions except financing the project. The project is owned by the government utility, but the responsibility for operations is contracted out to the private sector. The other variant is where the private entity designs, builds, arranges the financing, and operates the project. At some specified future date, the project is transferred to the government. This method is referred to as design, build, operate, transfer (DBOT or, sometimes, BOT or BOO). The experience in the United States has proven to be positive, as illustrated by the Seattle experience. However, some state laws still preclude this option for many utilities.

2.4   Franchise*

2.4.1   Leasing

This model is similar to the management contract arrangement, except that the lessee takes on additional functions. The local government owns the assets, but the private partner manages the facilities, provides working capital, and bills and collects from the customer. The lessee is responsible for maintenance and upgrading of facilities. Water rates are determined initially as a part of the bidding process, but the operator usually remits part of the revenue collected to the government as a lease payment.

2.4.2   Concession

When a publicly owned system requires both private management expertise and private capital, a concession contract can be arranged. Private firms typically bid for the contract, with the bids representing the level and structure of rates. Besides advancing all capital funds for new construction and working capital, the private firm manages the operation and maintains facilities, and bills and collects from the customer. Concessions tend to be granted for a long period (for example, as long as 30 years in some countries and 20 years or less in the United States, due to tax considerations) so that the concessionaire can recoup his or her investment. At the end of the contract, the assets would be acquired by the government and the contract likely would be rebid.

In general, fully utilizing the resources and capabilities of the private sector is a very strong argument in favor of these full privatization alternatives. Another, potentially more telling argument, is that a franchise can inject competition into a monopoly market. The competition is for the rights to the franchise. Since the bidding is stated in terms of rates and services to be provided, the franchisee has the incentive to reduce costs and employ the optimum technology. If the firm's managers perform poorly in terms of cost containment and profit maximization, the managers may be removed, or the firm itself may be subject to penalties if contract provisions are violated.

For many countries, using the franchise or concession model often can

---

*The terminology for both leases and concessions used here is based on Hanke and Walters (2000). Water consultants such as Tasman Asia Pacific (1997) also use it. The terminology is widely used in the literature.

yield positive results as Mark Dumol and Paul Seidenstat demonstrate for Manila, and Lillian Saade Hazin (Chapter 9) shows for Mexico's Federal District. The franchise model may fit many countries well and is widely employed currently in the developing world, as Kristin Komives points out in Chapter 7. However, the challenge is to develop a stable, responsive, and coherent regulatory system to induce international firms to bid and to accommodate reasonable rate adjustments, as economic circumstances require.

In the United States, leasing and concessions are used both as a means to cut costs and as a way to generate up-front fees. For example, Cranston, Rhode Island, executed a 25-year lease that included a $48 million up-front fee and a contractor agreement to make significant capital improvements. In New Brunswick, New Jersey, a 20-year franchise also was negotiated for both water and wastewater systems that included a $30 million payment.

2.5   Increased Use of Public–Private Partnerships (PPP)

Although private involvement in the water sector is still small relative to the public provision of water in developing countries, since 1990 private participation in developing countries has accelerated. The number of private water projects reaching financial closure increased more than tenfold between 1990 and 1997. At the end of 1997, private companies had invested US $25 billion in 97 projects in developing countries. The concession form of public–private partnerships is dominant, representing 49% of the projects and 80% of the total invested (Silva et al., 1998).

In the United States, as well, there has been a significant increase in the number of public–private partnerships. In 1997, the US Conference of Mayors (1997) conducted a survey of 261 cities that had over 40 million in population; the results revealed the extent of PPP involvement. With an involvement of 560 facilities, 40% of cities have some form of private sector involvement in their water services, and 39% have some involvement in wastewater services. Private participation as a percentage of systems was as follows:

Design and construction	71%
Meter reading	33%
Billing and collection	31%
Distribution system operation and maintenance (O&M)	25%

2.6   Factors Driving Local Government Use of Private Partnerships (Hudson Institute, 1999)

In the United States in the last decade, local governments looked more closely at enlisting the aid of the private sector in operating their water and wastewater systems. Financial issues and compliance problems with the federal Safe Drinking Water Act (SDWA) were key drivers for this effort.

Many local government-owned water systems facilities are well over 50 years old, and often have become obsolete, owing to the US Environmental Protection Agency (EPA) regulations.

The EPA estimates that, in the next 15 years, an expenditure of $138 billion will be necessary to install, upgrade, or replace infrastructure; 56% of that total will simply be necessary to replace or rehabilitate aging infrastructure. Much of that funding will have to come from the local and state government, as major increases in federal government funding are not expected. The financing situation is further aggravated by the requirements of the SDWA standards. The EPA estimates that water systems will have to spend $34 billion in the next 10 years just to meet the higher drinking water standards (US Environmental Protection Agency, 2001).

In spite of the potential ability to finance operations by user fees, many smaller government water systems, as well as a few larger ones are operating at a deficit.* Raising water and sewer rates at a rapid clip is a very difficult political step for many local government water and wastewater systems. The consequences of these operating deficits include antiquated facilities that require upgrading and insufficient maintenance and repairs.

Smaller community water systems in the United States especially are at risk for the financial and water quality issues that will develop. The requirements of raising capital and the inability to take advantage of economies of scale have left small systems (under 3000 people) vulnerable and consolidation or sale to private companies has been occurring. From 1982 to 1993, the number of small water systems fell by 8%, while the number of large water systems grew by 41% (Hudson Institute, 1999). It is increasingly likely that large investor-owned utilities will absorb these systems, will utilize partnerships, or will consolidate with other adjoining small government-owned systems.

---

*As late as 1997, for all US local governments, costs exceeded revenues by $4.1 billion for water systems and $2.57 billion for sewer systems. See testimony by Beecher (2001).

2.7 Potential Benefits from Public–Private Partnerships

A study of 29 public–private partnerships in the United States in the 1990s by the Hudson Institute (1999) for the National Association of Water Companies documents a number of benefits obtained by utilizing the private sector. The types of partnerships included:

Type	Percent of Total
Asset transfer (concession)	31
Outsourcing (billing)	6
Lease	14
Operations & management	49

The benefits received by the contracting government included:

1. *Capital Investment.* The governments involved received capital infusions (in millions) as follows:

Annual fees	$27.8
Concession fees	$34.6
Transfer (of assets) fees	$536.0
Capital expenditures	$55.0

2. *Cost Savings.* Costs savings of 10–40% were obtained on five of the projects.
3. *Water Quality.* Prior to engaging the private partner, 12 projects were out of compliance of the SDWA standards. Within a year after the private partner took over, all 12 were in compliance.
4. *Controlling Rate Increases.* Owing to infrastructure requirements and operating inefficiencies, many systems in the sample contemplated rates increases from 10 to 50%. The partnership model dampened or eliminated all proposed increases.
5. *Improved Customer Service.* The private partner, typically a subsidiary of a large parent company, was able to improve service by integrating the customers into its parent company's call-in center and billing and collection system.

3. PRIVATE OWNERSHIP SUBJECT
TO PUBLIC REGULATION

Under varying circumstances, governments have decided to rely solely upon private utilities to provide water and wastewater services. In some cases, private ownership developed the local market, as was the case of investor owned water utilities in the United States. There remains a viable private sector in the United States. In 1995, investor-owned water systems accounted for one-third of the nation's community water systems and served 13% of the US population, or more than 30 million people (Hudson Institute, 1999).

In other cases, the government owned the water or wastewater utility and decided to sell the assets to a private utility. Great Britain accomplished this "load shedding" in a dramatic fashion in the 1980s when it sold 10 government-owned and -operated English and Welsh water and sewerage companies.

Normally, the private company is awarded a monopoly right to serve a particular franchise area and, in some cases, a companion wastewater market. Since the private utility is a monopoly and is not subject to competitive constraints as to output or price, the government regulates the utility in terms of rates, quality of product, and service.

Operating a regulatory system that can enforce effectively an economically efficient level of rates, costs, productivity, product quality, and service is very difficult at best. Yogita Mumssen and Brian Williamson (Chapter 2) analyze the major regulatory issues. They argue that the various approaches to regulation—that is, rate of return, price cap, or New Zealand's regulation by threat—are not as distinct in practice as in theory. The dilemma of regulation is to provide incentives for cost reduction while ensuring that prices that are not too far out of line with costs. Also, the regulators should attempt to allow a reasonable chance of cost recovery without condoning excessively high costs. Moreover, the regulator must resist regulatory opportunism, that is, taking advantage of the regulated company if it earns high profits because of efficient operations or moving sluggishly to adjust rates in the face of rising costs. The authors cite the conditions for efficient regulation: commitment and stability of regulation, openness, transparency, consistency, and accountability.

During the 1990s, most countries in Latin America undertook major reforms to their water supply industry. Vivian Foster (Chapter 3) dissects this reform movement with special emphasis on the changes that were wrought in the regulatory framework. She stresses the importance of separating the functions of policy-maker, regulator, and service provider. Throughout the region efforts were made to insulate the private utility company or the corporatized public company from political interference and to ensure that it be business-like in providing its water service. Regulators used a rate-of-return system in the regulatory regime.

George Day (Chapter 5) reflects upon and analyzes the British regulatory system. In the price-cap type regime under the supervision of the Office of Water Services (Ofwat), the emphasis was on encouraging adequate investment to meet rising quality and environmental obligations, assessing comparative performance of companies, and the balancing of incentives with consumer benefits. After a decade and a half of experience, there is some momentum to injecting competitive elements in the system to improve its performance, especially in providing more benefits to consumers.

4. UNREGULATED PRIVATE OWNERSHIP

Although the presumption has been that the water sector is a classical monopoly and must be government owned or regulated, there may be circumstances in which neither choice of industry structure is necessary. Penelope Brook and Tyler Cowen (Chapter 20) argue that, in developing countries where tariffs are set well below costs, water utilities have no incentive to deliver services to large groups of people, typically poor. As a consequence, 30–60% of the population may not have access to a safe and reliable water supply. They argue that unregulated private monopoly potentially may represent a policy improvement. In attempting to maximize profits, the monopolist may connect as many customers as possible (and subsequently extract the maximum possible consumer surplus), be encouraged to minimize cost, and by effectively engaging in price discrimination, may maintain high product quality. Under some conditions, the unregulated monopoly may not be economically desirable but it should be analyzed as a viable choice in some cases.

Some commentators suggest that by treating the pipeline distribution or collection system as a common carrier, entry of adjoining water treatment plant competitors is feasible. Allowing entry could transform the retail step of the industry from monopoly to a structure with competitive elements, allowing the elimination of public regulation. However, there are some major technical and administrative problems that would have to be overcome.*

5. LESSONS LEARNED FROM THE EXPERIENCES OF VARIOUS OPERATING MODELS

Table 25.2 summarizes the various advantages of public and private models. In many cases, introducing the private sector will likely reduce costs, allow faster introduction of technology, and improve management. These results can be achieved with various models of public–private partnerships. To achieve the best results from a full privatization model of a franchise would require a comprehensive contract and a system of efficient regulatory oversight.

As mentioned above, many nations can benefit the most by a franchise if the environmental conditions for success can be introduced and enforced. Also, the immediate concern for some countries is to connect citizens as quickly as possible to a piped system of safe water (World Health Organization, 2000). All countries are striving for both safe and clean water at an affordable cost. Since the technology associated with wastewater treatment is more complex and more rapidly changing, private sector involvement would likely move forward more rapidly for that operation compared to the more technically simple and stable water treatment process.

Public water and wastewater utilities, in contemplating alternative delivery options that more efficiently meet their and their customers' needs, might derive some lessons by examining experiences of variously organized utilities in the world. Below are four areas that deserve attention given the potentially large improvements that they can bring to public systems.

5.1   Objectives

The objectives of the water utility should be made very clear if they are to be achieved. Providing clean and safe water on a consistent basis at a reasonable cost is an oft-stated goal. However, other objectives might be set that conflict with this overall goal. Especially in developing countries, extending connections as rapidly as possible to unserved people or pricing

---

*For a discussion of this option, see Webb and Ehrhardt (1999).

Table 25.2 Evaluation of Structural Models

Model / Mode of Operations	Strengths	Potential Weaknesses / Problems
Public Ownership		
Public	Direct public control	Lack of incentives No penalty for poor performance Operating budget issues Capital funding constraints Political aspects to rate setting Constrained personnel policy
Public–Private Partnerships	Competition among bidders Private funding Profit incentives Unconstrained personnel policy Absorbs risk of meeting water quality standards	Designing effective contract Enforcing contract Rate adjustment process Political opposition
Private Ownership		
With Public Regulation	Limited profit incentive Private funding Unconstrained personnel policy Absorbs quality standard risk	Inefficient regulatory rules Regulatory lag in adjusting rates
Unregulated	Full profit incentive Private funding Unconstrained personnel policy Absorbs quality standard risk May accelerate connection rate	Monopoly performance

water at less than cost to help the poor are important objectives that have to be included in operational plans but that may lead to higher costs, especially in the short run.

Also, for public systems, pursuing other objectives might interfere with the clean and safe goal as well. For example, a policy of maximizing voter support by suppressing necessary rate increases in the short run in the face of rising costs can result in underinvestment in facilities and maintenance.

5.2 Performance Indicators and Transparency

If a public agency's performance is not transparent, it is difficult to sustain pressure to improve performance. A water utility may be producing a generally accepted quality of product but at an elevated cost level, or the water quality could be lower than the level that the revenue base is capable of producing. Establishing generally accepted measures of performance and publicizing the results would be effective in pressuring efficiency. Such efforts as the American Water Works Association's initiative to collect performance data and publish the results can aid in establishing standards for water systems.

The British regulatory device of benchmarking has been instrumental in pointing the way toward establishing performance standards. Stephen Ramsey (Chapter 4) closely examines the use of benchmarking. He indicates that the regulatory agency, Ofwat, uses it as a regulatory stick, and that it has served to motivate companies to seek greater operational efficiencies and to improve standards of service to customers. Ramsey argues that success from benchmarking is not inevitable, but rather requires clear management support, strong staff participation, setting of clear priorities, adaptation of best practices to the company's particular situation, and effective implementation. When British companies followed these core guidelines, benchmarking had a real and positive impact on the financial and operational performance of the companies.

Using econometric models that adjust for differential values of independent variables can allow for reasonable performance comparisons of water companies. Both metric and process benchmarking can be employed to allow utilities to examine their performance and to assist external examiners in evaluating them. As the potential value of benchmarking to evaluate and improve performance is recognized, its explicit use has begun to spread. For example, in the United States, the Water Utility Benchmarking Association (WUBA™) has been established as a free association of public and private water utilities. It conducts benchmarking studies to identify practices that improve the overall operations of the members.

Benchmarking can be very useful for internal purposes in the utility. Benchmarked performance results available to the public can aid in public pressure to improve performance. In the United States, a transparency requirement was built into the Safe Drinking Water Act by amendments passed in 1996. Relating to drinking water quality, the Act was intended to inform the consumer about the quality of tap water. The law requires every water supplier to complete a Consumer Confidence Report that provides data on contaminants found in the drinking water and relates them to allowable standards. The Environmental Protection Agency makes a national database available on its Internet web site.

5.3   Governance Structures and Processes

As a typical government department, the water utility can be subjected to political interference and be restricted in following government rules relating to personnel issues and to incentive mechanisms. One approach to avoid some of these difficulties is to give a degree of independence to the utility and let it operate as an enterprise fund. The utility would be cut off from government budget support and would be required to fund its operations by generating its own revenue. At the same time, bonuses for performance and personnel decisions would be at the discretion of the utility management. If elected officials wished to subsidize poor people or favor large business users, explicit subsidies would be provided to the water utility. However, policy issues as to rate structure and allowable levels of profitability likely would remain in the hands of government decision-makers or those of a regulatory body (whose members are often appointed by elected officials).

An alternative way to change the organizational structure to accommodate a more business-oriented approach would be to establish a government-owned stock corporation. As explained above, this would entail an independent operation but would still be government owned. In the charter of this corporation explicit rules as to profitability, financing, personnel policy, and product quality and service could be established. The United States, for example, has organized its postal service along these lines. It is likely to increase accountability and offer benefits beyond the existing municipal structure, particularly for large water systems.

5.4   Optimizing Public–Private Partnerships

As indicated earlier, the experience of many water and wastewater systems is that the utilization of private producers can improve operations and reduce costs. Consequently, involvement of the private sector in operating the system should be an option. Various types of partnerships should be examined in terms of their applicability and political feasibility. A benefit-cost analysis could be undertaken to assess the options. Putting various aspects of the operation—from using vendors to read meters to hiring a private company to take full operational management responsibility—out to the bidding process can be a cost-effective approach.

Allowing the government agency to bid against the private bidders—termed "competitive contracting" or "managed competition"—can be considered as one operating option as well. However, it is essential that the bids are fair, that is, that all relevant costs are included in both the private bids and the government bids, that the bidding elements are clearly stated, that an independent advisory group is used in rating the bidders or in reviewing the decision, and that private firms should not be asked to bid unless the government is committed to letting the contract to the winning bidder.

6. THE FUTURE OF WATER AND WASTEWATER MARKETS

6.1   The Balance Toward Government Ownership

As a result of the present regulatory structures and tax regimes, government ownership of water and wastewater systems is likely to continue into the future. The sector's monopoly structure and associated regulatory regime make it difficult for market forces and technology to provide more opportunities for private ownership models. Moreover, strong forces are in place to preserve the status quo. The economic benefits enjoyed by elected officials, public unions and workers, public managers, and vendors will not lightly be surrendered. The philosophical bias toward public ownership of essential services also is well established in some countries.

6.2   Reforming Public Regulation of Investor-Owned Utilities

The advantages of private business in terms of incentives, flexibility, and external pressure to optimize performance are difficult to achieve in the case of a monopoly under private ownership. The policy of subjecting the private monopoly to explicit public oversight can short-circuit to an even greater extent the generally superior market and incentive forces of private enterprise. This is especially true of rate-of-return type regulation that is commonly employed. Although the British system of price capping offers the advantage of providing profit incentives, over time there is pressure on the regulators to translate efficiency accomplishment into rate-payer benefits rather than utility profits, such that regulation becomes similar to some form of straight rate-of-return regulation. As George Day's discussion on the role of regulation in Great Britain concluded, it is the development of competitive elements in the water sector that will likely lead the search for greater efficiency and that the role of regulation is "to provide an overall framework within which companies are free to innovate and seek the most effective methods."

Rebecca Rehal (Chapter 21) discusses the attempts at improving the British regulatory system. She points out that a major thrust is the amplified use of competition in water markets. Although the option was not aggressively used, the original regulatory framework had allowed for entry from adjacent water markets or for entrants displacing existing suppliers at specific sites. New legislation expands the competitive possibilities by allowing common carriage in distribution pipelines and allowing new forms of ownership and financial restructuring to reduce risk and allow greater efficiency. The law would add promoting competition as an additional primary duty of the regulator.

In the United States, a new look at regulation is being undertaken with a view to overcoming some of the weaknesses of the rate-of-return regime. The price-cap model is being considered. Various types of structured competition, such as relaxing regulatory requirements when public–private partnerships are used, allowing profit-sharing between privatizers and consumers, and allowing regulators to resolve contract disputes with private partners have been suggested. (See Beecher, 2000.)

6.3   More Extensive Use of Public–Private Partnerships

The growing evidence of the economic advantages of partnerships suggests that trend will continue as more governments become aware of the benefits and comfortable with the process of organizing private participation. To allay the difficulties of raising sufficient funds and the risks and technical complexities of water and wastewater systems, developing countries will continue to turn to the private contractor. In the United States, as the large bill for infrastructure replacement, upgrading, and expansion begins to be presented to local government systems, more consideration will be given to enlisting the private sector to ensure that capital is efficiently employed and to control operating costs so as to minimize rate increases. Any negative impact to the local economy will likely accelerate this trend as local governments become focused on stretching scarce public funds.

7. CONCLUSION: IMPROVING PERFORMANCE BY INJECTING COMPETITION

Water and wastewater systems operate in a monopoly environment that is slowing evolving to provide more and more opportunities to inject competition. Whether the systems are publicly or privately owned, overcoming the corrosive effects of monopoly on incentives and on resultant performance is the major challenge. Consequently, taking advantage of market forces should be a chief objective of policy-makers interested in improving the water and wastewater services.

With private ownership, fashioning a public regulatory regime that rewards good performance and penalizes poor results is the key challenge to optimizing performance. Countries across the globe are experimenting with various regulatory reforms. Great Britain and Australia are at the forefront of this effort, and various states in the United States have turned their attention to the issue. Moving from a rate-of-return model to some form of price-cap model is contemplated. More ambitiously, efforts to devise methods to introduce structural competition, and thus minimize the role of regulation and the necessary distortions it creates, are underway.

Worldwide experience demonstrates that the strongest and most reliable force for injecting competition in publicly owned utilities is the basic mechanism of organizing competition among private companies for the right to deliver a specific service. These public–private partnership models range from simple outsourcing of limited functions to a fully privatized franchise arrangement. In countries in all continents of the world, these partnerships have demonstrated that where contracts are properly designed and enforced, costs can be controlled, water quality improved, infrastructure expanded and maintained, and new management techniques and technology introduced.

Under public ownership and public operations, there are efforts in process to devise modes of operation to simulate private businesses by freeing agencies from the typical government budget constraints and by offering monetary incentives to managers and employees based on performance.

Some sort of "corporatization" or "quasi-independent" status of the public utility is an option that is working in practice. Additionally, regional mergers of smaller publicly owned utilities can achieve advantages of size, and if operated by autonomous organizations, can be operated with a measure of independence and accountability.

Competition is infused by private companies—both large multinational enterprises and small local businesses—bidding for the right to perform various functions or to operate the utility for a limited period of time. The pressure to perform is inherent with the threat of nonrenewal of the contract, the fashioning of performance-based contracts, and the concern about reputation that could affect future contracts. The challenge is for public authorities to devise effective contracts, aggressively monitor the contracts, and to develop sensible regulatory schemes for necessary long-term contract adjustments.

In the face of rising environmental standards and accompanying higher outlays of capital, and increasing consumer demand for better services but little appetite for higher rates, prudent public utilities must look beyond the limited resources of their organizations to tap those of the broader water service delivery marketplace. A systematic examination of the alternative modes of operations, as well as lessons of other countries, will provide a valuable starting point toward achieving long-term sustainable performance.

REFERENCES

Association of Metropolitan Sewerage Agencies and Association of Metropolitan Water Agencies (1998). *Thinking, Getting, and Staying Competitive: A Public Sector Handbook*. Washington, DC: Association of Metropolitan Water Agencies.

Baer, Walter S., Edmund D. Edelman, James W. Ingram, III, and Sergej Mahnovski (2001). *Governance in a Changing Market: The Los Angeles Department of Water and Power*. Los Angeles: The Rand Corporation.

Beecher, Janice A. (2000). "The Role of Economic Regulation in Water and Wastewater Utility Privatization," in Paul Seidenstat, Michael Nadol, and Simon Hakim, *America's Water and Wastewater Industries: Competition and Privatization*. Vienna, VA: Public Utility Reports.

Beecher, Janice A. (2001). Testimony on Water Infrastructure Needs, Subcommittee on Water Resources and the Environment of Committee on Transportation and Infrastructure, U.S. House of Representatives, Washington, DC, March 28.

Beito, D. (1993). "From Privies to Boulevards: The Private Supply of Infrastructure in the U.S. During the Nineteenth Century." In J. Jenkins and D. E. Sisk (Editors), *Development by Consent: The Voluntary Supply of Public Goods and Services.* Oakland, CA: ICS Press, pp. 23–48.

Economist, The (1990). "Of Wealth and Water," 317, 7675, October 6, pp. 69–70.

Hanke, Steve H., and Stephen J. K. Walters (2000). "H$_2$ Ownership: Privatizing Waterworks in Theory and Practice," in Paul Seidenstat, Michael Nadol, and Simon Hakim, *America's Water and Wastewater Industries: Competition and Privatization.* Vienna, VA: Public Utility Reports.

Hudson Institute (1999). *The NAWC Privatization Study: A Survey of the Use of Public-Private Partnerships in the Drinking Water Utility Sector.* Indianapolis, June.

Seidenstat, Paul (1996). "Privatization: Trends, Interplay of Forces, and Lessons Learned," *Policy Studies Journal,* Vol. 24, No. 3, pp. 464–477.

Silva, Gisele, Nicola Tynan, and Yesim Yilmaz (1998). "Private Participation in the Water and Sewerage Sector—Recent Trends." In: *Public Policy for the Private Sector.* Washington: The World Bank Group, September.

Tasman Asia Pacific (1997). *Third Party Access in the Water Industry.* A report prepared for the National Competition Council, Melbourne, Australia, September.

US Conference of Mayors (1997). "Status Report on Public/Private Partnerships in Municipal and Wastewater Systems: A 261 City Survey," Washington, DC, September.

US Environmental Protection Agency, Office of Water (2001). *Drinking Water Infrastructure Needs Survey,* Second Report to Congress, February.

Webb, Michael, and David Ehrhardt (1999). "Improving Water Services Through Competition." Washington, DC: World Bank Group.

Westerhoff, Garret P., Diana Gale, Paul D. Reiter, Scott A. Haskins, and Jerome B. Gilbert (1998). *The Changing Water Utility.* Denver: American Waterworks Association.

World Health Organization (2000). *Global Water Supply and Sanitation Assessment 2000 Report.* New York: United Nations, p. 1.

Index